ATLAS OF THE OCEANS: WIND AND WAVE CLIMATE

Related titles of interest

Forthcoming for 1996:

Marine Literature Review
(bibliographic CD-ROM Subscription)

Ocean Sciences Interactive Series (Elsevier):

YOUNG AND HOLLAND Atlas of the Oceans: Wind and Wave Climate, scientific and interactive CD-ROM

Books

EMERY AND THOMSON
Data Analysis Methods in Physical Oceanography

EMERY, BALDWIN AND TUCKER
The Advanced Very High Resolution Radiometer: Applications in Atmospheric Science, Oceanography and Terrestrial Physics

Journals

Continental Shelf Research
Deep-Sea Research, Part I, Oceanographic Research Papers
Deep-Sea Research, Part II, Topical Studies in Oceanography
Oceanographic Literature Review (incorporating Ocean Data News)
Progress in Oceanography

Pergamon is an imprint of Elsevier Science Ltd.

Full details of all Pergamon publications and a free sample copy of any Pergamon journal are available from your nearest Elsevier Science office.

The complete Elsevier Science Catalogue of books, journals, newsletters, electronic products and major reference works is available on the Internet:

Gopher: gopher.elsevier.nl
WWW: http://www.elsevier.nl

ELSEVIER SCIENCE

Catalogue on INTERNET

ATLAS OF THE OCEANS: WIND AND WAVE CLIMATE

IAN R. YOUNG

School of Civil Engineering
Australian Defence Force Academy
Canberra, Australia

and

GREG J. HOLLAND

Bureau of Meteorology Research Centre
Melbourne, Australia

PERGAMON

U.K. Elsevier Science Ltd, The Boulevard, Langford Lane, Kidlington, Oxford OX5 1GB, U.K.

U.S.A. Elsevier Science Inc., 660 White Plains Road, Tarrytown, New York 10591-5153, U.S.A.

JAPAN Elsevier Science Japan, Tsunashima Building Annex, 3-20-12 Yushima, Bunkyo-ku, Tokyo 113, Japan

---

First edition 1996

Library of Congress Cataloging in Publication Data

Young, I. R.
Atlas of the oceans: wind and wave climate / by I. R. Young
and G. J. Holland. – 1st ed. 1996
p. cm.
Uses data from the GEOSAT satellite
1. Marine meteorology–Charts, diagrams, etc. 2. Ocean waves–Charts, diagrams, etc. 3. Winds–Charts, diagrams, etc.
I. Holland, Greg.
G2801.C8Y8 1996 <G&M>
551.5'246'0223–dc20 R94-43942

British Library Cataloguing in Publication Data

A catalogue record for this book is available from the British Library

ISBN 0 08 042519 4

Printed and bound in Great Britain by BPC Wheatons Ltd, Exeter

CONTENTS

LIST OF FIGURES

Chapter 1

INTRODUCTION

1.1 Scope of Publication

There is a host of applications which require information on the wind and wave conditions likely to occur at various times and locations. Such applications include:

- Scientific requirements in areas such as climate research
- Determination of engineering design parameters
- The detailed planning of ship routes
- Recreational activities such as sailing or surfing

Naturally there will be many other possible applications not specifically mentioned in the above list.

To date, reliable data for use in such applications has only been available from *in situ* instruments such as wave buoys. Due to the expense of routinely operating such equipment, the number of locations where such data are available is extremely small. This is particularly true when the vastness of the world's oceans is considered.

The only previous alternative to such point specific data has been to use visual observations compiled by ships. This form of data is, however, relatively unreliable. This lack of reliability is partly due to the inherent inaccuracy of visual estimates and partly due to the poor spatial coverage provided by ships, which tend to follow standard shipping routes.

With the advent of oceanographic satellites, it has now become possible to obtain global data of wind and wave conditions on a continuous basis and with relatively high accuracy. This Atlas presents results obtained from three years of global data obtained from the GEOSAT satellite. Rather than merely presenting the data, an attempt has also been made to provide some interpretation of the observed results. The volume of data is, however, so large that this interpretation cannot be complete. Individual readers, depending on their specific interests and applications, will undoubtedly find many points of interest and importance not raised by the authors. The discussion provided within the Atlas is intended only as an initial aid to interpretation for the reader.

1.2 Intended Readership and Reading Pattern

The intended readership for this work is potentially very diverse, ranging from research scientists and design engineers to ships' captains and recreational sailors and surfers. Naturally the interests and requirements of these groups will be very different and hence the intended reading patterns will also differ. The content of the various chapters and intended readership is outlined below.

- **Chapter 2** presents an introduction to global meteorology. It is intended to introduce readers to basic concepts such as the processes which produce winds and to some common meteorological systems. As such, the level of presentation is relatively basic: the chapter is intended for readers with little background in meteorology. It has been included because this type of basic knowledge provides a sound basis for the interpretation of the subsequent data.
- **Chapter 3** provides an introduction to ocean waves in a similar fashion to that provided for wind in Chapter 2. The reader is introduced to

the basic mechanisms by which the wind generates ocean surface waves. In addition, the concept of significant wave height, by which we can represent an apparently random wave field by a single representative value, is introduced. This concept is important as it is used in all subsequent wave data presented. As with Chapter 2, the level of material presented in this chapter is relatively elementary and is intended for readers with little or no background in oceanography or ocean engineering.

- **Chapter 4** describes the GEOSAT satellite mission in detail, indicating the spatial coverage of the satellite and the resolution of the radar altimeter used to measure wind and waves. The techniques used to extract wind and wave data from the altimeter radar returns are also discussed in detail. The material within this chapter is relatively involved and is included for readers who wish to know the potential limitations imposed on the data by the resolution and accuracy of the instrumentation.

- **Chapter 5** describes the detailed processes used to construct the global wind and wave statistics presented in the Atlas from the raw GEOSAT data. Again, the value of this information is that it provides details of the data processing (such as averaging periods) which may be important in some scientific and engineering applications.

- **Chapter 6** provides a validation of both the wind and wave data presented in the Atlas against *in situ* buoy measurements. These results are essential to demonstrate the credibility of the data processing used to produce the Atlas results.

- **Chapter 7** presents the actual data and interpretations of many of its features. The interpretation section is included as a guide for the reader. As such, it provides a valuable introduction to the data and is intended to speed the development of familiarity with the data.

- **Chapter 8** contains the detailed set of references for the text.

- **Chapter 9** presents a Glossary of commonly used terms relating to both winds and waves. In addition, conversion factors between different unit systems are presented.

Chapter 2

GLOBAL METEOROLOGY

2.1 Major Surface Wind Patterns

The atmosphere is a complex cauldron of interacting systems and circulations ranging from local wind gusts and convective clouds with lifetimes of minutes, to huge global circulations which evolve on seasonal and longer scales. Many of the interactions are chaotic so that small errors can grow rapidly and the atmosphere's "memory" can be very short. The scale and rapidity of these interactions are remarkable. For example, consider the information required to make a local forecast: for the next half hour this can be accomplished reasonably by little more than extrapolation of visible trends within eyesight; tomorrow's forecast requires analyses and information on the scale of the Australian continent; five days into the future requires information from the whole globe; and monthly or seasonal forecasts require detailed information on the oceanic circulations as well. Fortunately, there are energetic and physical processes that both drive and constrain the global weather and ensure reasonably consistent patterns similar to those for surface flow in Fig. 2.1. The equatorial regions are generally characterized by low winds and high rainfall, whereas the mid-latitudes have a zone of high pressure and anticyclonic flow (the subtropical ridges) which have little rainfall. Extending equatorward of the subtropical ridge are the trade wind easterlies. In some regions the trade flow converges into long bands of cloud, the Intertropical Convergence Zone (ITCZ). In Asia/Australasia and Africa, the air flows across the equator from the winter to the summer hemisphere to form broad belts of westerly flow converging into zonal troughs of low pressure and high rainfall. This is the gigantic monsoonal circulation that sustains two thirds of the world's population. Poleward of the subtropical ridges are belts of westerlies and further poleward are cyclonic regions of low pressure and belts of easterlies circling the poles.

Whilst there are complex processes contributing to the development of the flow patterns in Fig. 2.1, the essential features can be understood from a few basic principles. It will be shown that the major energetics and constraints on the global circulation are provided by solar heating, the Earth's rotation, the presence of water substance in the air, and the distribution of oceans and their contents. But first, a few basic meteorological principles and laws are described.

2.2 What Drives the Wind?

The wind field at the surface is governed by Newton's second law:

$$F = ma \tag{2.1}$$

where F is the total force, m is the mass and $a = \mathrm{d}U/\mathrm{d}t$ is the acceleration. If the acceleration of a parcel of air of unit mass in a coordinate system attached to the parcel (the "natural" coordinate system, Fig. 2.2) is considered, the following relationships can be derived:

$$\begin{aligned} \frac{U^2}{r} - fU + \frac{1}{\rho}\frac{\partial p}{\partial n} &= 0, \\ \frac{\mathrm{d}U}{\mathrm{d}t} + \frac{1}{\rho}\frac{\partial p}{\partial s} &= F, \end{aligned} \tag{2.2}$$

where U is the wind speed, f is the Coriolis parameter ($2\Omega \sin\phi$, where $\Omega = 7.292 \times 10^{-5}\ \mathrm{s}^{-1}$ is the

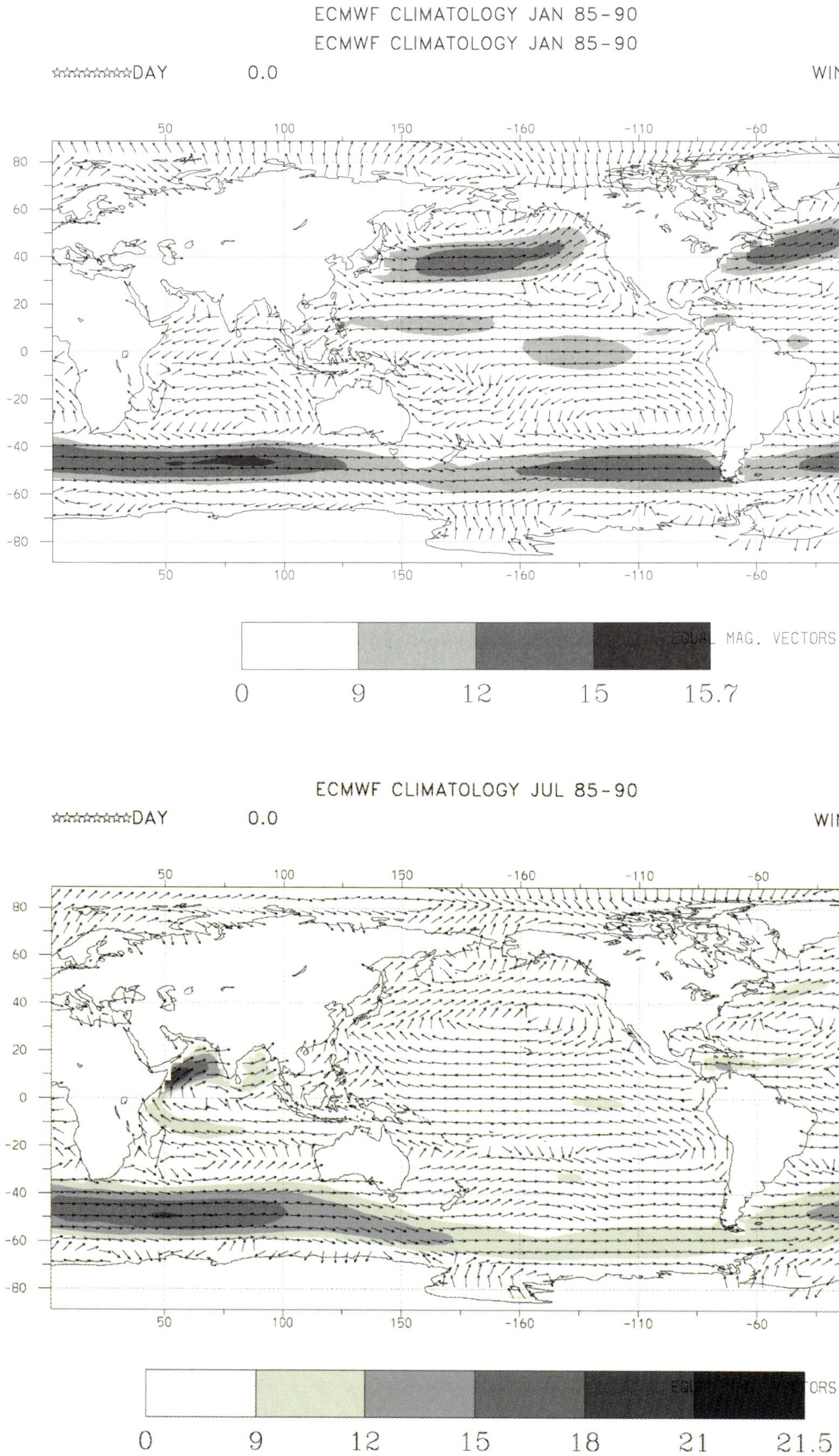

Fig. 2.1 Mean surface wind field for (a) January and (b) July.

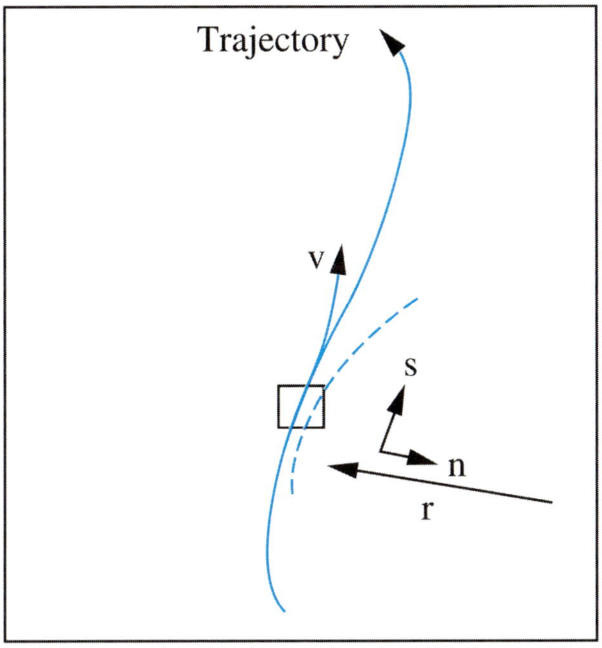

Fig. 2.2 The natural coordinate system that follows the motion of an idealized air parcel.

Earth's rotation and ϕ is latitude), ρ is the air density, p is the surface pressure, and F the frictional dissipation at the Earth's surface. The $\partial p/\partial n$ term is the component of pressure gradient normal to the direction of motion of the parcel and the $\partial p/\partial s$ term is the pressure gradient along a streamline. The U^2/r term is called the centrifugal force, as it relates to the acceleration experienced by a parcel moving around a curved trajectory of radius r. The Coriolis term is often referred to as a "force" because of its role in turning the flow. However, this term arises from the movement of an air parcel on a rotating spherical Earth, which is a non-inertial frame of reference. Thus, the trajectory of a parcel that has no forces acting upon it and is straight and steady to an observer in outer space would appear to curve to an observer on the rotating Earth.

Suppose now that an air parcel initially at rest has a pressure gradient imposed across it. Then Eq. (2.2) shows that the parcel will accelerate towards low pressure. If friction is neglected, as the speed increases the Coriolis term (fU) will turn the parcel to the right (left) in the Northern (Southern) Hemisphere. After a period of time of around $1/f$ the parcel will have turned to move along the isobars, and a balanced state governed by the first part of Eq. (2.2) will ensue. A number of different balanced states are possible, depending on the relative magnitudes of each of the terms in Eq. (2.2). The most general flow is when all terms are equally important and gradient balance results:

$$U = -\frac{fr}{2} \pm \sqrt{\frac{f^2r^2}{4} - \frac{r}{\rho}\frac{\partial p}{\partial n}}. \qquad (2.3)$$

Note that the Coriolis acceleration is always towards the left (right) in the Southern (Northern) Hemisphere, and the pressure gradient force is always towards low pressure, but the centrifugal force is directed outwards from the curved flow (see Fig. 2.3). Thus a low pressure system, or cyclone, has both Coriolis and centrifugal terms in the same direction and opposing the pressure gradient. Very strong winds and sharp curvature can thus be sustained. However, a high pressure system has both the centrifugal and pressure gradient forces opposing the Coriolis term, so that anticyclones are forced to be weak, with only slightly curved flow. This is a partial explanation for the observations that anticyclones tend to be broad and weak, whilst cyclones can be very intense and small.

Frictionless motion along straight isobars ($F = 0$, $U^2/r = 0$) is called geostrophic flow:

$$U = -\frac{1}{\rho f}\frac{\partial p}{\partial n}. \qquad (2.4)$$

This balance is typical of much of the large-scale flow in the atmosphere (see Fig. 2.4).

When the Coriolis term is small compared to the centrifugal force ($F = 0$, $fU \ll U^2/r$) we have cyclostrophic flow:

$$U = \pm\sqrt{\frac{r}{\rho}\frac{\partial p}{\partial n}}, \qquad (2.5)$$

which occurs in regions of sharply curved high winds, such as near the centre of a tropical cyclone (see Fig. 2.5).

When the Coriolis term is negligible, the flow cannot turn to form the balanced state given by the first part of Eq. (2.2) and the only balance is between pressure gradient acceleration and friction. This is called antitiptic flow and occurs near the equator or in local wind circulations, such as sea breezes (see Fig. 2.6). Other potential types of balance and flow conditions are described in Holton (1992).

Introducing friction slows the winds at the Earth's surface. Since the Coriolis and centripetal terms in Eq. (2.2) are functions of the wind speed, but the pressure gradient is not, the effect is to turn the flow towards low pressure (see Fig. 2.7). Thus air at the surface tends to flow out of anticyclones and towards cyclones. The influence of surface friction decays rapidly with height and is generally negligible by about 1 km altitude. This lowest atmospheric

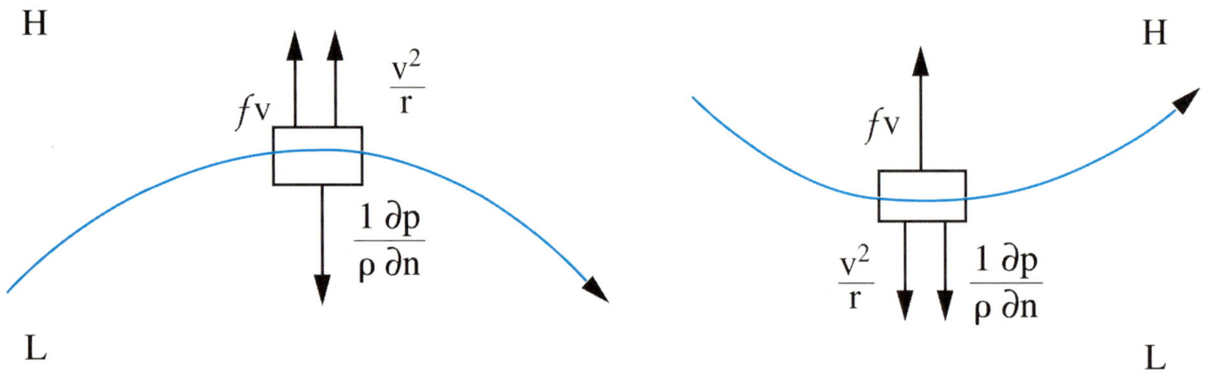

Fig. 2.3 Force balance in Eq. (2.3).

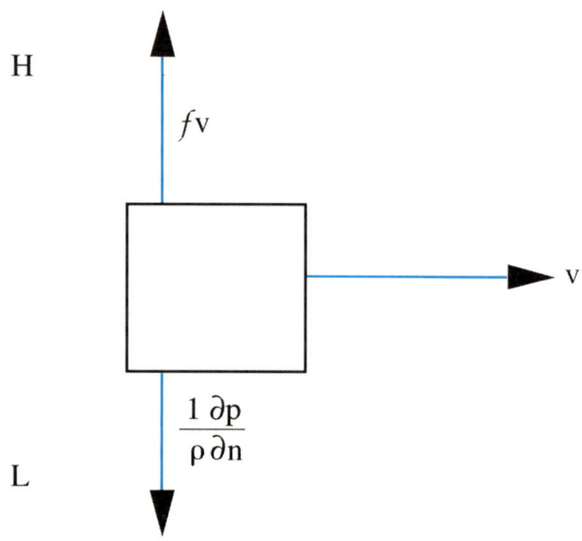

Fig. 2.4 Force balance in Eq. (2.4).

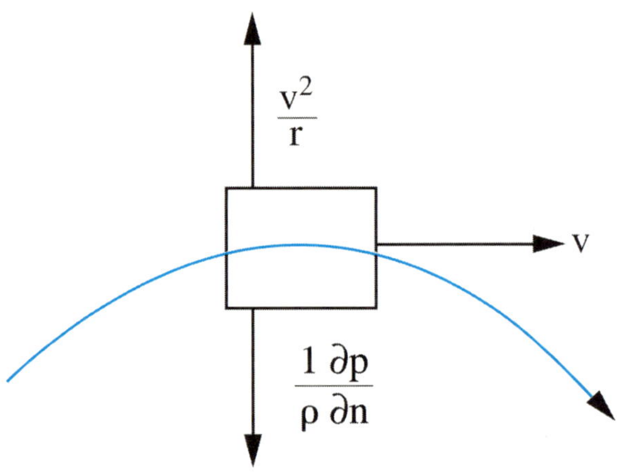

Fig. 2.5 Force balance in Eq. (2.5).

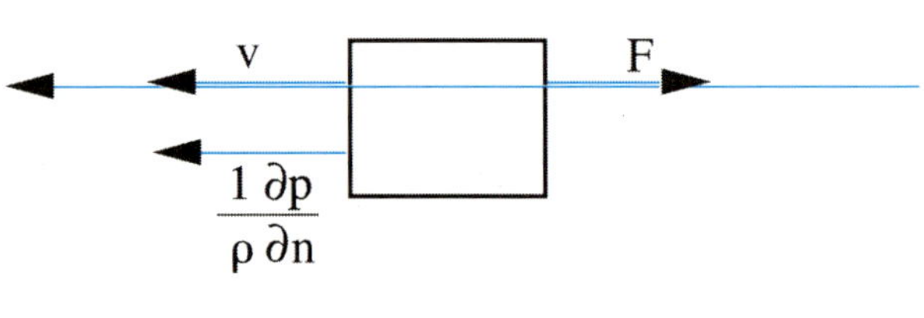

Fig. 2.6 Force balance in antitiptic flow.

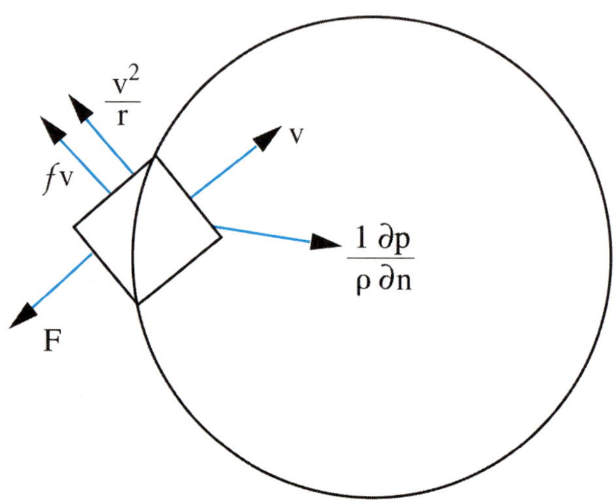

Fig. 2.7 Force balance with the inclusion of friction.

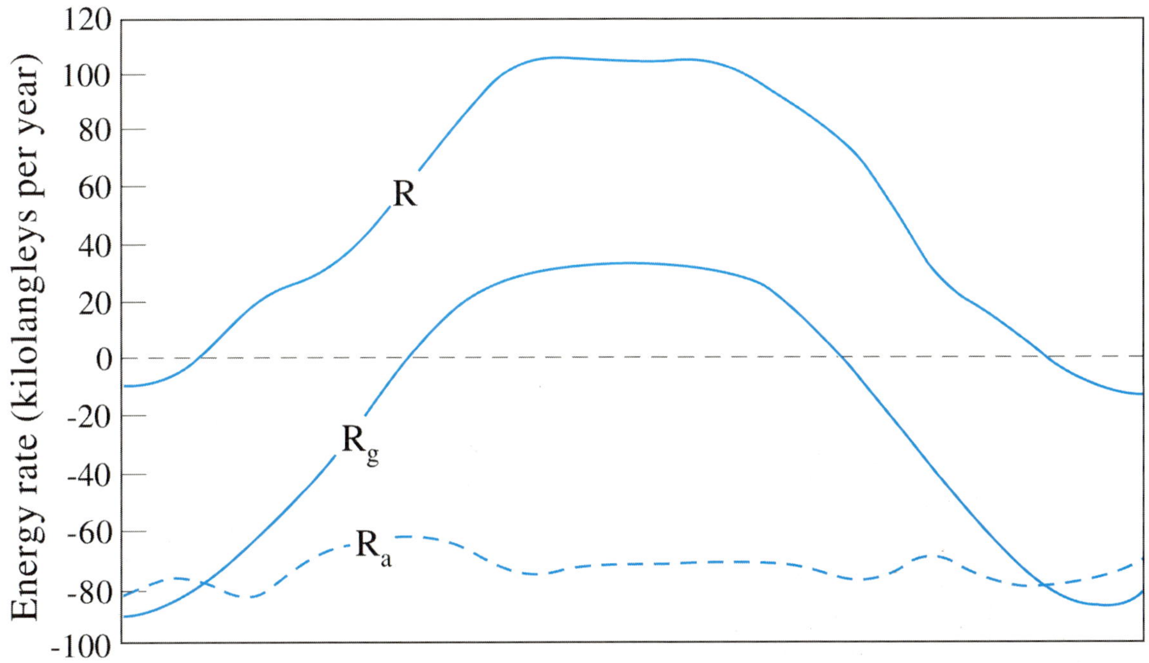

Fig. 2.8 Incident solar radiation on the spherical Earth (Sellers, 1965).

layer is often referred to as the boundary, or friction layer and the 1 km level is called the gradient level to reflect the balance in Eq. (2.3). Conservation of atmospheric mass then requires that there be upward motion out of the friction layer in cyclones and downward motion in anticyclones. This partially explains the observations that anticyclones are generally cloud free, whilst cyclones contain many cloud varieties.

2.3 The Global Meridional Circulation

The primary energy source for the global circulation is heating by solar radiation. The incoming radiation is nearly incident to the surface in tropical regions and tangential in polar regions (Fig. 2.8), so that the net heating in the tropics is much higher than that at the poles. Some of this incident radiation is reflected back to space by clouds and surface features (Fig. 2.9) and a small amount is absorbed in the atmosphere, but the majority is absorbed at the Earth's surface, where it contributes to both heating and evaporation of water vapour. The net atmospheric cooling from radiation balance is around 1–2°C per day. This maintains conditions suited to growth of clouds, which will be shown to be fundamental to the maintenance of the global circulation. Clouds also warm the atmosphere by a combination of latent heating and local circulation. Maintenance of a steady state requires a global mean rainfall of approximately 3 mm per day, together with an equivalent evaporation from the Earth's surface. The actual rainfall occurs in local, intense weather systems, however these must average 3 mm over the globe.

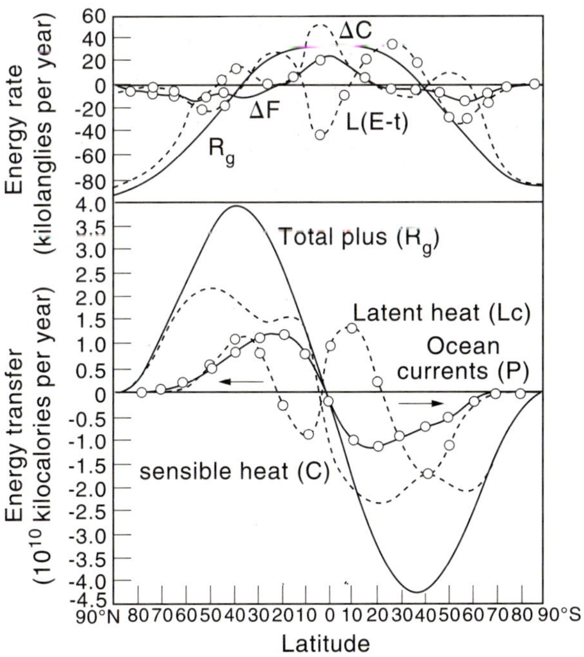

Fig. 2.9 The radiation budget for the Earth–atmosphere system (Sellers, 1965).

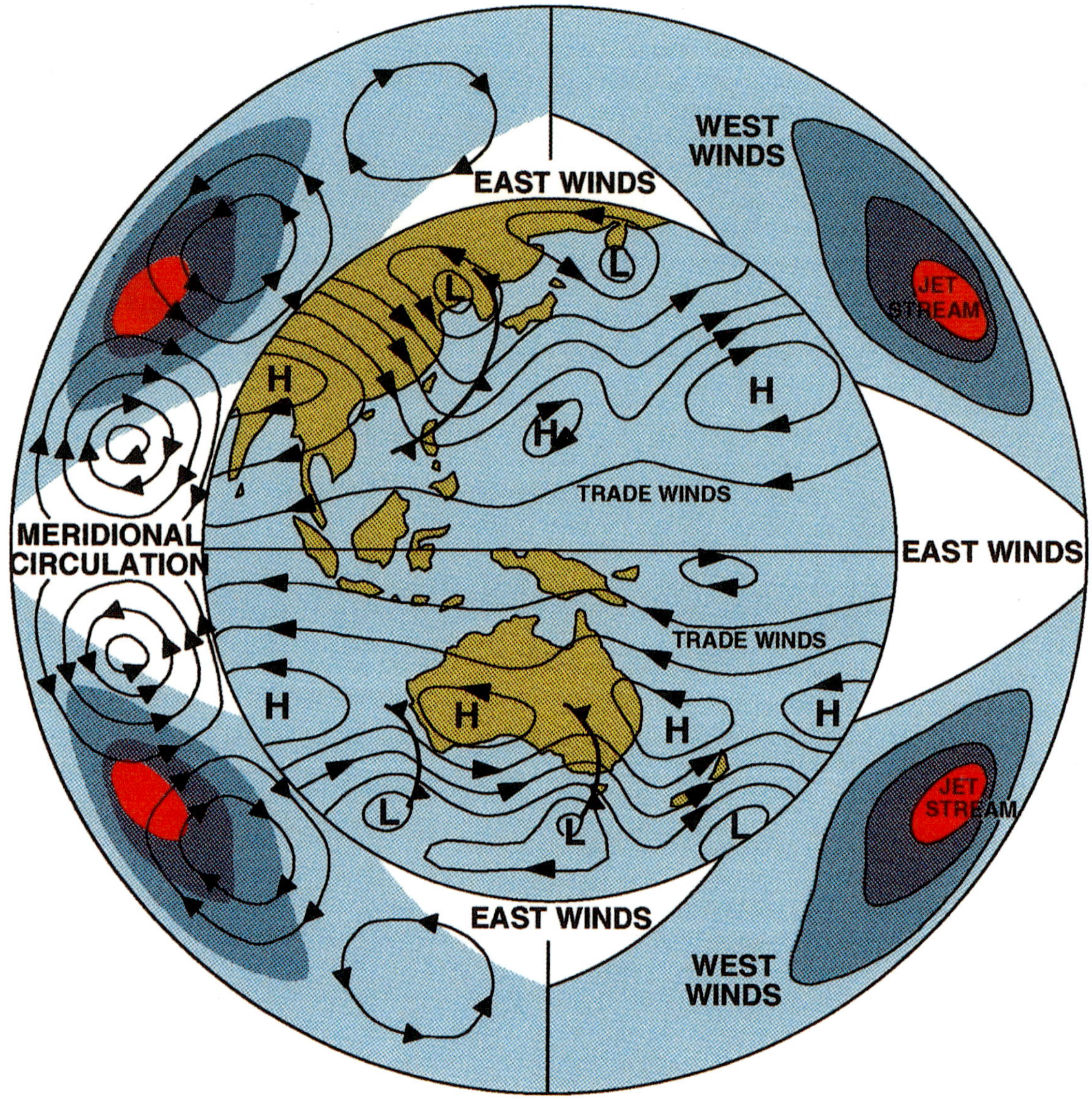

Fig. 2.10 Features of the mean meridional circulation of the atmosphere (provided by the Australian Bureau of Meteorology).

The net radiation budget provides excess heating to the tropics and a cooling of the higher latitudes. This maintains the meridional temperature gradient that supports growth of mid-latitude weather systems. Thermal balance also requires a transfer of heat energy from the tropics to the polar regions, which is accomplished by a combination of oceanic and atmospheric circulations.

The atmospheric circulation is most easily conceptualized by a combination of mean meridional circulations and weather-system type deviations, known as eddies. The mean circulation (Fig. 2.10) consists of a large Hadley circulation extending from tropical to subtropical latitudes, a smaller Ferrell circulation in mid-latitudes and generally negligible circulation at high latitudes.

The second powerful constraint on the global circulation is the rotation of an approximately spherical Earth. As air parcels move meridionally, their distance from the axis of rotation of the Earth changes. Conservation of angular momentum (Ur =constant) requires that their speed changes and these changes are in the opposite sense to those of the Earth's surface, which is in solid-body rotation (U =constant r).

For example, an air parcel at rest relative to the Earth's surface (Fig. 2.11) is actually moving at an azimuthal speed given by,

$$U = \Omega a \cos \phi, \tag{2.6}$$

where $\Omega = 7.292 \times 10^{-5}\ \mathrm{s}^{-1}$ is the angular rotation of the Earth, $a = 6370$ km is the mean Earth radius and ϕ is the latitude. For $\phi = 30^\circ$, $U = 400\ \mathrm{ms}^{-1}$ or over 800 kt. If the parcel is moved to the equator in the absence of any external force, conservation of angular momentum will reduce its absolute speed to 335 ms^{-1}. Since the Earth's surface at the equator is moving at approximately 455 ms^{-1}, an easterly wind of 120 ms^{-1}, or 250 kt, would be generated. These wind speeds do not develop fully because of frictional losses at the surface, but this is the primary mechanism responsible for the easterly trade wind flow that dominates much of the tropics (Fig. 2.1). The essence of this process was first described by Hadley (1735), after whom the overall circulation is named.

The opposite process occurs in the reverse circulation at high levels, leading to generation of westerly flow in subtropical latitudes. Here, lack of direct frictional effects enables the very high winds of occasionally over 250 kt observed in the subtropical jet stream.

The Hadley circulation is thermally direct, in that it arises from the differential heating–cooling between the tropics and mid-latitudes. The upward motion in the tropics can only be accomplished by the release of latent heat following condensation/freezing of water vapor in tropical clouds. The required moisture is provided largely by evaporation from the warm ocean. In particular the trade winds evaporate huge amounts of water. This moisture is constrained beneath the trade wind inversion, a stable atmospheric layer that develops at an altitude of 1–3 km. A result of this inversion is that the trade wind regime is relatively free of all but small "fair weather" cumulus clouds. The moisture is transported equatorward to support the relatively narrow band of tropical clouds that make up the Intertropical Convergence Zone (ITCZ).

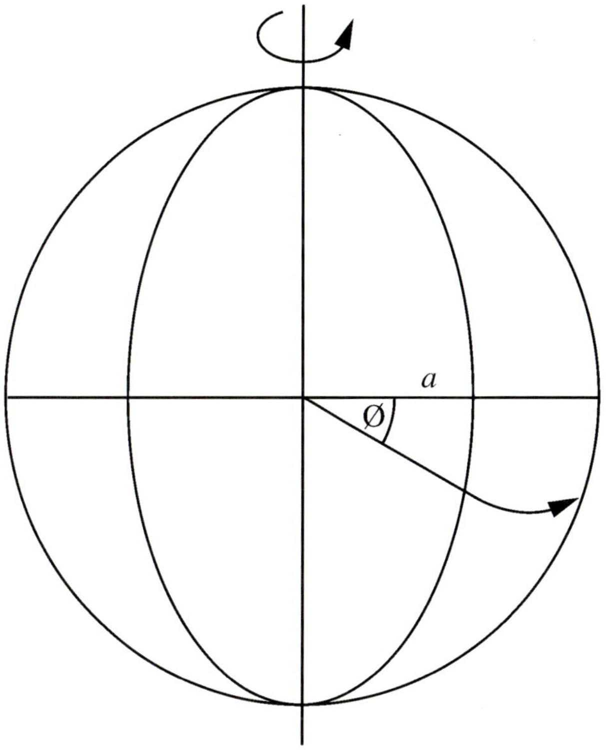

Fig. 2.11 Geometry of an air parcel located on the spherical Earth.

The steady nature of the easterly winds and the lack of major weather systems were effectively exploited by early mariners who named these the "trade winds":

> The fair breeze blew, the white foam flew,
> The furrow followed free[1]

By comparison, the light winds and high humidity in the vicinity of the ITCZ could lead to long periods of idleness in sweltering conditions, leading to this region being named "the doldrums":

> Day after day, day after day,
> We stuck, nor breath nor motion;
> As idle as a painted ship
> Upon a painted ocean[1]

[1]Coleridge, The Rime of the Ancient Mariner.

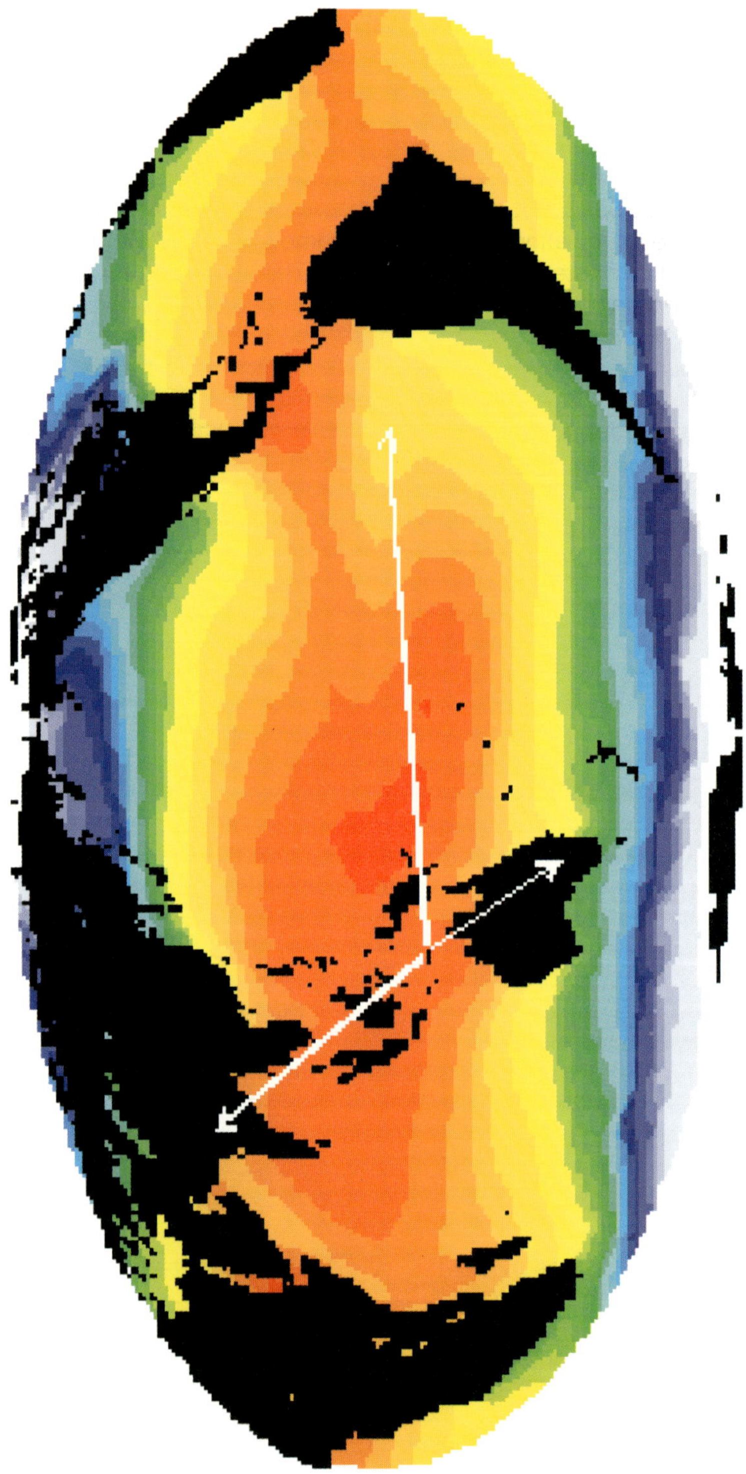

Fig. 2.12 Sea surface temperature deviations from the zonal mean, showing the strong gradients across the equatorial Pacific Ocean (provided by the Australian Bureau of Meteorology.)

The intense tropical convection and frictional convergence into the low-pressure zone [Eq. (2.2)] provides a narrow upward branch of the Hadley circulation. By comparison, the subsiding branch in mid-latitudes is heated by the compression required to descend to the dense lower atmosphere. This heating is compensated by radiative losses to space, which are much weaker than latent heating. As a result, the substance is spread over a large area and at the surface is manifested in the broad belt of anticyclones known as the subtropical ridge.

The subtropical ridge and subtropical jet mark the poleward extent of the Hadley circulation. Further poleward lies the broad belt of westerly winds, a dangerous region of high winds and storms, referred to as the "roaring forties" by early mariners in the Southern Hemisphere;

> The ship drove fast, loud roared the blast,
> And southward aye we fled[1]

In this region the eastward momentum gained by the atmosphere in the equatorward branch of the Hadley circulation is returned to the Earth by frictional dissipation.

The westerlies occupy a region of strong horizontal temperature gradients, or baroclinity, known as the "polar front". The baroclinity maintains increasing westerly flow with height, through the thermal wind relationship, leading to the polar front jet stream (Fig. 2.10). Some of the baroclinic energy is also utilized to drive the small, thermally indirect Ferrell circulation.

The statistical concept of mean meridional circulations breaks down at very high latitudes. Here cyclonic storms (or eddy circulations) provide the meridional heat transport by bringing cold air equatorward to the west and warm air polewards to the east.

2.4 Zonal Circulations

Whilst the major temperature gradients occur between the polar and tropical regions, significant zonal gradients of surface temperature and cloudiness can occur. The most important of these lies across the near-equatorial Pacific Ocean and is named the Walker circulation after Sir Gilbert Walker who discovered the changes in surface pressure associated with fluctuations of the circulation, which he named the Southern Oscillation (Walker, 1924). An excellent review of the Walker circulation and its consequences may be found in Philander (1990). In essence, the easterly trade flow off South America (Fig. 2.1) drives surface water away from the coast by frictional coupling. The compensating upwelling of cool subsurface water results in a region of sea surface temperatures that are typically 5°C or more below the zonal average (Fig. 2.12). The upwelling and cold surface temperatures are maintained along the equator by Coriolis drift of the ocean surface water, which is to the left (right) in the Southern (Northern) Hemisphere.

At the western end of the equatorial Pacific Ocean lies the pool of warmest ocean on Earth (the "Warm Pool") and the myriad of tropical islands known as the "Maritime Continent" (Ramage, 1968). These provide for sustained upward motion and, together with the region of suppressed cloudiness and generally subsiding atmospheric motion in the eastern Pacific, drive the zonal Walker circulation (Fig. 2.13). The zonality of the Walker circulation means that there are no significant changes of angular momentum, so that the large speed changes associated with the Hadley circulation do not occur. Notice in Fig. 2.13 that the Hadley and Walker circulations converge on the Maritime Continent/Warm Pool region. This is the region of greatest cloudiness and rainfall anywhere on Earth. Extending southeastward from this region is a secondary zone of convergence and cloudiness known as the South Pacific Convergence Zone (SPCZ). A direct result of the combined Hadley and Walker circulations is the presence of surface easterly winds across the entire tropical Pacific Ocean that are sustained throughout the year.

Occasionally the eastern Pacific warms up, the Walker circulation weakens and moves eastward and an El Nino cycle commences. The mechanisms involve complex ocean and atmospheric interactions, some of which are not well understood [see Philander (1990), for a detailed description]. However, the El Nino/Southern Oscillation cycle is recognised as a major influence on the world weather, being associated with changes to the Indian summer monsoon, floods in California and tropical cyclones in Tahiti.

2.5 Storms

A wide variety of storms is encountered over the oceans of the world. In severe cases high winds of up to 150 kt, driving rain and phenomenal seas of up to 34 m peak height (according to the Guinness Book of Records) may be encountered. For discussion purposes these storms are grouped into the three main families experienced in maritime regions: tropical cyclones, mid-latitude cyclones, and

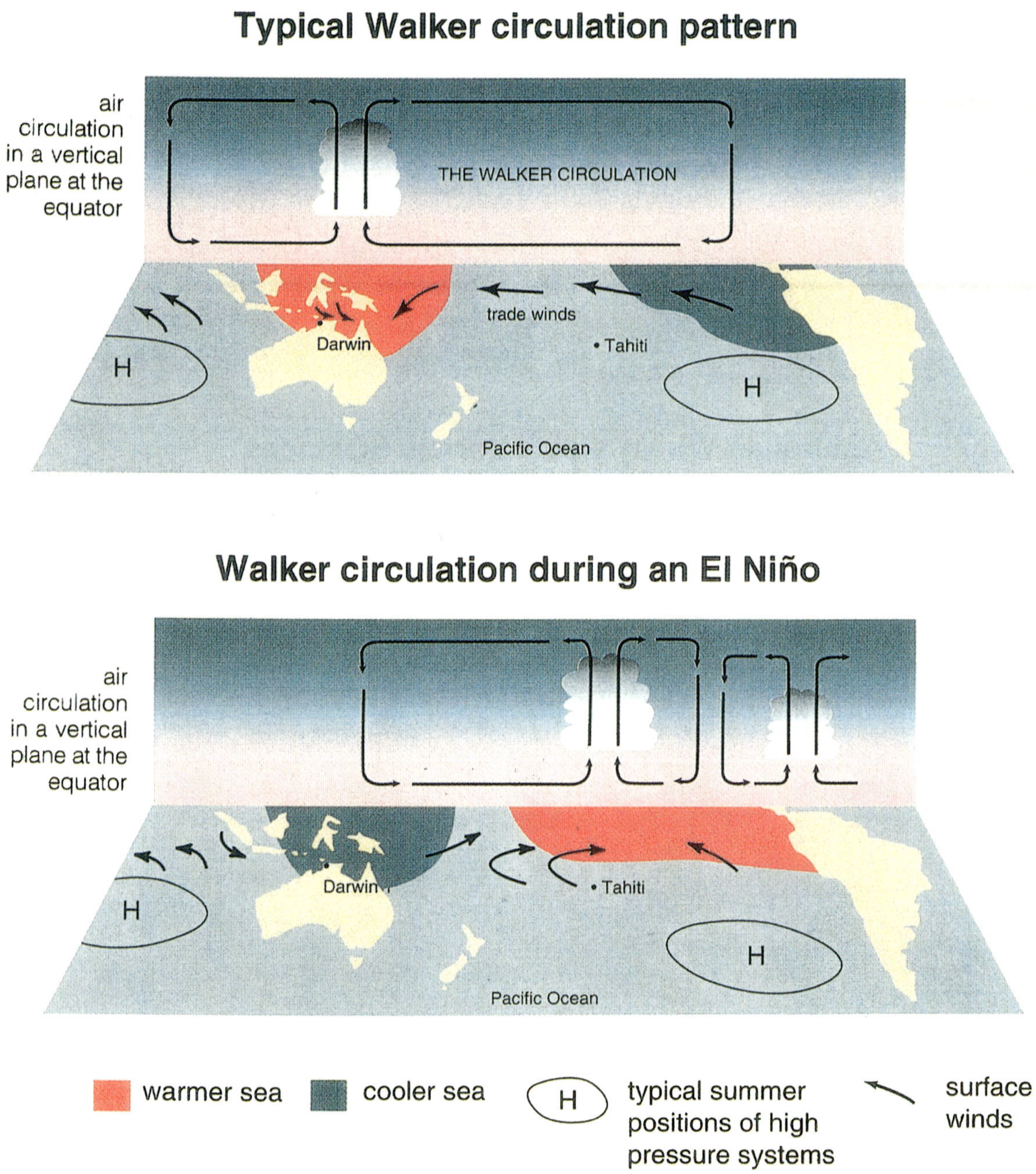

Fig. 2.13 Features of the Walker and Hadley circulations across the Pacific Ocean (provided by the Australian Bureau of Meteorology).

Fig. 2.14 Satellite view of Hurricane Allen in the Gulf of Mexico (provided by NASA).

hybrid systems. Further discussion on local storms, such as wind squalls, convective squall-lines and severe thunderstorms may be found in Cotton and Anthes (1989).

The energy to drive these storms comes from three sources: latent heat released by condensation/freezing of water vapor in clouds, together with oceanic interactions; release of the baroclinic energy stored in horizontal temperature gradients; and barotropic energy contained in special types of horizontal wind variations. This latter source of energy is of relatively minor importance and the reader is referred to specialist literature such as Holton (1992) for further information.

2.5.1 Tropical Cyclones

> When that cloud begins to move apace, you may expect the Wind presently. It comes on fierce, and blows very violently at N.E. Twelve hours more or less... When the wind begins to abate it dyes away suddenly, and falling flat calm, it continues for an hour, more or less: then the wind comes about to the S.W. and it blows and rains as fierce from thence, as it did before at N.E. and as long.[2]

Release of latent heat, and complex interactions with the ocean surface are the primary mechanisms driving the development and intensification of tropical cyclones. These dangerous systems have a characteristic satellite signature (Fig. 2.14) of large cloud bands spiralling into a central core of dense, high cloud with a relatively cloud free eye.

For intense tropical cyclones the lowest pressure is found at the centre and the maximum winds occur near the edge of the eye (Fig. 2.15). The central pressure values can be converted to maximum sustained wind by the approximate relationship (Atkinson and Holliday, 1977):

$$U_m = 1.74\left(p_{env} - p_c\right)^{0.644}, \qquad (2.7)$$

where the speed is in ms^{-1}, p_{env} is the environmental pressure, usually taken as the outer closed isobar on a surface pressure analysis, and p_c is the pressure at the cyclone centre.

Generally there is a rapid increase of pressure and decrease of wind speed with increasing radius, but sustained strong winds can be maintained over large areas, and secondary wind maxima are often observed. Whilst the destructive winds are located close to the centre, tropical cyclones can vary enormously in size. Extreme examples include Tropical Cyclone Tracy at Darwin in 1974, which had a radius of maximum winds of 6 km and gale-force winds extending to 50 km, and Supertyphoon Tip in the western North Pacific in 1979, which extended a cyclonic circulation over much of the ocean basin.

The strongest winds are typically found in the front left (right) quadrant for cyclones in the Southern (Northern) Hemisphere. This is the dangerous semi-circle feared by early mariners. However, some cyclones can develop their strongest winds in other quarters.

Tropical cyclones are children of the warm tropical oceans (Fig. 2.16). They require warm surface sea temperatures (SST)$>$ 26°C to develop and they interact with the ocean surface to provide additional energy for the intensification process.

The degree of intensification is limited by the ocean temperature and near environmental conditions. Theoretical and empirical evidence can be combined to provide the curve of maximum potential intensity as a function of SST in Fig. 2.17 (Holland, 1995; DeMaria and Kaplan, 1995). The solid curve in Fig. 2.17 indicates the intensity that can be reached by the developing tropical cyclone. Only a very few tropical cyclones reach the maximum intensity indicated in Fig. 2.17. A wide variety of adverse conditions normally combine to arrest development before this stage is reached.

The rate of intensification varies considerably. It often takes several days for a cyclone to reach maximum intensity, but most severe cyclones experience a period of rapid intensification. Record pressure falls of over 100 hPa, or increase of 130 kt, in 24 hrs have been observed.

The movement of tropical cyclones can be complex and difficult to forecast, as indicated by the wide variety of movement in Fig. 2.16. The climatological tendency is for westward movement in low latitudes followed by poleward then eastward movement in high latitudes, a process known as recurvature. There are, however, many exceptions to the rule. In the western South Pacific cyclones tend to track continuously eastward, and erratic movement, sharp speed changes and even loops are common.

2.5.2 Subtropical Cyclones

These are the large cyclones with trailing cold fronts found on weather charts for extratropical latitudes (Fig. 2.18). These cyclones gain most of their energy from horizontal temperature gradients, a

[2] Abstract from the log of William Dampier on 4 July 1687, in the South China Sea (Shaw, 1926).

(a)

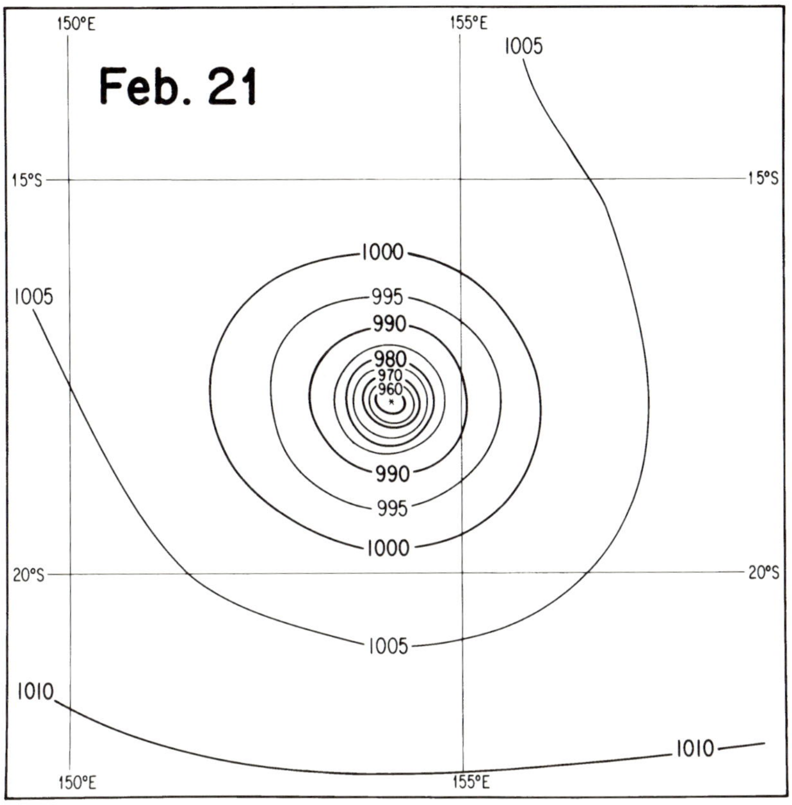

(b)

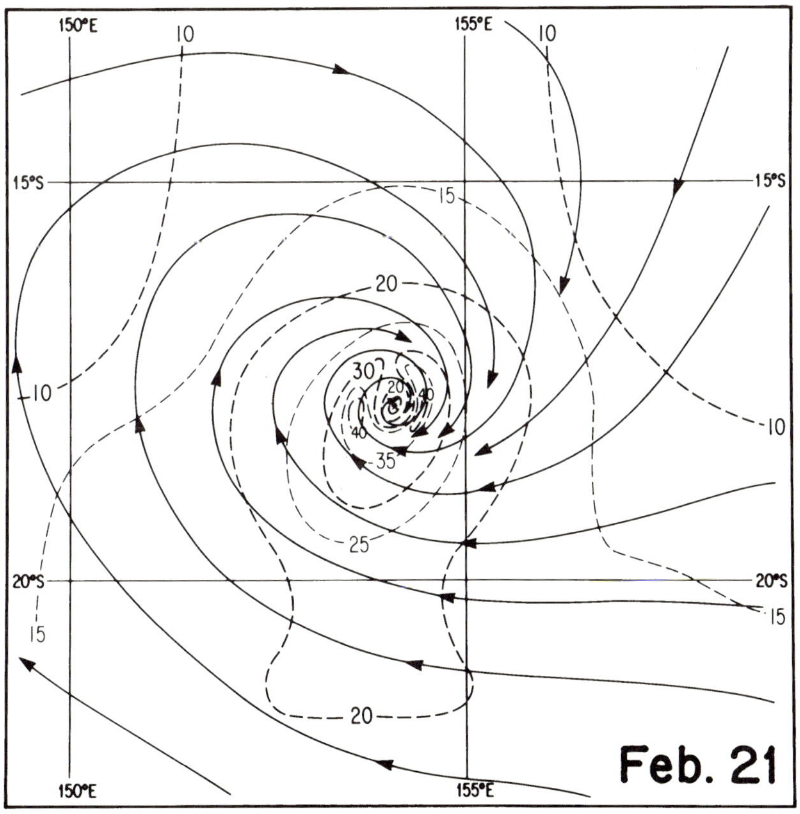

Fig. 2.15 (a) Surface pressure and (b) wind fields for Tropical Cyclone Kerry in the Coral Sea (Black and Holland, 1995).

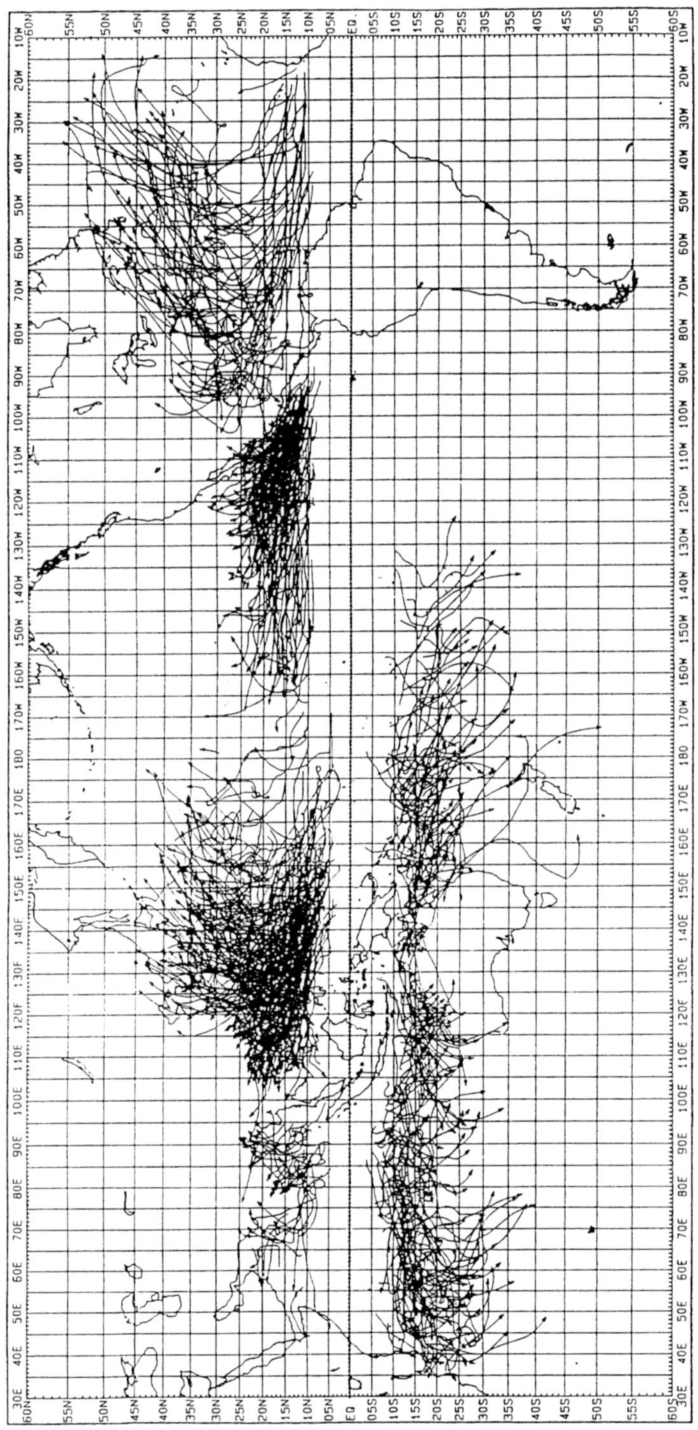

Fig. 2.16 Tracks of tropical cyclones (maximum winds > 34 kt) for the period 1979–1988 (Neumann, 1993).

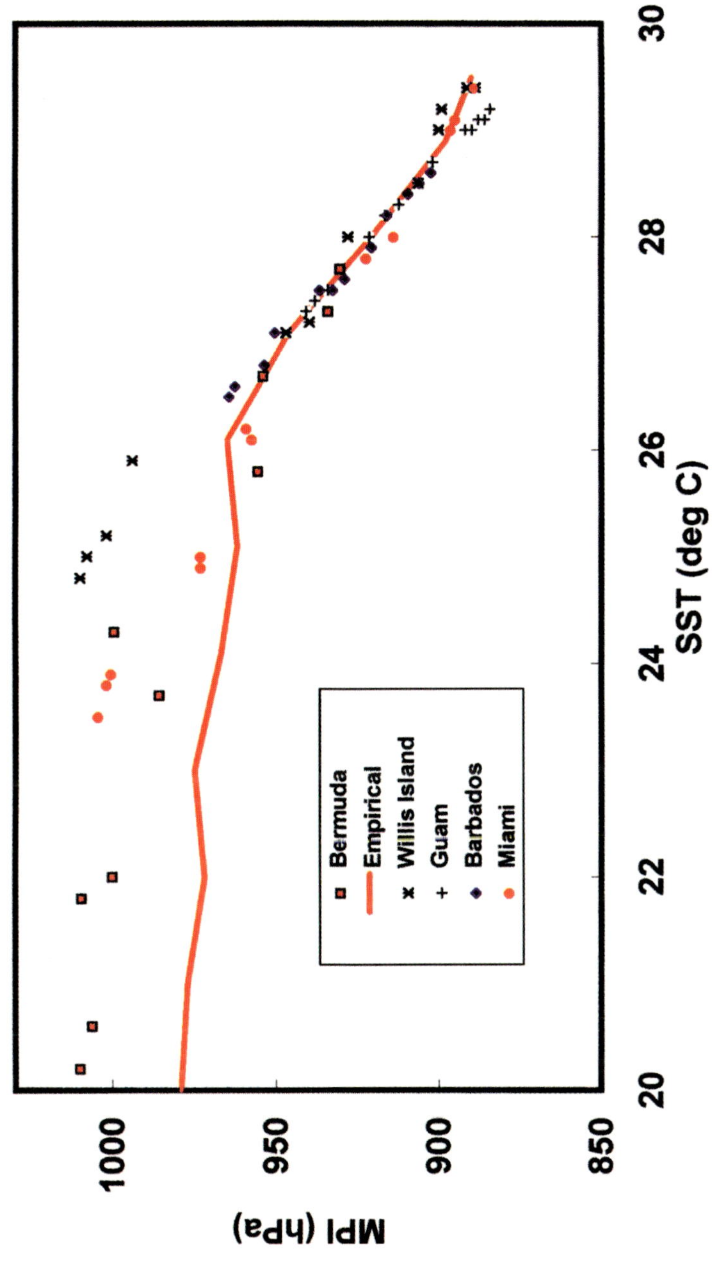

Fig. 2.17 Approximate relationship between ocean temperatures and the maximum possible intensity of tropical cyclones.

Fig. 2.18 Satellite imagery of a large subtropical cyclone in the North Pacific Ocean.

process known as release of baroclinic instability. Some energy is also derived from latent heat release in the broad bands of cloud that develop (Fig. 2.18), however, in contrast to tropical cyclones this is not the major energy source. These cyclones have cold air near the centre in the lower atmosphere with warm intrusions of stratospheric air aloft. To maintain a balance between the mass and wind fields, a process known as "thermal wind balance" as described, for example, by Holton (1992), the cyclonic winds must increase with height up to the level of the polar front jet stream. The maximum wind belt near the surface is also a long way from the cyclone centre, which is generally relatively benign, in strong contrast to the dangerous centre of a tropical cyclone.

Subtropical cyclones are highly asymmetric (Fig. 2.18). Large belts of cold air typically sweep equatorward on their western side, and warm air moves poleward to the east. Where the cold air undercuts the warmer air a cold front forms, with sharp gradients of temperature and long bands of cloud (Fig. 2.18). Here the surface winds can change rapidly from southerlies (northerlies) to strong, gusty northwesterlies (southwesterlies) in the Northern (Southern) Hemisphere. Aloft there is a complex interplay of jet streams and rising/subsiding air.

The baroclinic energetics require a westward slope with height for energy to be transferred to the cyclone; when cyclones become vertical, or slope eastward they are at the end of their lifecycle and moving to a decaying mode.

2.5.3 Hybrid Systems

The advent of satellite observations of the oceans provided evidence for a further group of cyclones that develop over the subtropical and polar oceans. These cyclones come in a variety of types and have been called "Polar Lows" (Rasmussen, 1979), "Arctic Hurricanes" (Emanuel and Rotunno, 1989), and "Maritime Cyclones" or "Meteorological Bombs" (Sanders and Gyakum, 1980). Their energetics are a combination of the aforementioned baroclinic and latent heating/ocean interaction processes responsible for tropical and extratropical cyclones. In particular, such cyclones are often associated with the movement of cold air over a warmer ocean and/or with strong temperature gradients in the lower atmosphere. Such conditions are common near the eastern coasts of the major continents and the edge of polar icepacks, and these regions can be prolific breeders of hybrid cyclones.

The cyclones can develop very quickly, which leads to the "bomb" description by Sanders and Gyakum. Systems have been observed off the east Australian coast that have developed to hurricane force in less than 10 hrs. The systems are also often very small, sometimes only a few tens of km across, so that unwary mariners could rapidly sail into very difficult conditions. A general discussion of their characteristics may be found in Businger (1991) and more technical descriptions in Twitchell *et al.* (1989).

Chapter 3

OCEAN WIND WAVES

3.1 Basic Ocean Wave Properties

As in many scientific areas, surface waves on the ocean can be represented by the simple sinusoidal form

$$\eta = a \sin(\omega t + \phi) \tag{3.1}$$

where η is the water surface elevation, $\omega = 2\pi/T$ is the frequency, T is the wave period and ϕ is a phase angle. The wave form represented by Eq. (3.1) is shown in Fig. 3.1. Based on the relatively simple wave form represented by Eq. (3.1), Linear Wave Theory (Airy, 1845; Dean and Dalrymple, 1991), predicts a number of useful properties of ocean waves. In particular, the phase speed or speed of propagation of the wave form, C is

$$C = \frac{L}{T} = \frac{gT}{2\pi} \tanh \frac{2\pi d}{L} \tag{3.2}$$

where L is the wave length (see Fig. 3.1), T is the wave period or time for two successive wave crests to pass a fixed point, d is the water depth and g is gravitational acceleration. The wave length of the waves, L, is predicted by linear wave theory to be

$$L = \frac{gT^2}{2\pi} \tanh \frac{2\pi d}{L}. \tag{3.3}$$

Hence, if any two of the quantities, C, L or d are known, the other is immediately defined. There is a direct analogy here between water waves and optics. The water depth, d acts in a similar manner to the refractive index of the medium through which light travels. As the medium and hence the refractive index changes, the speed of propagation changes. Similarly, for water waves, as the water depth decreases, the phase speed of the waves decreases.

For the case of deep water (i.e. large d), Eqs (3.2) and (3.3) can be simplified to

$$C = C_0 = \frac{L}{T} = \frac{gT}{2\pi} \tag{3.4}$$

and

$$L = L_0 = \frac{gT^2}{2\pi} = C_0 T. \tag{3.5}$$

The term "deep" is a relative one and depends on the scale of the waves, as measured by the wave length. For the present application, however, it can be interpreted as meaning all oceanic areas with the exclusion of the continental shelves. With this simplification it can be seen that specification of the wave period immediately defines the wave length and phase speed. These relationships are shown in Fig. 3.2. It is clear from Fig. 3.2 that both the wave length and the phase speed increase with wave period. Equations (3.4) and (3.5) indicate that the wave length and phase speed are not functions of the wave height, $H = 2a$. This is a characteristic of Linear Wave Theory. Higher order wave theories which assume a non-sinusoidal wave form predict a weak dependence on the wave height. Although there are some engineering applications where these higher order effects are important, for the present application they may be neglected.

Since the wave period is independent of the wave height, it is, in principle, possible to have any wave height for a given period or wave length. In practice, however, there is a limit to the maximum wave height which can exist for a given wave length. For a fixed wave length L, as the wave height H increases, the wave slope H/L increases. Eventually, the wave becomes unstable and begins to break. This breaking is characterized by the commonly termed "white-caps" or "white-horses" observed on the wind-roughened sea. The turbulence

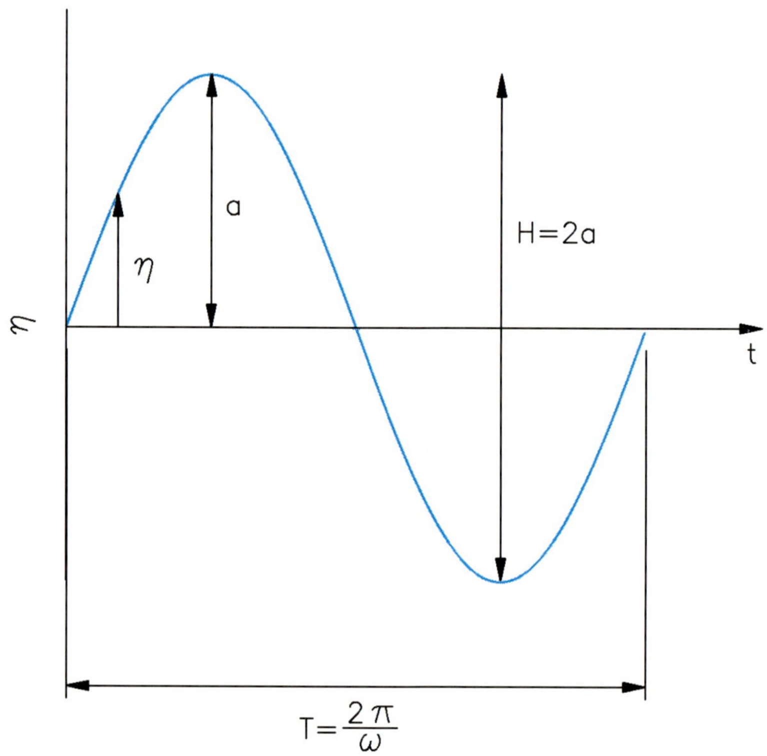

Fig. 3.1 A sinusoidal or linear wave as defined by Eq. (3.1).

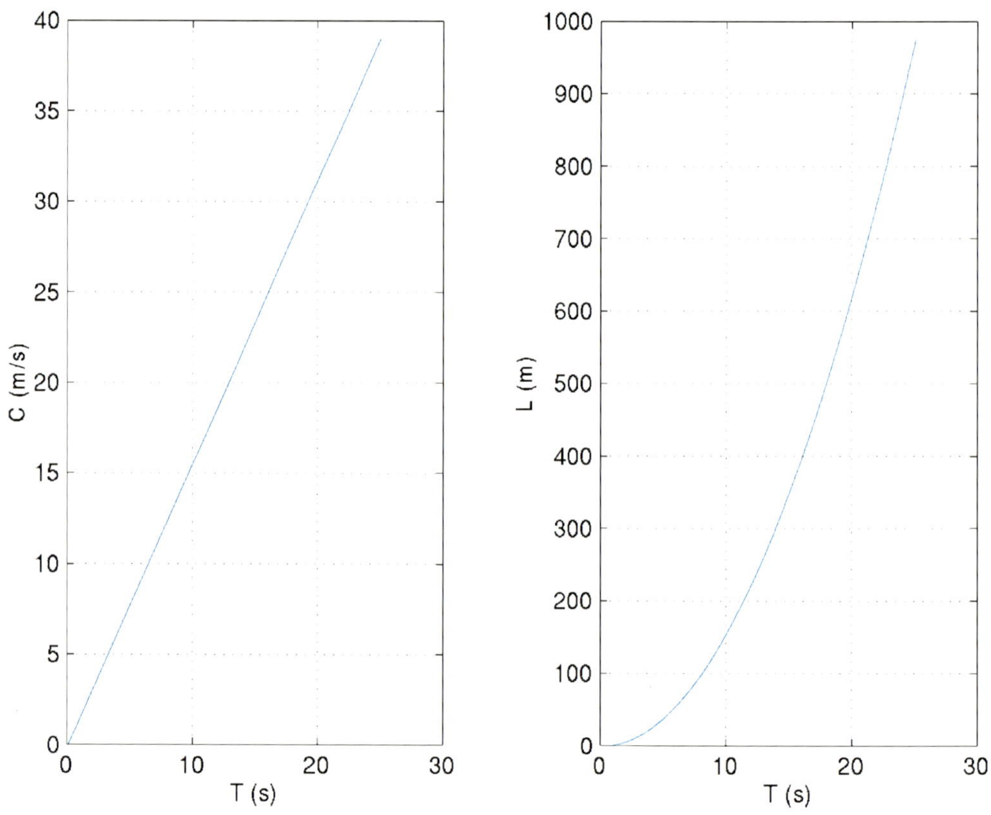

Fig. 3.2 The wave length *L*, and wave phase speed *C*, as a function of the period *T*, for deep water conditions.

induced by the breaking process is a very effective energy dissipation mechanism and reduces the wave height. The process of "white-cap" dissipation is mathematically very challenging, being a strongly non-linear process, and a full understanding is yet to be obtained. Nevertheless, observations indicate that there is an upper limit to the wave slope given by

$$\frac{H_0}{L_0} = 0.14. \tag{3.6}$$

Figure 3.3 shows this maximum limiting wave height as a function of both wave length and period.

Typical ocean wave periods are in the range 10–14 s and hence from Fig. 3.3, limiting wave heights would be between 22 m and 43 m. Clearly, waves of this magnitude are rarely reported and hence it can be concluded that steepness is seldom the criterion which determines the wave height of such long period waves.

Before examining the quantities which govern the wave height, it is pertinent to consider exactly what is meant by the term "wave height". The definition applying in the sinusoidal form defined in Fig. 3.1 is clearly the vertical distance between the crest and the trough of the wave. It makes no difference which wave is selected as they are all identical. Even the most casual observer, however, will have noted that the sinusoidal form represented by Eq. (3.1) and Fig. 3.1 is a gross simplification of reality. Typical ocean waves have a variety of heights, periods and directions. There is an apparent randomness about the sea surface which, on initial inspection, defies explanation. A typical example of an open ocean wave record is shown in Fig. 3.4, the record having been recorded at the Southern Ocean site of Macquarie Island ($-54°30'$S, 159°E). The irregular nature of the water surface is clear. Both the height and period of the individual waves vary. The record shown in Fig. 3.4 is an extreme case, with the largest wave in the approximately 3 min interval, having a height of 20 m.

3.2 The Random Sea Surface

The representation of apparently random processes, such as ocean wind waves, is commonly encountered in a range of scientific and mathematical endeavours. The most common approach to the problem is through Fourier analysis. Fourier analysis assumes that any time or space dependent property can be represented by a sum of sinusoidal waves of the form defined by Eq. (3.1) and Fig. 3.1. Rather than having a single amplitude, frequency and phase angle as specified by Eq. (3.1), these properties will vary for each sinusoid in the sum. The mathematical form for this Fourier series is

$$\eta = \sum_{i=0}^{\infty} a_i \sin(\omega_i t + \phi_i)\,. \tag{3.7}$$

Equation (3.7) assumes that any wave profile, no matter how complex, can be represented by this infinite series of sinusoidal waves, each with its own amplitude, a_i, frequency, ω_i and phase, ϕ_i. An example of the potential power of this approach is shown in Fig. 3.5, where a rather complex water surface is created by the summation of only three Fourier components.

The advantage of adopting this seemingly cumbersome mathematical representation is that the mathematical formulations of Eqs (3.4) and (3.5) remain valid for each of the individual components in the summation [Eq. (3.7)]. Although this type of representation is clearly a far better approximation to the ocean wave surface than the simple form represented by Eq. (3.1), it is far less simple to use. For practical purposes, an estimate of single representative values of the wave height and period is often required. The Fourier model [Eq. (3.7)], however, produces an infinite number of heights and periods, one for each component in the summation. Clearly, this is a serious deficiency.

Although ocean waves are not simply sinusoidal as represented by Eq. (3.1), they do have a dominant height and period. For instance, if an observer were asked to estimate the height and period of waves passing a point, they would have no difficulty interpreting the quantities they were being asked to report. It should be noted, however, that visual observations of wave properties are notoriously inaccurate. Clearly, the human observer is determining some average wave height for the random wave surface. Such a process can also be performed mathematically. Firstly, however, it is necessary to define what is meant by an individual wave in the random record, as each wave will have a different height and period.

There are a number of systematic ways in which individual waves can be defined for a measured wave profile. The most widely adopted approach is the Zero Up-crossing Method (Goda, 1985). Individual waves are defined by the region between successive crossings of the zero elevation by the water surface, η, in which the water surface changes from negative to positive. This process is shown schematically in Fig. 3.6.

A wave height, H_j, can then be defined for each of the individual waves (defined by the counter j) in the record. This wave height is defined as the vertical distance from the crest to the trough (see Fig. 3.6). As all waves are not of the same height, there will be a range or distribution of wave heights,

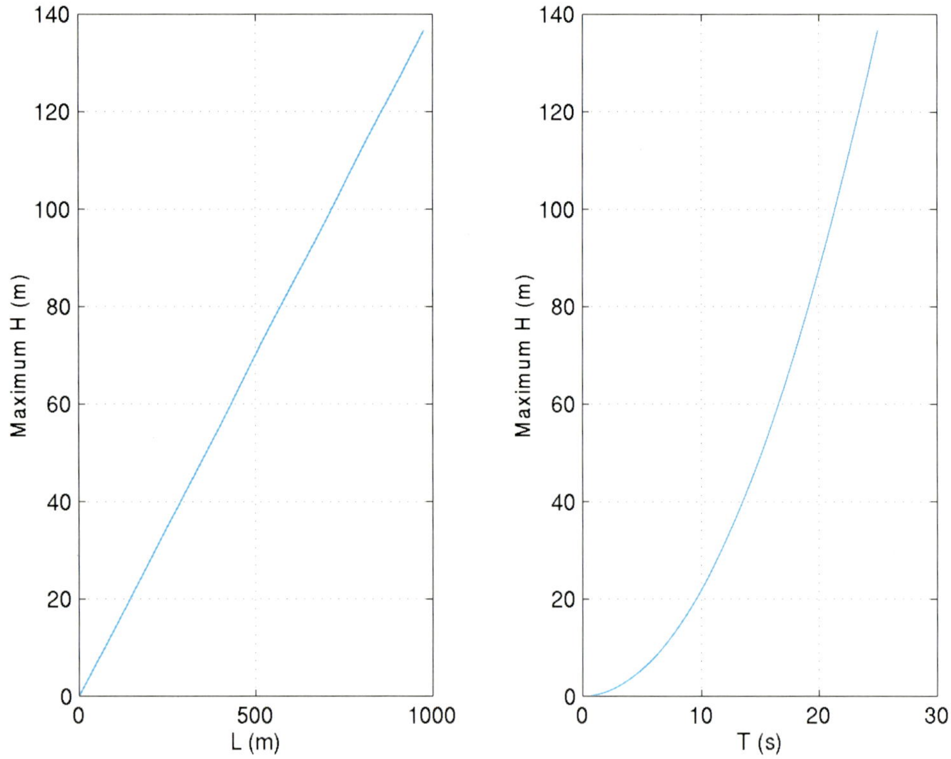

Fig. 3.3 The maximum limiting wave height, H_{max}, as a function of wave length, L, and wave period, T. These figures consider that the wave height is limited only by steepness.

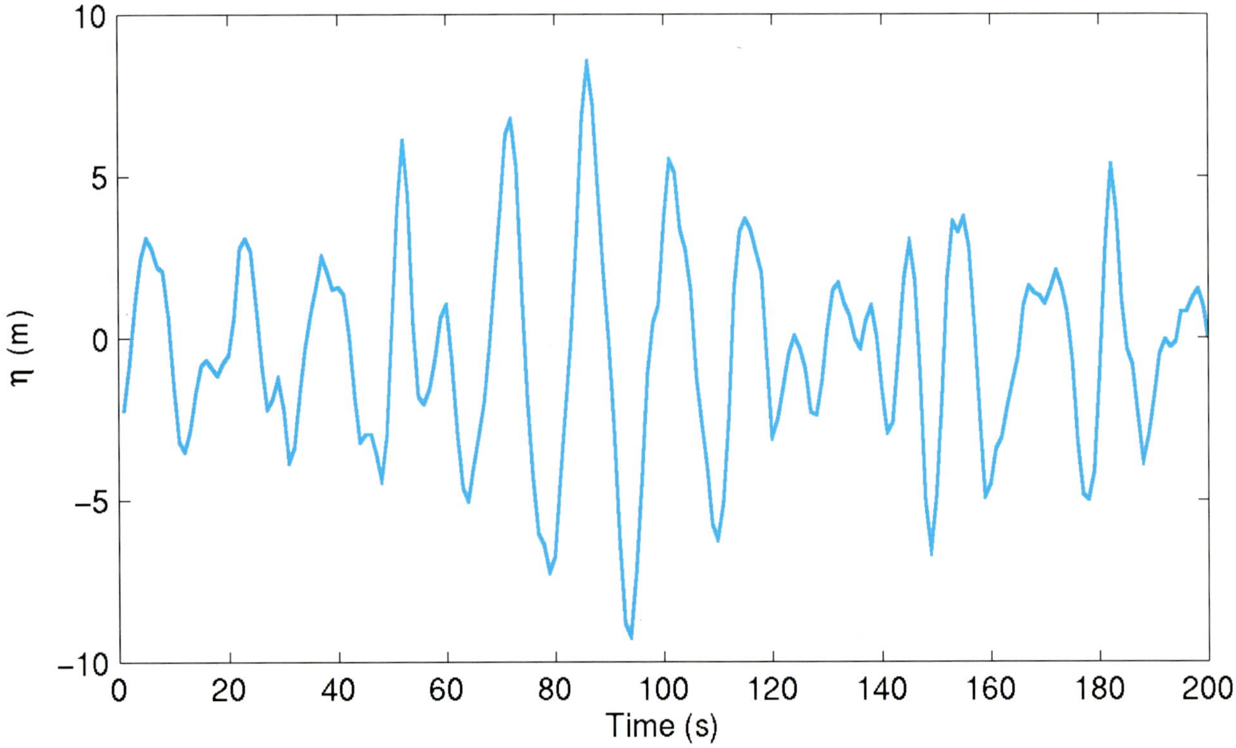

Fig. 3.4 An example of an open ocean wave record recorded at the Southern Ocean site of Macquarie Island ($-54°30'$S, $159°$E) on 13 May 1989.

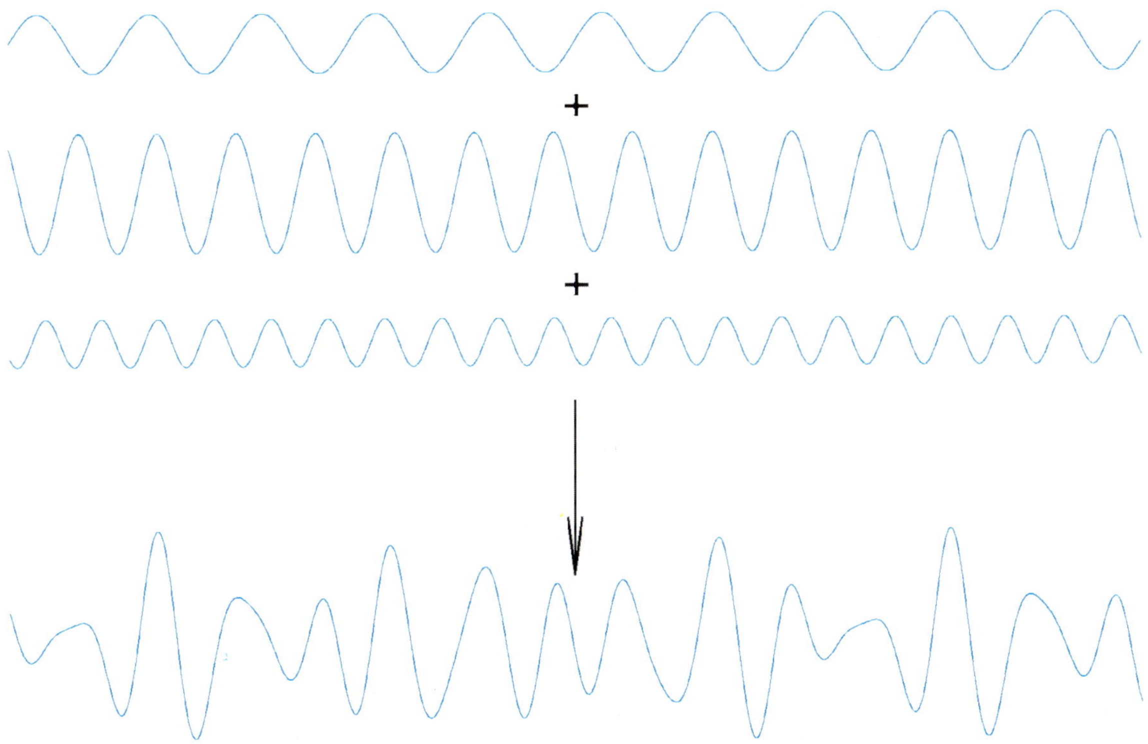

Fig. 3.5 The complex water surface in the above figure is created by the summation of the three simple sinusoidal (Fourier) components shown.

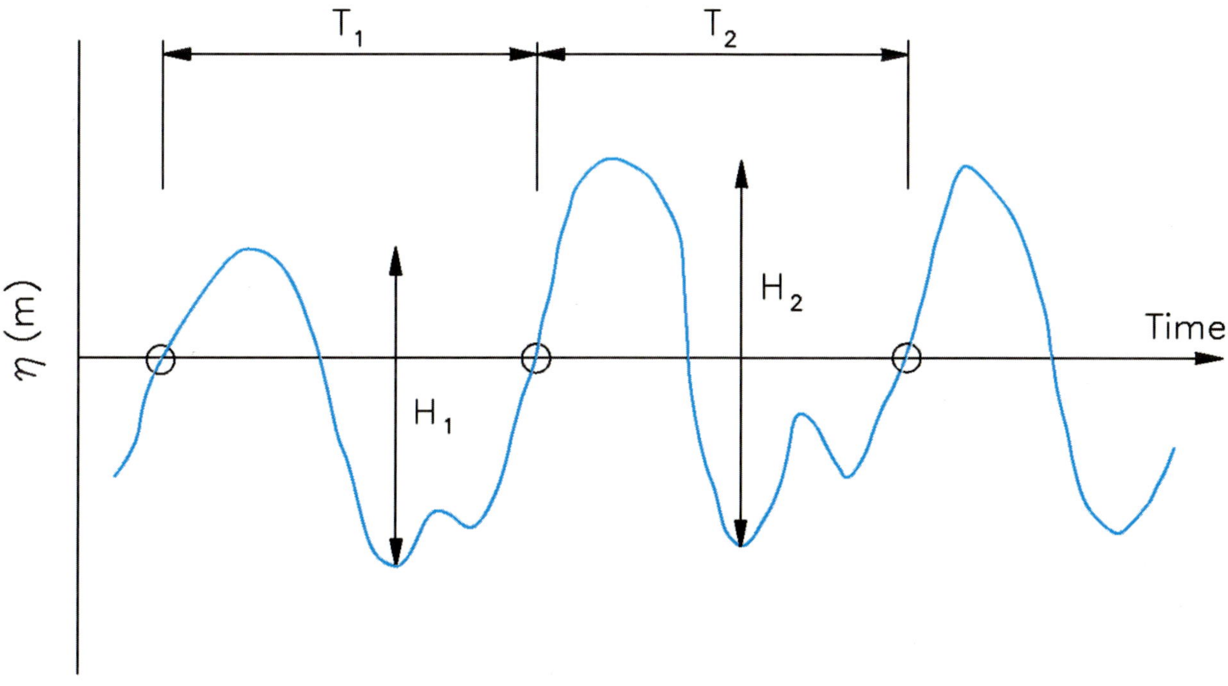

Fig. 3.6 Decomposition of a wave record into individual waves using the Zero Up-crossing Method. The period, T, and height, H, of the individually identified waves are shown.

H_j. The average of these wave heights, $\overline{H}$, represents one method of parameterizing the sea state of the random wave field in terms of a single value. For most practical purposes (e.g. engineering design, shipping), interest is centred on the larger waves in the record. Further, comparisons of visual observations of the wave height with measurements show that observers consistently estimate the wave height to be significantly larger than the mean. The observer's estimate is biased by the larger waves. For these reasons, a wave height parameter related to the larger waves in the record is commonly utilized, particularly for engineering purposes.

The wave heights can be sorted in descending order from the largest to the smallest. It has been found that visual estimates of the wave height are well represented by the average of the highest third of the waves in this list. This quantity is termed the significant wave height, H_s and will be utilized throughout this atlas.

Although the significant wave height is a convenient single parameter to characterize the wave field, it should be remembered that there will be individual waves in any record both larger and smaller than this value. Extensive measurements have shown that the variability in the heights of ocean waves can be well represented by the Rayleigh probability distribution (Goda, 1985). Adopting this distribution, it is possible to relate quantities such as those listed below to the significant wave height, H_s:

$$\overline{H} = 0.63H_s \tag{3.8}$$

$$H_{10} = 1.27H_s \tag{3.9}$$

$$H_1 = 1.67H_s \tag{3.10}$$

$$H_{0.2} = 1.91H_s \tag{3.11}$$

where $\overline{H}$ is the mean wave height, H_{10} is the average of the highest 10% of waves, H_1 is the average of the highest 1% of waves and $H_{0.2}$ is the average of the highest 0.2% of waves.

As the Rayleigh probability distribution has no upper bound, there is no way of estimating the maximum wave height for a given value of H_s. The result shown in Eq. (3.11), however, indicates that as a crude estimate, the maximum wave height, H_{max} will be approximately twice H_s.

$$H_{max} \approx 2H_s. \tag{3.12}$$

3.3 Dependence on Wind Speed and Fetch

The relationship between wind speed and wave height is obvious. Wave height is, however, not solely dependent on the wind speed. Other quantities, such as the duration or period for which the wind has blown and the distance or fetch of water over which the wind blows also influence the resulting wave height. The physical processes by which waves are generated by the wind are quite complicated and, despite considerable research, a full understanding is still lacking. It is beyond the scope of this Atlas to consider this work in detail. An understanding of the basic concepts is, however, important for the interpretation of the results presented in the body of the Atlas.

The simplest possible case of wave generation by the wind is termed Fetch Limited Growth. In Fetch Limited Growth, a wind of constant speed and direction blows perpendicular to a long and straight coastline as shown in Fig. 3.7. In addition, the wind is assumed to blow for a long enough period such that the duration of the wind has no influence on the resulting waves.

For this relatively simple situation, it has been observed that both the significant wave height, H_s and the wave period increase with distance from shore (termed the fetch, F). Insight into wave development in this situation was first obtained during the Joint North Sea Wave Project (JONSWAP, Hasselmann *et al.*, 1973). Based on data observed during JONSWAP, the following empirical relationships were developed.

$$\frac{gH_s}{U_{10}^2} = 1.6 \times 10^{-3} \left(\frac{gF}{U_{10}^2}\right)^{1/2} \tag{3.13}$$

$$\frac{gT}{U_{10}} = 2.857 \times 10^{-1} \left(\frac{gF}{U_{10}^2}\right)^{1/3} \tag{3.14}$$

where g is gravitational acceleration (g = 9.81 ms^{-2}), F is the fetch length (see Fig. 3.7), U_{10} is the wind speed measured at a reference height of 10 m and T is the wave period. These relationships are shown in non-dimensional form in Figs 3.8 and 3.9 and in dimensional form in Figs 3.10 and 3.11.

It is clear from Figs 3.8–3.11 that, in addition to the wind speed, the fetch also has a significant role in determining the resulting wave height. Equations (3.13) and (3.14) predict unbounded growth. An infinitely long fetch will yield infinitely large waves. Clearly, such a result is unrealistic and there must be some upper limit to growth where a further increase in the fetch has no influence on the resulting waves. Observations (Pierson and Moskowitz, 1964) confirm that such a limit exists. Known as the Pierson–Moskowitz limit, it is given by

$$\frac{gH_s}{U_{10}^2} = 2.433 \times 10^{-1} \tag{3.15}$$

$$\frac{gT}{U} = 8.134. \tag{3.16}$$

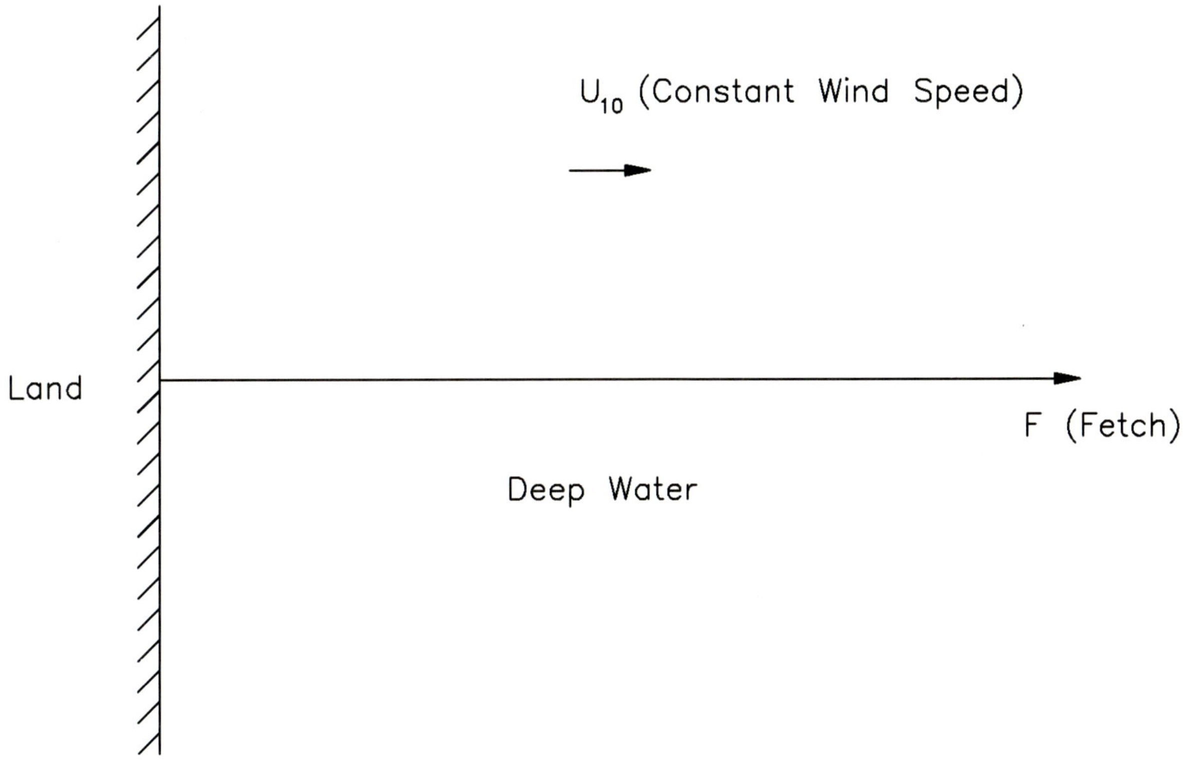

Fig. 3.7 A schematic representation of the idealized situation of fetch limited wave growth.

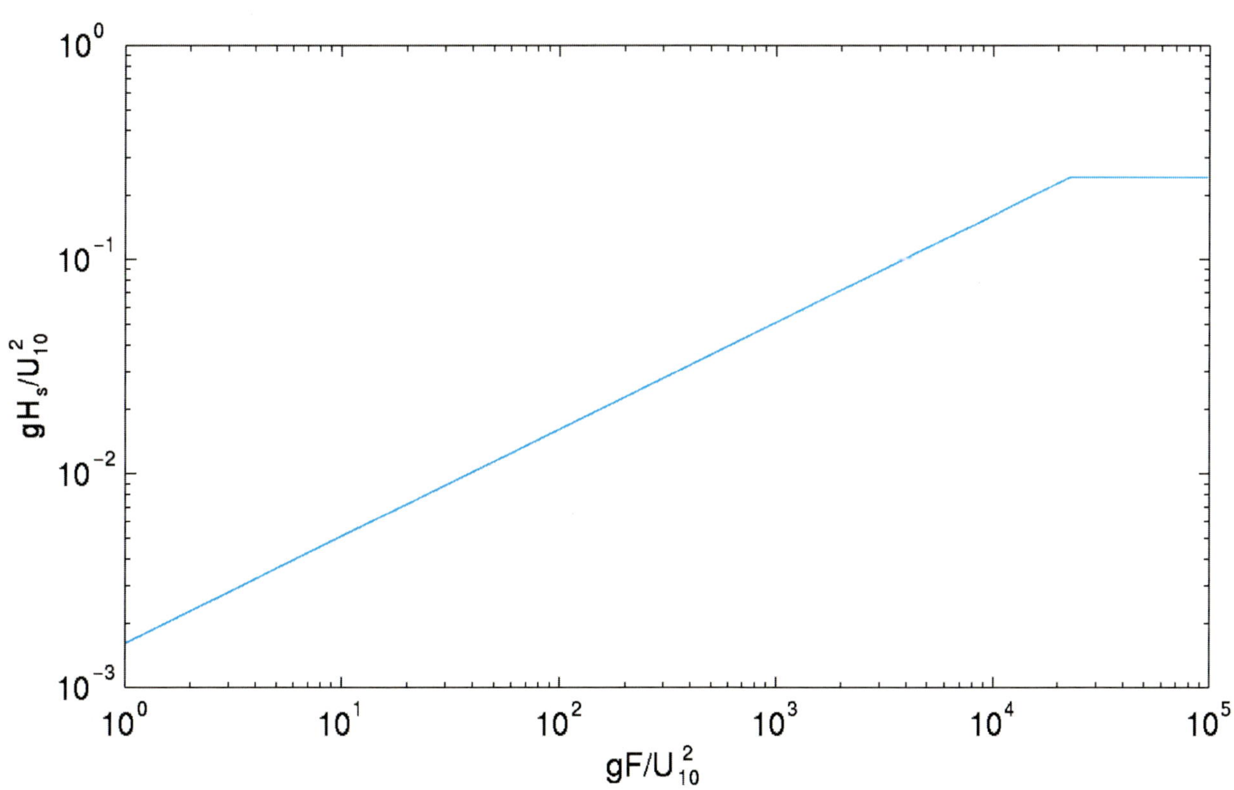

Fig. 3.8 Development of the non-dimensional wave height gH_s/U_{10}^2 as a function of non-dimensional fetch gF/U_{10}^2 as predicted by the JONSWAP relationship, Eq. (3.13). The Pierson–Moskowitz limit [Eq. (3.15)] is represented by the horizontal section at large non-dimensional fetch.

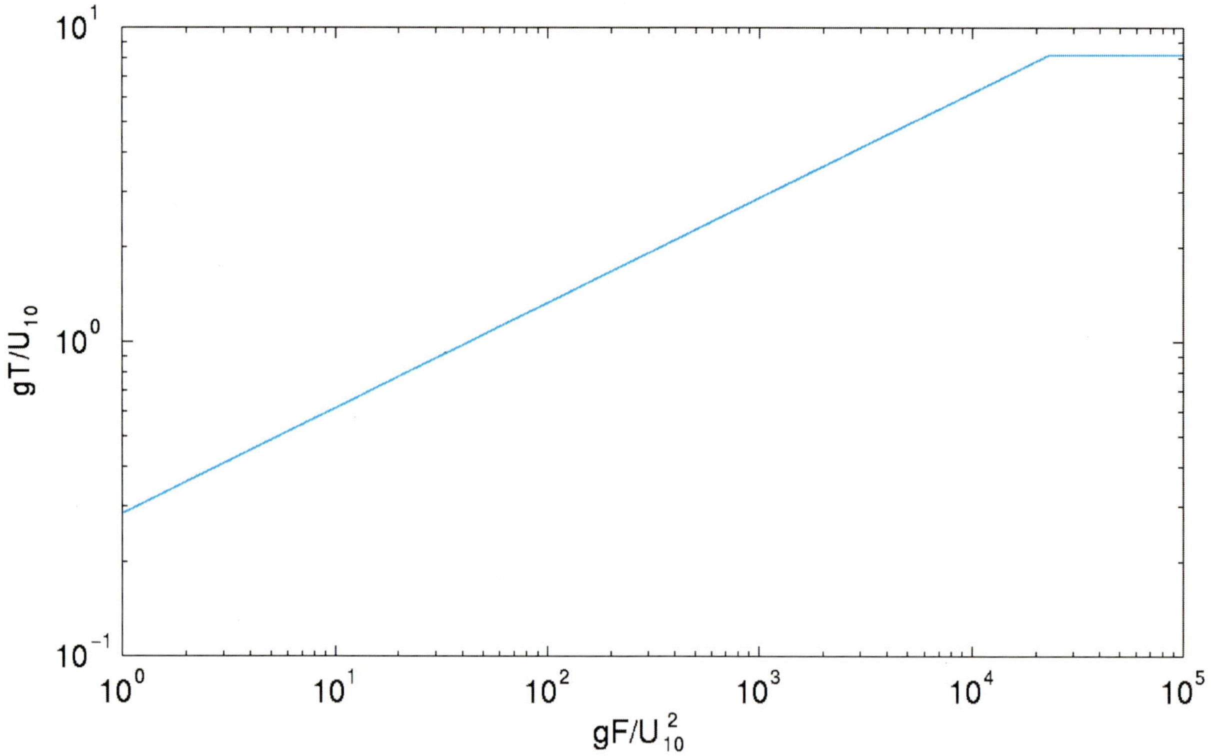

Fig. 3.9 Development of the non-dimensional wave period gT/U_{10} as a function of non-dimensional fetch gF/U_{10}^2 as predicted by the JONSWAP relationship, Eq. (3.14). The Pierson–Moskowitz limit [Eq. (3.16)] is represented by the horizontal section at large non-dimensional fetch.

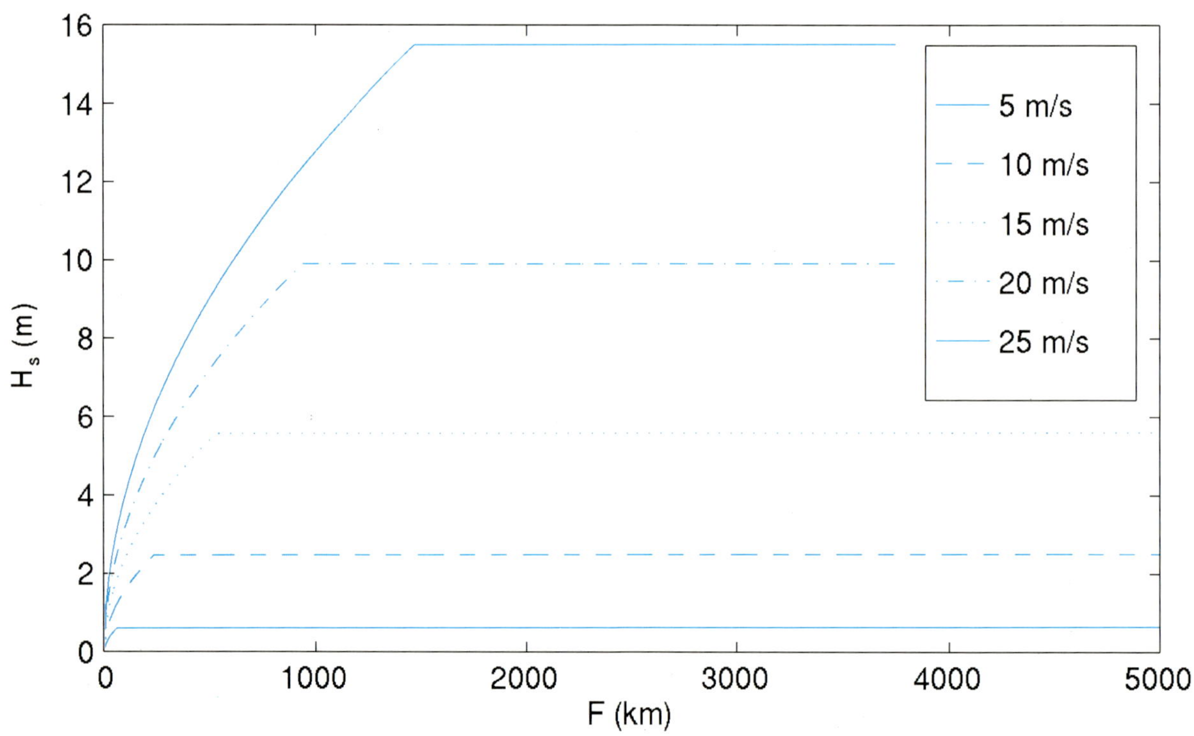

Fig. 3.10 Development of the significant wave height H_s as a function of fetch F as predicted by Eq. (3.13). The results for wind speeds, U_{10} of 5, 10, 15, 20 and 25 ms^{-1} are shown.

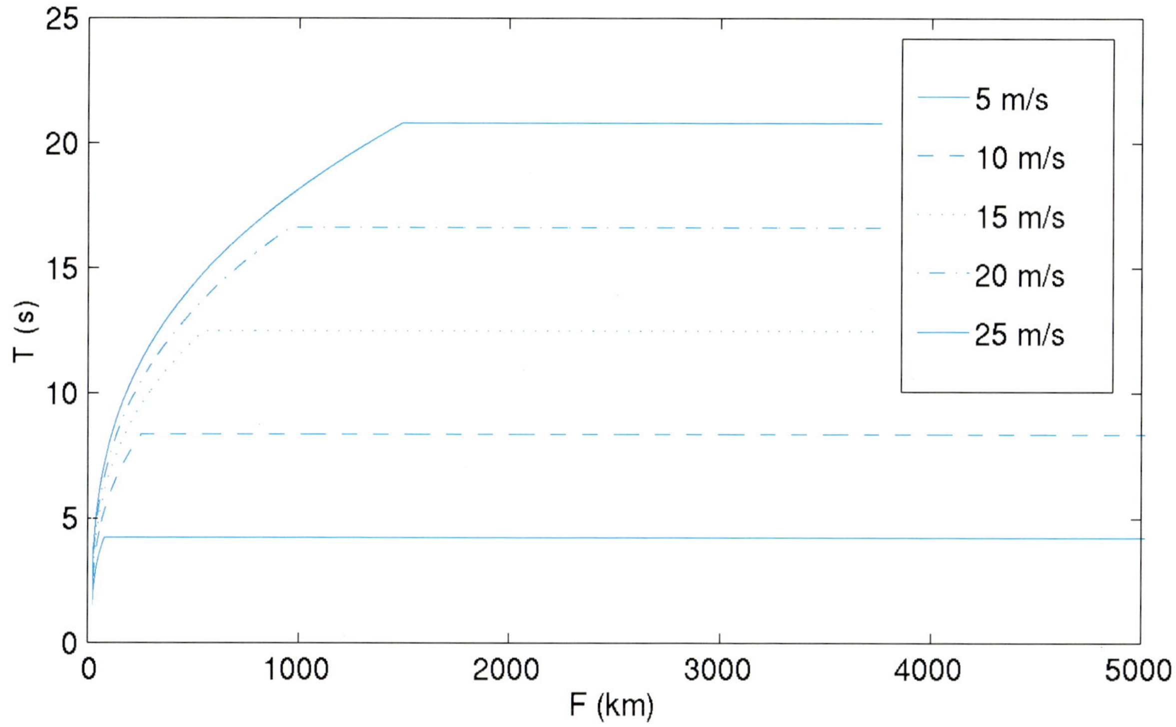

Fig. 3.11 Development of the wave period T as a function of fetch F as predicted by Eq. (3.14). The results for wind speeds, U_{10} of 5, 10, 15, 20 and 25 ms^{-1} are shown.

These relationships have been included in Figs 3.8–3.11 where they appear as the horizontal "plateau" regions.

The physical reason for the existence of such a limit can be explained in the following manner. As shown by Eqs (3.13) and (3.14), the increase in wave height is accompanied by an increase in the wave period. As the wave period increases, the speed at which the waves propagate (the phase speed, C) also increases as shown by Eq. (3.4). Eventually, the point will be reached where the wave phase speed is equal to the wind speed. As the waves are now propagating at the same speed as the wind, the relative wind speed sensed by the waves is zero. Consequently, the wind can transfer no energy to the waves and all growth ceases.

3.4 Wave Propagation and Swell

The description which appears above concerns actively wind generated waves. When the wind stops or decreases, these wind seas do not immediately decrease. Once generated by the wind, waves can propagate huge distances across the globe with very little decrease in their height. Indeed, waves generated in the region of New Zealand have been tracked as they propagated across the Pacific Ocean to finally break on the shores of Alaska (Snodgrass *et al.*, 1966). There was almost no measurable decay of these waves even over this enormous distance. Once waves cease to receive active energy input from the wind, they are termed swell. As illustrated above, swell decays very slowly and hence it is ubiquitous in the open oceans.

A wind generated sea can be described in terms of a summation of sinusoidal terms [Eq. (3.7)]. Each component in this summation will propagate at a phase speed given by Eq. (3.4). Hence, the longer period components will propagate at a faster speed than the shorter period components. This process means that as swell propagates away from its generation region, the waves will tend to disperse, based on their period. The longer period components will propagate fastest and hence propagate ahead of shorter period components. As a result of this dispersive property of the wave field, the swell at any point will be composed largely of only a single period component. Consequently, it will appear almost sinusoidal in form [see Eq. (3.1)]. In contrast to an active wind sea, which appears random in nature, swell is well ordered with each wave of similar height and period. Once generated by the wind, waves propagate along Great Circle paths across the globe.

Chapter 4

THE GEOSAT SATELLITE MISSION

4.1 The GEOSAT Data Coverage

Before the advent of satellite based measurement systems such as GEOSAT, the only reliable means for the acquisition of quality wind and wave data was through the use of *in situ* instruments. Visual observations from ships considerably increase the quantity and spatial coverage of data, but their reliability is highly questionable. As a result, available ocean data tend to be confined to the major continental shelves. Even here, however, the record is far from complete. The expense of deep water, long term oceanographic moorings means that such data are almost non-existent in the deep oceans.

The first significant satellite based oceanographic measurements were made in the 1970s using the GEOS 3 and SEASAT satellites. These measurements were, however, of only limited success. GEOS 3 provided relatively poor coverage and SEASAT operated successfully for only a few months.

The U.S. Navy satellite GEOSAT was launched in March 1985. For the first 18 months of its operation, the satellite performed a classified Geodetic Mission for the U.S. Navy. This mission was completed in November 1986 at which time the unclassified Exact Repeat Mission commenced. The satellite continued to operate in this mode until January 1990, when it ceased to function. The satellite began to malfunction, however, in early 1989 and there was a significant loss of data after March 1989. It is the data from this approximately three year period of the Exact Repeat Mission which has been used to derive the results in this Atlas.

In its Exact Repeat Mission, the GEOSAT satellite was placed in orbit at a height of approximately 800 km. The orbit was retrograde and inclined at 108°, covering latitudes between ±72°. The satellite repeated the same ground tracks to an accuracy of approximately 1 km every 17 days (termed a 17 day repeat cycle). The resulting ground track separation at the equator was 150 km, this separation decreasing towards the poles. The satellite orbit is shown schematically in Fig. 4.1 and the resulting ground track pattern for a typical 17 day repeat cycle is shown in Fig. 4.2. In Fig. 4.2, the ground track of the satellite is only shown where reliable ocean data were recorded.

GEOSAT was designed and built by the Johns Hopkins University Applied Physics Laboratory and employed a pencil beam nadir looking (directly below) altimeter to measure wind speed, significant wave height and sea surface height along its ground track. A sketch of the satellite appears in Fig. 4.3. The altimeter operated at a single frequency of 13.9 GHz and transmitted 1000 pulses per second. These pulses were averaged to produce data values at a frequency of one per second. The ground speed of the satellite was approximately 6.5 $km\,s^{-1}$ and hence one data value was obtained every 6.5 km along the ground track. Excluding land areas, this yields approximately 50,000 observations of both wind speed and significant wave height per day. Over the full duration of the Exact Repeat Mission this corresponds to approximately 45 million observations of these two quantities.

4.2 Wave Measurement Using the GEOSAT Radar Altimeter

As stated above, the altimeter transmitted 1000 pulses per second. These pulses radiate away from the satellite altimeter antenna as a spherical shell as shown in Fig. 4.4. The pulse reflects from the

Fig. 4.1 Schematic diagram showing the orbit of the GEOSAT satellite.

water surface and the return pulse is sampled by the satellite. Rather than the return pulse being continuously sampled, it is measured in discrete intervals called "range gates". Hence it is convenient to consider the pulse during each time increment corresponding to each range gate (Dobson and Monaldo, 1996). As shown in Fig. 4.4, during the first range gate, the pulse illuminates a small circular region directly below the satellite. In subsequent time increments, the surface is illuminated by a series of expanding annular rings. The area of each of these annular rings is constant.

If the ocean surface was perfectly flat, the return pulse would exhibit a rapid rise, corresponding to the return from the inner circular region. This would be followed by a flat section resulting from each of the constant area annular rings. If the ocean surface is roughened by the presence of waves, the leading edge of the transmitted pulse will interact with the crests of the waves a short time before intersecting the troughs. As a result, the leading edge of the return will be broadened in comparison to the flat ocean case. As the wave height increases, this broadening of the leading edge of the return pulse also increases. Therefore, the slope of the leading edge of the return pulse can be used as a measure of the wave height. An example of this process is shown in Fig. 4.5 (Dobson and Monaldo, 1996). Figure 4.5 shows the one second average return pulse for the case of a relatively calm sea (H_s = 0.36 m) and a rough sea (H_s = 7.47 m). The slope of the leading edge of the pulse clearly decreases as the significant wave height increases.

A typical example of a satellite ground track and the GEOSAT altimeter values of significant wave height along this ground track are shown in Fig. 4.6.

The ground track in Fig. 4.6 is shown as a smooth line with the significant wave height and wind speed as the shaded region adjacent to the ground track. The perpendicular distance from the ground track to the top of the shaded region is proportional to either H_s or U_{10}. The case shown in Fig. 4.6 occurred on 25 February 1989 at 17.8 hrs UT. The pass is south going over the Indian Ocean. The satellite crosses the Asian coast in the Bay of Bengal and tracks across the central Indian Ocean before crossing the Antarctic ice sheet. At approximately 20°S the satellite passes directly over Tropical Cyclone Hanitra which at the time had a central pressure of 950 HPa. The altimeter recorded a maximum significant wave height of 10.9 m and a maximum wind speed of 42 m s^{-1} as it passed over the tropical cyclone. In addition, the rapid fall off in both wind speed and wave height as the ground track passed through the calm eye of the storm can be seen in the figure. Significant storm events can also be seen both to the north and south of Hanitra. In particular, the event at approximately 35°S has wind speeds of approximately 20 m s^{-1} and significant wave heights of 5.8 m. As the satellite tracks

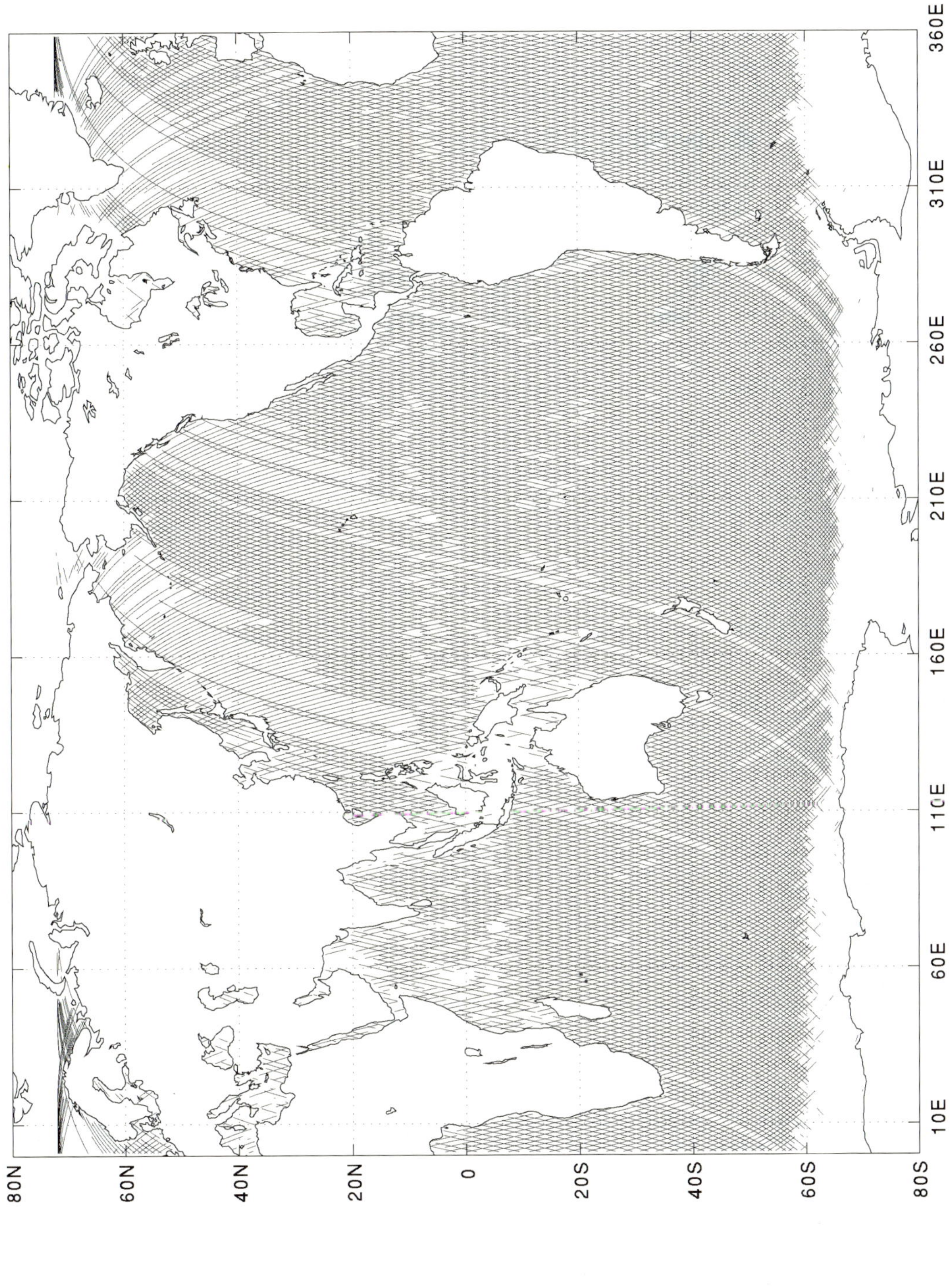

Fig. 4.2 Ground track pattern for the 17 day repeat cycle of the GEOSAT satellite between 8 November 1986 and 24 November 1986.

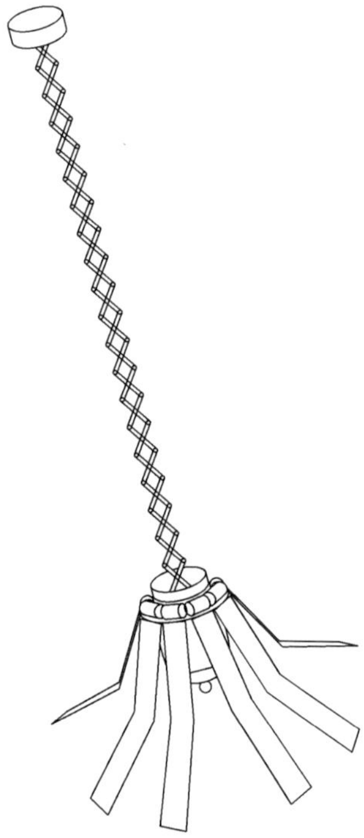

Fig. 4.3 Sketch of the GEOSAT satellite.

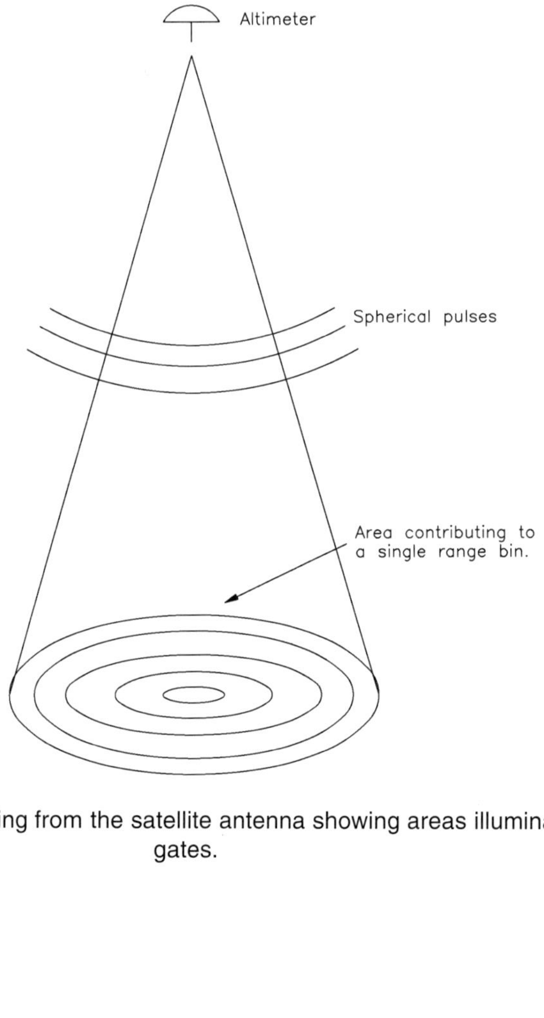

Fig. 4.4 Spherical altimeter pulses radiating from the satellite antenna showing areas illuminated in successive range gates.

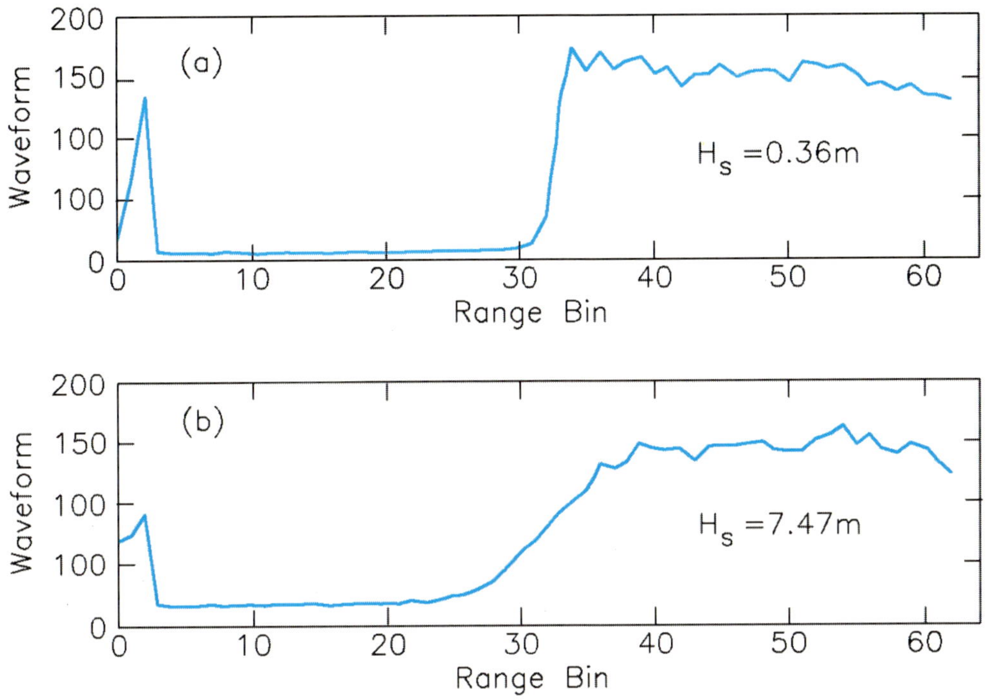

Fig. 4.5 Examples of the one second average return pulse from the GEOSAT altimeter for (a) $H_s = 0.36$ m and (b) $H_s = 7.47$ m [after Dobson and Monaldo, 1996].

over the Southern Ocean it reveals persistent wind speeds of 15 m s^{-1} and resulting significant wave heights of 5 m over a large area, consistent with the year-round extreme conditions at high latitudes.

4.3 Wave Height Measurement Validation

Although the quantity of data obtained from an orbiting platform is a number of orders of magnitude larger than could be obtained from *in situ* instruments such as buoys, the accuracy of the data must be established. The obvious means of obtaining such a validation is through the comparison of GEOSAT values of H_s with corresponding measurements made from buoys. There are, however, a number of practical limitations with this approach. Firstly, the overflight of the satellite rarely corresponds exactly with either the measurement times or locations of buoys. There is almost always a temporal and spatial mismatch between paired GEOSAT and buoy observations. The potential error introduced by such a mismatch will obviously decrease as the mismatch decreases. A quantitative estimate of the magnitude of such errors is, however, difficult to determine. There are also inherent differences in the manners in which the significant wave height is determined from the altimeter and buoy. The altimeter measures a spatial average of the approximately 6 km diameter radar foot print. In contrast, buoy values of significant wave height are determined from time series, generally of a length of approximately 30 min, obtained at a single point (see Chapter 3). Buoy data will also have some instrumentation and sampling errors. Young (1986) has shown that *in situ* buoy measurements of significant wave height can have errors of the order of $\pm 10\%$, simply due to the length of the finite time series from which H_s is derived.

The pre-launch specifications for significant wave height measurements with the GEOSAT altimeter were for an accuracy of ± 0.5 m or 10% of the significant wave height, which ever was greater. A number of validation experiments have shown the altimeter has obtained this goal. Dobson *et al.* (1987) have compared GEOSAT altimeter derived values of H_s with values from U.S. National Ocean Data Center (NODC) buoys. They limited data to cases where there was a spatial mismatch of less than 50 km and a temporal mismatch of less than 30 min. With these colocation limitations, the data set was reduced to 116 comparisons obtained from 43 buoys. Their comparison is shown in Fig. 4.7 along with the pre-launch specifications. The rms (root mean square) difference for their data is 0.49 m and the mean difference 0.36 m, indicating the GEOSAT H_s values are lower than the buoy data.

Based on a larger data set, again obtained from

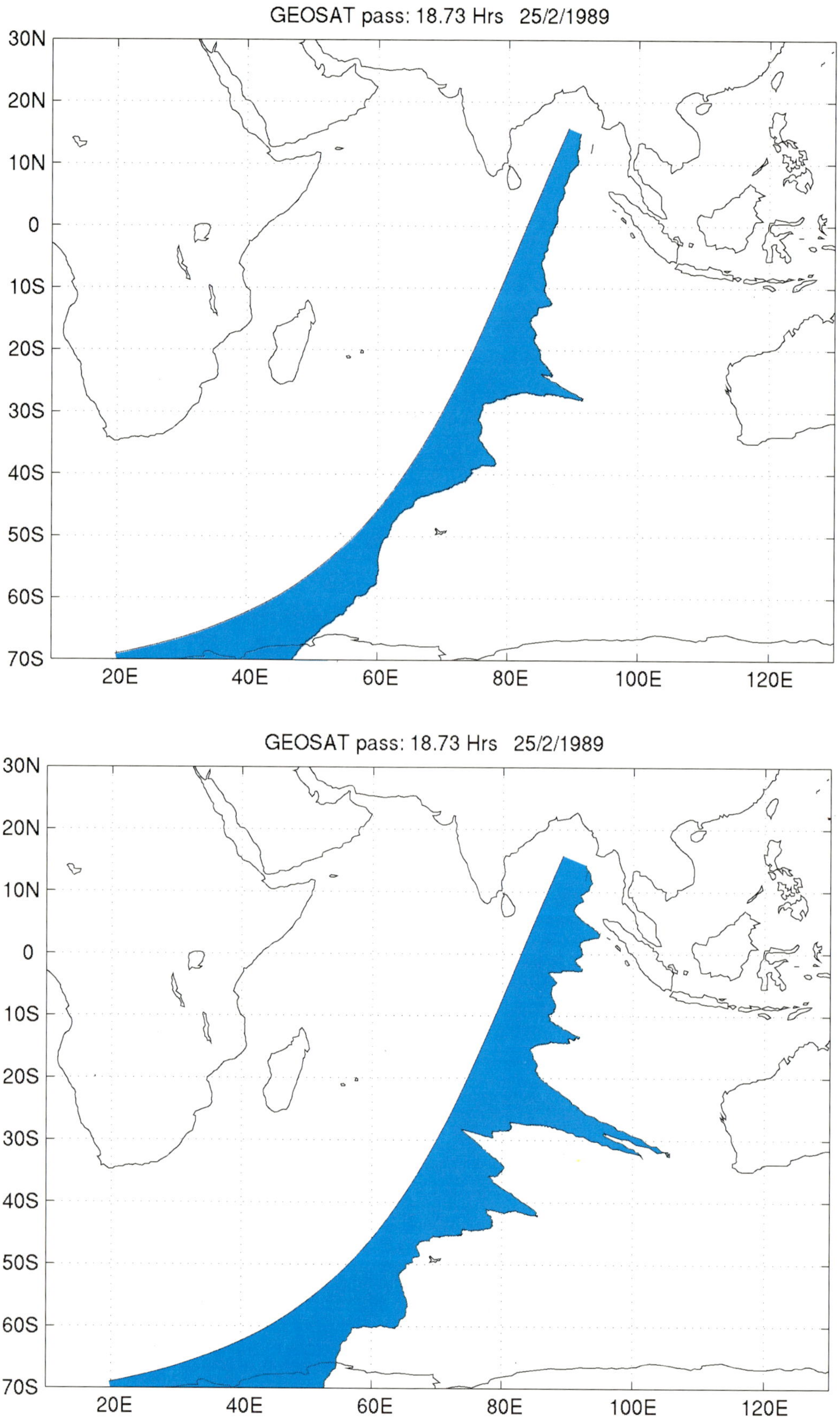

Fig. 4.6 A GEOSAT ground track over the Indian Ocean recorded on 25 February 1989 at 18.7 hrs UT. Also shown are (a) the variation in significant wave height along the track and (b) the variation in wind speed along the track.

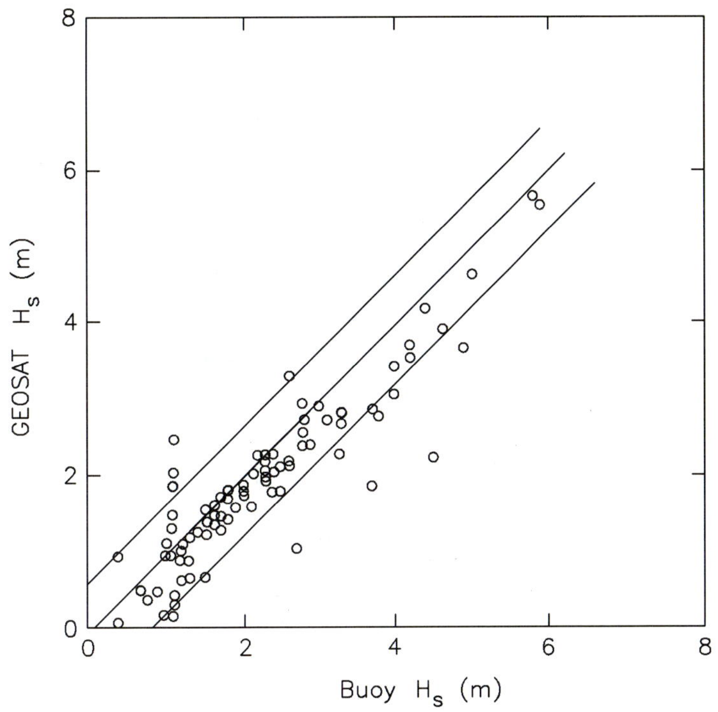

Fig. 4.7 Comparison between GEOSAT derived values of significant wave height and NODC buoy measurements as presented by Dobson *et al.* (1987). The pre-launch specifications are shown by the solid lines.

NODC buoys, Glazman and Pilorz (1990) found that GEOSAT measurements were 0.4 m smaller than buoy measurements. Further analysis of this same data set by Carter *et al.* (1992) showed that GEOSAT measurements were 13% lower than buoy data.

Guillaume and Mognard (1992) have compared GEOSAT values of H_s with the METRO FRANCE VAG numerical wave model (Guillaume, 1987). Since a model was used, a significantly larger data set could be obtained than for buoy comparisons. A total of 2260 data points were obtained in the range $H_s = 1$–9 m, with 84 greater than 5 m. Data at high sea states had not previously been available from the buoy comparisons. Again, they show that GEOSAT yields values lower than the wave model, the error increasing with wave height. Their mean error is shown as a function of significant wave height in Fig. 4.8. They find that the difference becomes significant above $H_s = 5$ m, where a difference as large as 24% exists. Although these results are consistent with the buoy comparisons, they are indirect and questions as to the accuracy of the numerical model must be considered, particularly at high sea states.

In conclusion, GEOSAT derived values of H_s are within the pre-launch specifications of ± 0.5 m or 10%. There is, however, evidence that GEOSAT values may, in the mean, be lower than buoy data.

4.4 Wind Measurement Using the GEOSAT Radar Altimeter

The determination of the significant wave height utilizes only the slope of the leading edge of the return pulse. The actual strength of the return pulse is not utilized. The normalized radar cross section, σ_0 (the strength of the return of the near-nadir altimeter transmission) can, however, be used to determine wind speed. The radar cross section depends upon the amount of energy returned by the sea surface from the transmitted radar pulse. This returned energy will obviously depend on the physical properties of the surface, particularly the roughness. The smoother the ocean surface, the more energy is returned to the satellite. In contrast, a rough surface scatters energy from the facets on the surface and the resulting returned energy or σ_0 is decreased.

As indicated in Chapter 3, the wave height is not only a function of wind speed, but also depends on quantities such as fetch and duration. It is however, not the large energy containing waves which constitute the bulk of the roughness elements which cause the radar scattering. The very short wavelength components (centimetre scale) are very steep and make the major contribution to the mean squared surface slope (mss). In contrast to the large energy containing waves, the short wave length components are almost always in equi-

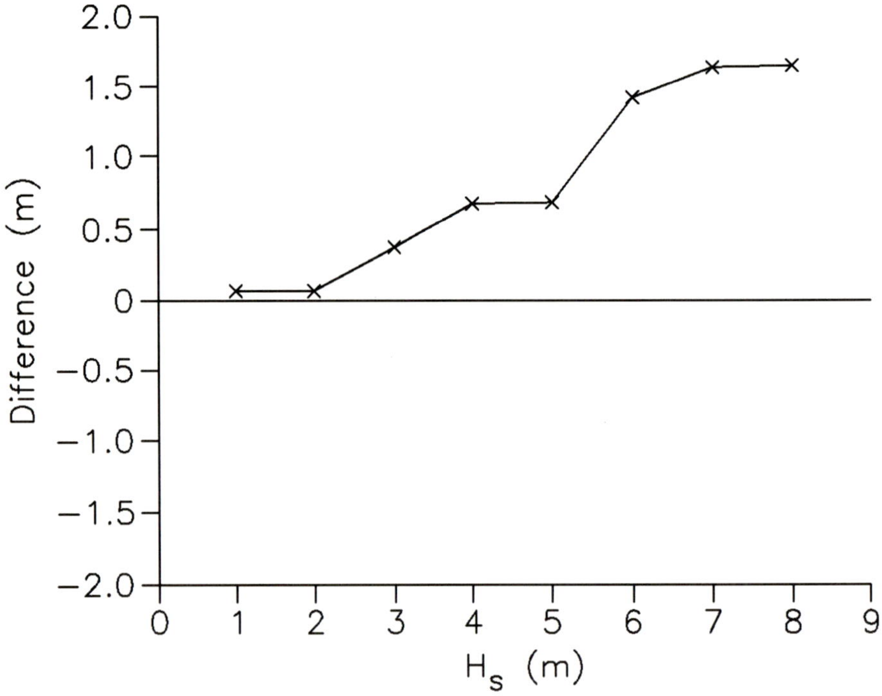

Fig. 4.8 Mean differences in H_s between the VAG numerical wave model and GEOSAT as a function of significant wave height.

librium with the local wind. Hence it could be expected that these short wavelength components would be strongly related to the local wind. Therefore, the radar cross section could be expected to be inversely related to the wind speed. As the wind increases, the surface roughness increases, causing enhanced radar scattering and a reduction in the radar cross section.

Clearly the relationship between wind speed and radar cross section is much more indirect than, for instance, the relationship between the significant wave height and the slope of the leading edge of the return pulse. Hence, the accuracy of wind speed measurements from the altimeter is generally poorer than corresponding measurements of significant wave height. In addition, there is considerably more controversy as to which, of a number of proposed relations between wind speed and radar cross section, is the most appropriate.

4.5 Wind Speed Measurement Validation and Choice of Algorithm

There is a wide variety of algorithms which have been proposed for the determination of wind speed from the radar cross section. The major contributions are discussed below.

4.5.1 Brown Algorithm

Brown *et al.* (1981) compared a total of 184 values of wind speed derived from buoy measurements with values of GEOS 3 σ_0. The data were in the range from 1 to 18 m s^{-1}. The resulting algorithm, commonly called the Brown three-branch algorithm, is

$$U_{10} = \begin{cases} \sum_{n=1}^{5} a_n W^n & W \leq 16\text{m s}^{-1} \\ W & W > 16\text{m s}^{-1} \end{cases} \qquad (4.1)$$

where

$a_1 = 2.087799$, $a_2 = -0.3649928$,
$a_3 = 4.062421 \times 10^{-2}$, $a_4 = -1.904952 \times 10^{-3}$,
$a_5 = 3.288189 \times 10^{-5}$,

and

$$W = \exp[(10^{-(0.21+\sigma_0/10)} - B)/A] \qquad (4.2)$$

with σ_0 being the radar cross section expressed in decibels and

$A = 0.080774$, $B = -0.124651$: $\sigma_0 < 10.12\text{dB}$
$A = 0.039893$, $B = -0.031996$: $10.12\text{dB} \leq \sigma_0 < 10.9\text{dB}$
$A = 0.01595$, $B = 0.017215$: $\sigma_0 \geq 10.9\text{dB}$

Because the GEOS 3 and SEASAT missions overlapped, Fedor and Brown (1982) attempted to cross-calibrate the altimeters for the two satellites. They compared 14 simultaneous (within one hour) measurements from the two platforms and found that within the approximate wind speed range from 1 to 10 $\mathrm{m\,s^{-1}}$, the SEASAT altimeter was 1.6 dB higher than the GEOS 3 altimeter. Hence they concluded that the Brown *et al.* (1981) algorithm could also be applied to SEASAT, provided the SEASAT values of σ_0 were firstly reduced by 1.6 dB. This was then confirmed by comparison of SEASAT altimeter derived winds with a total of 87 buoy measurements in the range 1–12 $\mathrm{m\,s^{-1}}$.

4.5.2 Chelton and McCabe Algorithm

Chelton and McCabe (1985) examined the full three months of altimeter data from SEASAT and represented it as a temporal average on a 2° × 6° grid. They concluded that the discontinuities in the Brown three-branch algorithm lead to an unrealistic multimode probability distribution of wind speeds. To formulate a smooth function, they compared spatial and temporal averages from SEASAT altimeter and scatterometer. Although the regions of the sea surface imaged by altimeter and scatterometer differ by approximately 200 km, Chelton and McCabe (1985) argued that the averaging process employed would ensure that errors due to this spatial separation would have zero mean. A total of 1947 averaged values were considered in the wind speed range 3–14 $\mathrm{m\,s^{-1}}$. The resulting algorithm was

$$U_{10} = 0.943 \times 10^{[(\sigma_0/10 - G)/H]} \tag{4.3}$$

where $G = 1.502$, $H = -0.468$, σ_0 has units of decibels, and the factor of 0.943 has been included here to convert the original algorithm from $U_{19.5}$.

Chelton and McCabe (1985) noted that there was an indication in their data that Eq. (4.3) overestimated U_{10} at high winds (greater than 12 $\mathrm{m\,s^{-1}}$) and that although Eq. (4.3) and Eq. (4.1) agreed well over the range of wind speeds for which data were available, they diverged significantly at high wind speeds.

4.5.3 Goldhirsh and Dobson Algorithm

In order to develop an algorithm for use with GEOSAT until more detailed calibration data became available, Goldhirsh and Dobson (1985) reviewed the algorithms proposed by both Brown *et al.* (1981) and Chelton and McCabe (1985). Rather than adopt the Chelton and McCabe (1985) algorithm, which is based on indirect measurements of wind speed from scatterometer, they smoothed the Brown three-branch algorithm with a fifth-order polynomial.

$$U_{10} = \sum_{n=0}^{5} a_n \sigma_0^n \tag{4.4}$$

where $a_0 = -15.383$, $a_1 = 16.077$, $a_2 = -2.305$, $a_3 = 9.896 \times 10^{-2}$, $a_4 = 1.8 \times 10^{-4}$, $a_5 = -6.414 \times 10^{-5}$, and σ_0 has units of decibels. The relationship is valid only over the range from 2 to 18 $\mathrm{m\,s^{-1}}$ for which the polynomial fit was applied.

4.5.4 Chelton and Wentz Algorithm

Chelton and Wentz (1986) pointed out that there were a number of weaknesses in the wind speed algorithm developed for SEASAT by Chelton and McCabe (1985). Principally, the scatterometer wind speed algorithm was erroneous. They reanalyzed the SEASAT altimeter and scatterometer data and developed a new algorithm based on data in the wind speed range from 0 to 21 $\mathrm{m\,s^{-1}}$. Their result was expressed in a tabular form and is reproduced in Fig. 4.9.

4.5.5 Witter and Chelton Algorithm

Witter and Chelton (1991) attempted to cross-calibrate SEASAT and GEOSAT altimeter data in order to develop a wind speed algorithm for GEOSAT. As noted earlier, there was no overlap between SEASAT and GEOSAT. Hence this cross-calibration was achieved by matching the probability distributions from the two instruments. Witter and Chelton (1991) argued that this process was valid since variations in such distributions between years would be minor. The results were again presented in a tabular form (reproduced in Fig. 4.9) covering the wind speed range between 0 and 20 $\mathrm{m\,s^{-1}}$.

4.5.6 Comparison of Existing Algorithms

Figure 4.9 shows a comparison of the various algorithms described in the preceding sections. There is reasonable agreement for wind speeds ranging from 3 to 15 $\mathrm{m\,s^{-1}}$, the region for which the bulk of the data exists upon which the algorithms were based. At higher wind speeds the various algorithms diverge widely. The fact that many of the algorithms have been developed using indirect techniques such as scatterometer winds and the comparison of wind statistics between years clearly demonstrates the difficulty of obtaining such data, even at these relatively low wind speeds. Recent independent comparisons by Ebuchi *et al.* (1992)

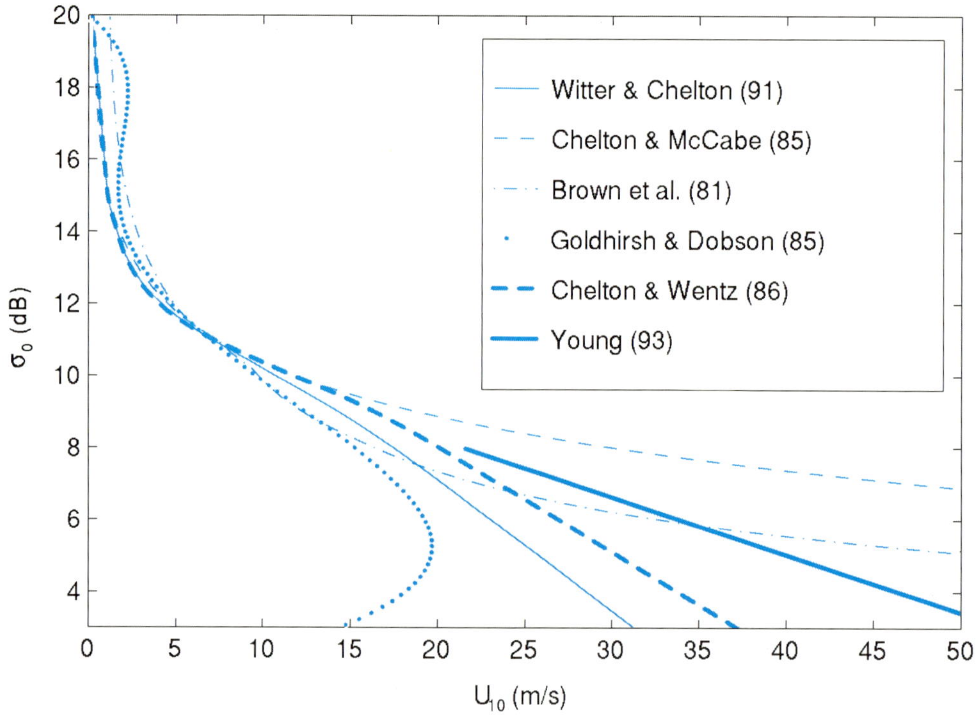

Fig. 4.9 A comparison between the various algorithms which have been proposed for the determination of wind speed, U_{10} from the altimeter radar cross section, σ_0.

and Guillaume and Mognard (1992), however, confirm the performance of these algorithms at low to moderate wind speeds.

The possibility of obtaining buoy data to extend these relationships to higher wind speeds is most unlikely. The instrumentation problems under such extreme conditions are considerable. In addition, it is necessary to impose tight spatial and temporal tolerances between buoy and satellite measurements to avoid the introduction of significant errors, as highlighted by Brown *et al.* (1981). Hence progress must rely on indirect means to determine the wind speeds for these severe conditions.

4.5.7 Young's Algorithm for High Wind Speed

Although the various altimeter wind speed algorithms differ for wind speeds between 1 and 20 m s^{-1} (Fig. 4.9), these differences are small compared to the differences which develop when they are applied to higher wind speeds. This is hardly surprising since they are all empirical fits to data obtained for wind speeds less than 20 m s^{-1}. As there are almost no data available at higher wind speeds, Young (1993) applied an indirect method to estimate the relationship between wind speed and altimeter radar cross section at higher wind speeds.

From the full three year GEOSAT data base, Young (1993) selected all passes of the satellite which tracked directly over the centres of mature tropical cyclones (a total of five passes). Although there were no direct *in situ* measurements of wind speed in these storms, the well-formed structure of these meteorological systems means that it is possible to predict the wind speed along the respective satellite ground tracks with reasonable accuracy. Young (1993) applied the tropical cyclone model of Holland (1980) to estimate the wind field for each of the storms. The model predicted wind speeds in the range 5–45 m s^{-1}. The relationship between these predicted values of wind speed and the GEOSAT altimeter radar cross section was consistent with the existing algorithms at low wind speed. This provided confidence that the Holland (1980) model yielded a reasonable estimate of the tropical cyclone wind fields. Although the data scatter increased at higher wind speeds, there was a clear relationship between wind speed and radar cross section for winds as high as 40 m s^{-1}. For wind speeds in the range 20–40 m s^{-1}, it was proposed that the following algorithm be adopted

$$U_{10} = -6.4\sigma_0 + 72 \tag{4.5}$$

where U_{10} has units of m s^{-1} and σ_0 units of decibels.

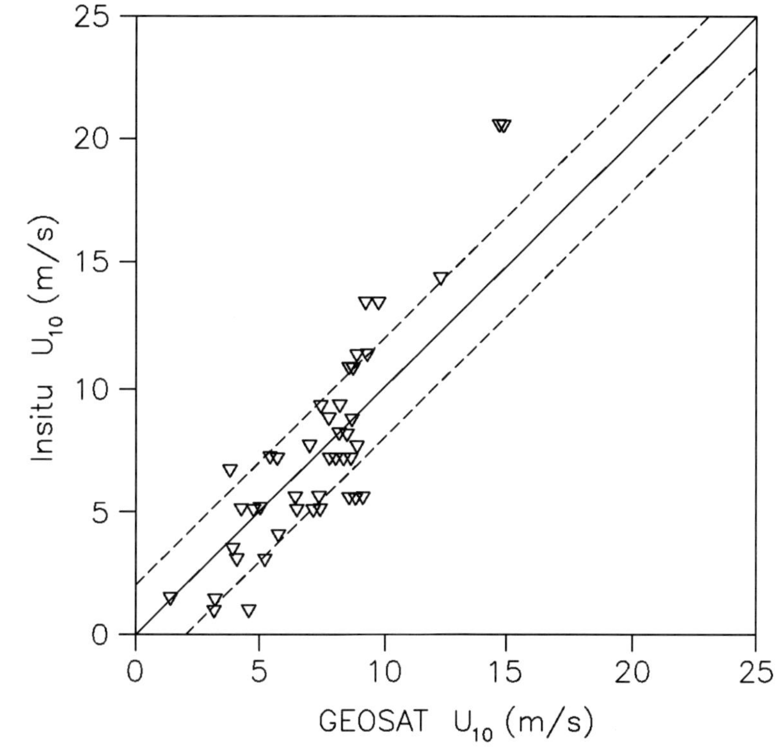

Fig. 4.10 Comparison of GEOSAT derived values of wind speed with *in situ* data from the North Sea. The dashed lines represent a ± 2 m s^{-1} deviation, after Tournadre and Ezraty (1990).

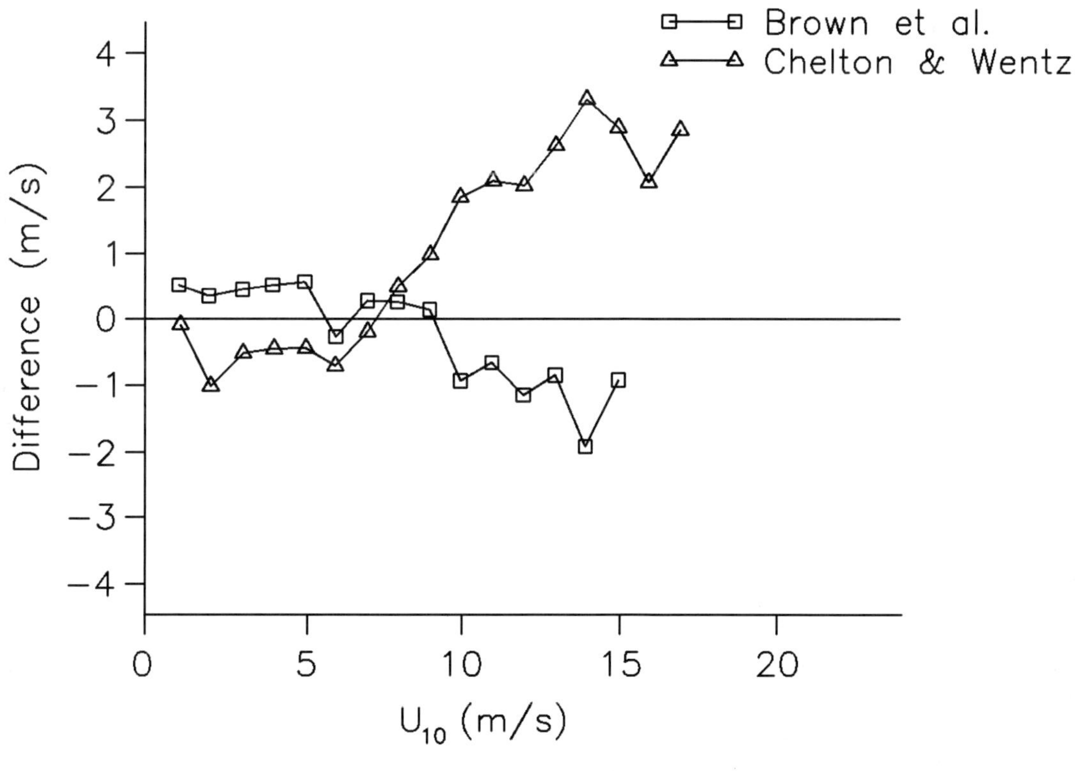

Fig. 4.11 Mean wind speed differences between the algorithms of Brown *et al.* (1981) and Chelton and Wentz (1986) and ECMWF as a function of wind speed.

4.5.8 Wind Speed Validation

Potential differences between buoy and GEOSAT derived values of significant wave height have been discussed above. A major difficulty arises due to the mismatch in space and time between the measurement systems. Similar problems exist in the validation of GEOSAT wind speed estimates against buoy data. Wind fields, however, are far less homogeneous than wave fields and hence potential differences due to the lack of colocation are greater. Rapid spatial and temporal variations in wind fields are quite common. An example is frontal systems. In such cases, wind speed measurements made at two relatively close locations can be quite different. Waves are generated by winds, and once generated, they propagate away from their generation source. As a result, wave fields are far more homogeneous than wind fields. Rapid changes in significant wave height similar to atmospheric fronts do not occur. Despite these potential difficulties, a number of attempts have been made to validate GEOSAT wind speed measurements against *in situ* buoy data. Tournadre and Ezraty (1990) compared GEOSAT estimates of wind speed obtained by Dobson *et al.* (1987) using a smoothed version of the Brown algorithm (Eq. 4.1) with data from the North Sea. Figure 4.10 shows the resulting comparison for colocation distances of less than 50 km.

Their data yield a mean deviation of 0.10 $m\,s^{-1}$ and a rms deviation of 1.92 $m\,s^{-1}$ for data with colocation distances less than 50 km. As expected, the difference increases with increasing colocation distance with an rms difference of 2.63 $m\,s^{-1}$ at 200 km. These values are consistent with the comparisons of Dobson *et al.* (1987) (mean—0.1 $m\,s^{-1}$, rms—1.9 $m\,s^{-1}$) and Shuhy *et al.* (1987) (rms—1.54 $m\,s^{-1}$).

Guillaume and Mognard (1992) compared the algorithms of Brown *et al.* (1981) and Chelton and Wentz (1986) with the wind analysis of the European Centre for Medium Range Weather Forecasts (ECMWF) atmospheric model. Although this is not a direct comparison, it greatly enhances the quantity of well colocated data for comparison. They find that the differences between the model results and both algorithms increase with wind speed. The Chelton and Wentz (1986) algorithm produced larger values than the model, and the Brown *et al.* (1981) algorithm smaller values. (see Fig. 4.11).

Although not evaluated by Guillaume and Mognard (1992), the algorithm of Witter and Chelton (1991) would produce wind speed values between those of Brown *et al.* (1981) and Chelton and Wentz (1986) for winds in the range 10–20 $m\,s^{-1}$ (see Fig. 4.9). As this is the region for which the algorithms deviate from the ECMWF predictions, it could be expected that the Witter and Chelton (1991) algorithm would yield a better agreement with the model data.

Although conclusive validations of GEOSAT derived winds, from which estimates of their accuracy can be obtained are lacking, the evidence suggests that in the wind speed range 0–15 $m\,s^{-1}$, wind speeds can be estimated with an accuracy of approximately ± 2 $m\,s^{-1}$. At wind speeds above 15 $m\,s^{-1}$, the error increases. Examination of Fig. 4.9 shows that the relationship between wind speed and radar cross section becomes progressively flatter with increasing wind speed. Hence, relatively small errors in σ_0 translate into large errors in U_{10}. This is confirmed by the high wind speed data of Young (1993). The scatter in these data indicates that at wind speeds above 20 $m\,s^{-1}$, the accuracy is of order $\pm 20\%$.

For the wind speed values used in this atlas, the algorithm of Chelton and Wentz (1986) has been used for values of radar cross section less than 8 dB. For radar cross section greater than 8 dB the algorithm of Young (1993) has been adopted.

Chapter 5

DATA PREPARATION

5.1 GEOSAT Data Format

The raw data utilized for the production of the atlas consisted of Geophysical Data Records (GDRs), distributed by the U.S. National Oceanographic and Atmospheric Administration (NOAA) (Cheney *et al.*, 1987). The raw GDR data consist of records recorded at a frequency of approximately one per second. Each record contains a large number of parameters, including:

- Measurement date and time
- Measurement position (latitude, longitude)
- Satellite height
- Sea surface height
- Significant wave height, H_s. Each second, 10 values of significant wave height are determined. These values are averaged to produce the value reported in the GDRs
- The standard deviation of the 10 values used to calculate H_s (σ_{H_s})
- The radar cross section, σ_0
- A flag indicating whether the satellite is over land or water according to a $1/12^\circ$ mask
- Satellite attitude, θ_s. The off-nadir orientation angle of the altimeter

5.2 Data Quality Control

Although a number of data quality checks were performed on-board the satellite, data contained in the GDRs are far from error free. Such errors occur for a number of reasons (Challenor *et al.*, 1990).

1. To produce reliable results, the altimeter return signal must be centred in the tracking loop window of the satellite.
2. The return signal should not be distorted by returns from ships, sea ice, exposed reefs, small islands and other objects, or attenuated by rain cells.
3. After a period of bad returns, data may continue to be faulty until the altimeter return has been reacquired correctly and centred in the tracking loop window.
4. Similar errors to (3) can occur when the satellite commences measurements again after it crosses a coastline.

Such errors are readily detected by visual inspection of the data (as demonstrated by Cheney *et al.*, 1987). Examples of typical errors are shown in Figs 5.1–5.4.

Figure 5.1 shows a descending pass of the satellite which occurred on 17 November 1986 at 21.6 hrs UT. The satellite track passes over North America crossing the California coast. The satellite ground track is shown as the smooth line crossing the coast and the significant wave height, H_s, as the shaded region adjacent to the ground track. The height of the shaded region measured perpendicular to the ground track is proportional to H_s. As the satellite passes from land to water it takes a finite period of time for the altimeter return pulse to be reacquired. The result is the large spike in H_s near the coastline. Once the signal is reacquired, the values of H_s become realistic. They show an increase in H_s as the satellite tracks offshore, the significant wave height reaching a value of approximately 1.5 m offshore.

Figure 5.2 shows an ascending pass recorded on 13 July 1988 at 13.1 hrs UT. The pass is over the Southern Ocean to the south of Australia. The most southern section of the pass shown in Fig. 5.2 is seawards of the Antarctic continent. As this pass occurs during the Southern Hemisphere Winter, however, there will be a significant region of sea ice in this area. This is clearly seen by the area of erroneously high values of H_s south of 65° S. Towards the edge of the sea ice the satellite obviously passes over a region of relatively clear water and the values of H_s momentarily drop to realistic values of approximately 2 m. As the satellite continues to track north it finally enters open water and records stable values of H_s of approximately 5 m.

Figure 5.3 shows a descending pass recorded on 18 July 1988 at 21.7 hrs UT. The pass is over the Coral Sea to the north-east of Australia. This is an area where there are numerous small islands and isolated reefs. At the northern end of the pass shown in Fig. 5.3, values of H_s are low at approximately 0.5 m, consistent with most data from the relatively calm equatorial regions. As the ground track passes over the region of the Solomon Islands to the east of New Guinea, large "spikes" are recorded in the record of significant wave height. Once free of this Island group the altimeter records increased wave heights of approximately 2 m as it moves south into regions where strong south-easterly trade winds would exist at this time of year. The recorded values of H_s are not, however, particularly stable. Large isolated "spikes" occur as the ground track passes over isolated islands and partially submerged reefs. Finally, before it passes over the Australian coastline another region of erroneously large data is recorded. This is not a "land" error as shown in Fig. 5.1, the erroneous data commences approximately 100 km seaward of the coastline. These data errors are caused by passage of the satellite ground track over the partially submerged reefs of the Great Barrier Reef.

Figure 5.4 shows the ground track and derived values of wind speed, U_{10}, which occurred for an ascending pass over the Southern Ocean on 18 July 1988 at 13.8 hrs UT. The data shown here begin at approximately 64° S, at the edge of the sea ice as shown in Fig. 5.2 which occurred over the same area but one week earlier. Wind speeds at the southern end of the pass are approximately 12 m s^{-1} and gradually increase to 20 m s^{-1} by latitude 55° S. At this point the satellite attitude begins to exceed one degree ($\theta_s > 1°$), the radar cross section, σ_0, drops and hence the wind speed erroneously increases. The nature of the erroneous increase in wind speed due to the attitude error is significantly different to the wave height errors shown in Figs 5.1–5.3. Attitude errors are not manifested as data spikes but as a section of erroneously high wind speeds. In this case U_{10} gradually increases from 20 m s^{-1} to as high as 70 m s^{-1}.

Clearly, the visual inspection of the full three year data set is not a feasible task. An automated scheme utilizing the following steps was developed to ensure data quality.

1. Each value of H_s was constructed from the average of 10 individual observations, taken at a rate of 10 per second. If any of these 10 values were anomalous, it will be evident in the standard deviation of the 10 values, σ_{H_s}. Only records for which

$$\frac{\sigma_{H_s}}{H_s} < 0.1 \tag{5.1}$$

were retained.

2. The land/water flag was checked. Any records which were over land according to the 1/12° mask used for this flag were discarded.

3. The data were sub-divided into blocks of 50 records each. Care was taken to ensure the block represented consecutive points, unbroken by passage over land. Close to land, where a continuous 50 point block did not exist, shorter blocks were utilized. The mean significant wave height of the block, $\overline{H}_s$ (block) and the standard deviation of the 50 values of H_s making up the block, σ_{H_s} (block) were formed.

4. The entire 50 records making up the block were discarded if

$$\frac{\sigma_{H_s}(\text{block})}{\overline{H}_s(\text{block})} > 0.5 \tag{5.2}$$

If the standard deviation of the values making up the block was larger than the value indicated by Eq. (5.2), it generally indicated that there was a number of data spikes and hence the entire block was discarded.

5. Individual records within the block were discarded if

$$\frac{|H_s - \overline{H}_s(\text{block})|}{\sigma_{H_s}(\text{block})} > 3 \tag{5.3}$$

Equation (5.3) flags individual points within the block whose significant wave height, H_s, deviate from the mean of the block by more than three standard deviations.

6. It was found that wind speed estimates, in particular, were very unreliable if the satellite attitude, θ_s, exceeds 1° (see Fig. 5.4). Such records were discarded.

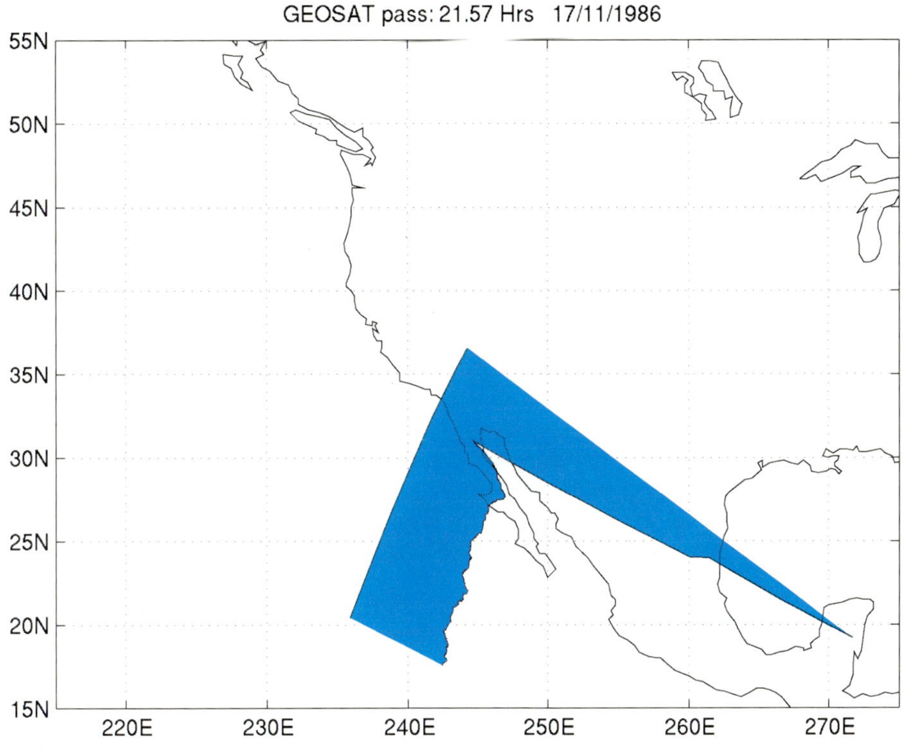

Fig. 5.1 Variation in the significant wave height along the "south going" GEOSAT ground track recorded on 17 November 1986 at 21.6 hrs UT. Note the data "spike" which occurs on the transition from land to water.

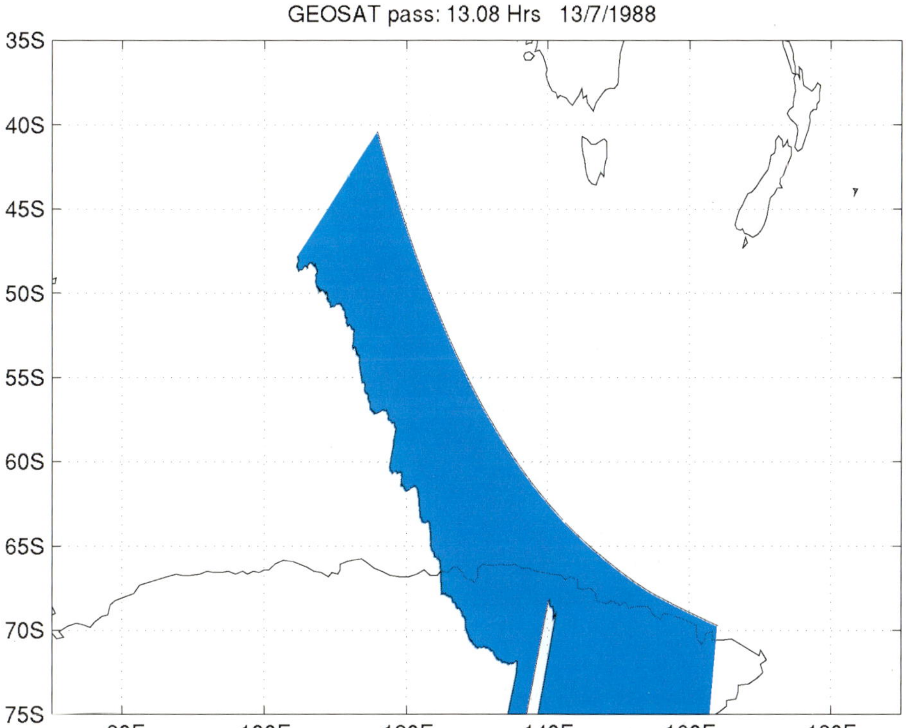

Fig. 5.2 Variation in the significant wave height along the "north going" GEOSAT ground track recorded on 13 July 1988 at 13.1 hrs UT. Note the data errors which occur as the satellite passes over the sea ice near the Antarctic coastline.

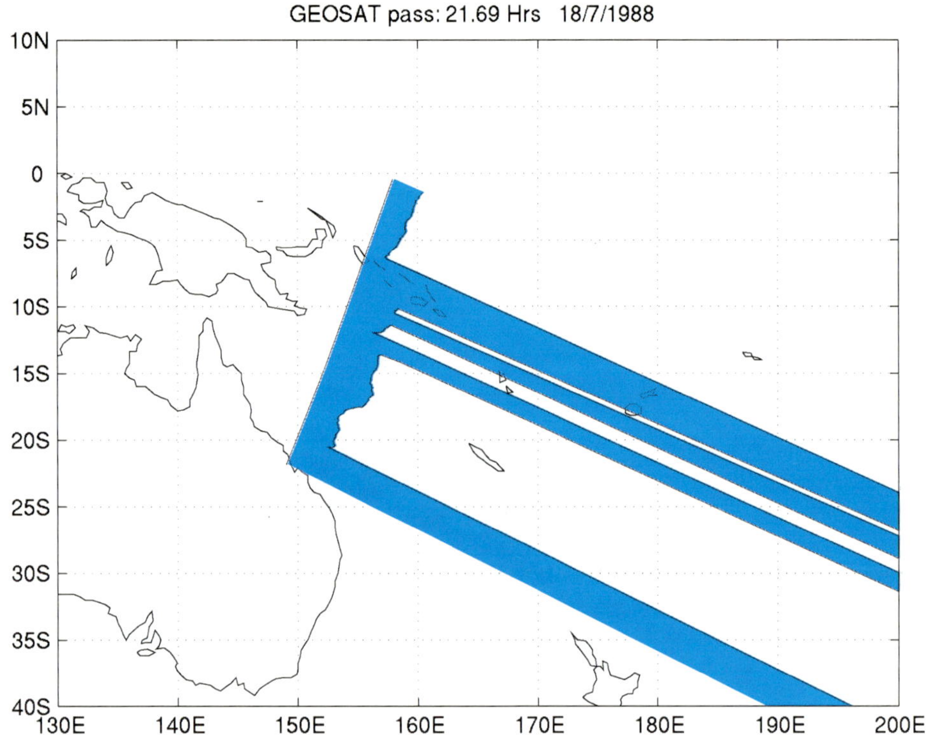

Fig. 5.3 Variation in the significant wave height along the "south going" GEOSAT ground track recorded on 18 July 1988 at 21.7 hrs UT. Note the data "spikes" which occur as the satellite passes over isolated islands and reefs in the Coral Sea.

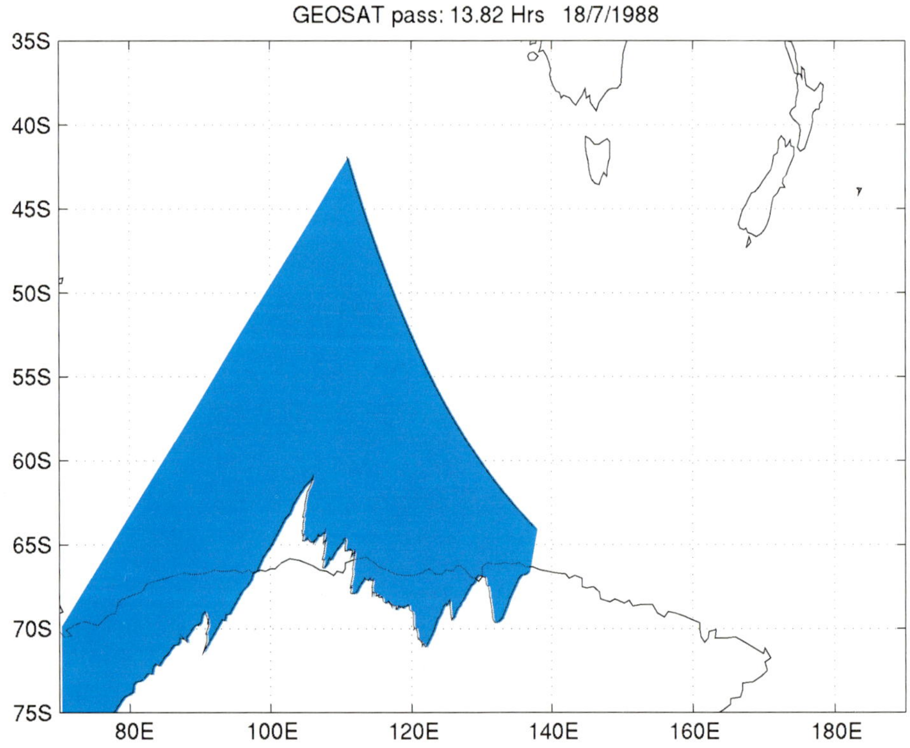

Fig. 5.4 Variation in the wind speed along the "north going" GEOSAT ground track recorded on 18 July 1988 at 13.8 hrs UT. Note the erroneously large values towards the northern extreme of the pass which occur due to an unacceptably large satellite attitude angle.

7. Finally, records containing highly improbable or unrealistic data were discarded. Records with data satisfying the following inequalities were discarded:

$$H_s > 20 \text{ m} \tag{5.4}$$

$$U_{10} > 40 \text{ ms}^{-1} \tag{5.5}$$

$$\sigma_0 < 0 \text{ dB} \tag{5.6}$$

Relatively large quantities of data were visually inspected to ensure the quality control process described above produced reliable data without unnecessarily discarding good data. The process proved highly reliable. As an example of the reliability of the quality control process, the same four examples shown in Figs 5.1–5.4 are shown in Figs 5.5–5.8 after checking by the quality control process. In all cases, the erroneous data has been successfully identified and removed. Taken over the full three year GEOSAT data set, the quality control process discarded approximately 12% of the data as erroneous.

5.3 Partitioning of Data

In order to construct global wind and wave statistics from the quality controlled data set it is necessary to divide the globe using a mesh. GEOSAT data can then be allocated to the appropriate grid element. In order to obtain maximum resolution, this discrete grid should be as fine as possible. If the grid is made too fine, however, there will be relatively few passes of the satellite ground track through each element of the grid and hence the resulting statistics for each grid element will be unreliable. A number of different grid resolutions were investigated. Finally, a $4^\circ \times 4^\circ$ resolution mesh was adopted. The grid extended over latitudes of $\pm70^\circ$ and over the full longitude range of 0°–360°. Hence, the full grid consisted of 90×36 values, a total of 3240 elements.

An indication of the potential reliability of the statistical results which will be presented in the later sections of the Atlas can be obtained by an examination of the number of satellite ground track passes through each of the grid squares. Figure 5.9 shows a contour plot of the total number of ground tracks through each $4^\circ \times 4^\circ$ grid square. The average number of passes through each grid square during the approximately three year duration of the GEOSAT mission is 230.

As shown in Fig. 5.9, the global coverage of the data is not uniform. The polar regions are affected by the presence of sea ice during their respective winters, thus reducing the available data compared to areas at lower latitudes. Available data also decreases close to land due to data corruption from spurious radar returns (see Section 5.2). This is true, not only near the coastlines of continents but also in areas where there are numerous small islands and reefs. An example of this is the region of the Greater Sunda Islands which make up Malaysia, Indonesia and New Guinea where the data density is very low (50–100 passes per grid square). There is also a significant fall off in data density in enclosed seas such the Mediterranean Sea and the Gulf of Mexico. The spatial grid resolution of $4^\circ \times 4^\circ$ is clearly larger than optimal for such regions.

Even in the deep oceans, far away from land, the data density is not constant. There is a clear banding along lines of constant latitude apparent in Fig. 5.9. This is caused by the ground track pattern and the incomplete data return for the near polar orbit of GEOSAT. Figure 5.10 shows the ground track pattern for repeat cycle 37 of the satellite (12 July 1988–28 July 1988). A repeat cycle is a 17 day period, after which the satellite retraces the ground tracks. The repeated passage of north and south "going" orbits, traces out a diamond shaped pattern on the Earth's surface. The north and south "going" passes cross (cross-over points) along lines of approximately constant latitude. This is more clearly apparent in the expanded region of the North Pacific Ocean shown in Fig. 5.11. Figure 5.11 also shows the $4^\circ \times 4^\circ$ mesh which was utilized. As shown in Fig. 5.10, data return from the satellite was not perfect. In any repeat cycle there were regions of the globe which received poorer coverage than others. Due to the ground track pattern these occur as the "diamond" shaped patches in Fig. 5.10. As can be seen, they tend to fall along lines of approximately constant latitude. For instance, note the areas of poor data return along latitudes $+20^\circ$, -20°, $+55^\circ$ and -55°. Towards the end of the GEOSAT mission, satellite performance declined significantly and the frequency of such data gaps increased. When averaged over the whole mission, these data losses yield the banded and intermittent pattern shown in Fig. 5.9.

Data density also varies on a seasonal basis. Figures 5.12–5.15 show the global data density for each month of the year. Again, values are cumulative for the full GEOSAT mission. For example, all passes which occurred during the month of April for the three years have been summed to form the total passes for the month of April shown in Fig. 5.13. The average number of passes for each grid square and each month are summarized in Table 5.1.

Figures 5.12–5.15 clearly show the seasonal variability of the sea ice extent around Antarctica. Data are available much closer to the Antarctic con-

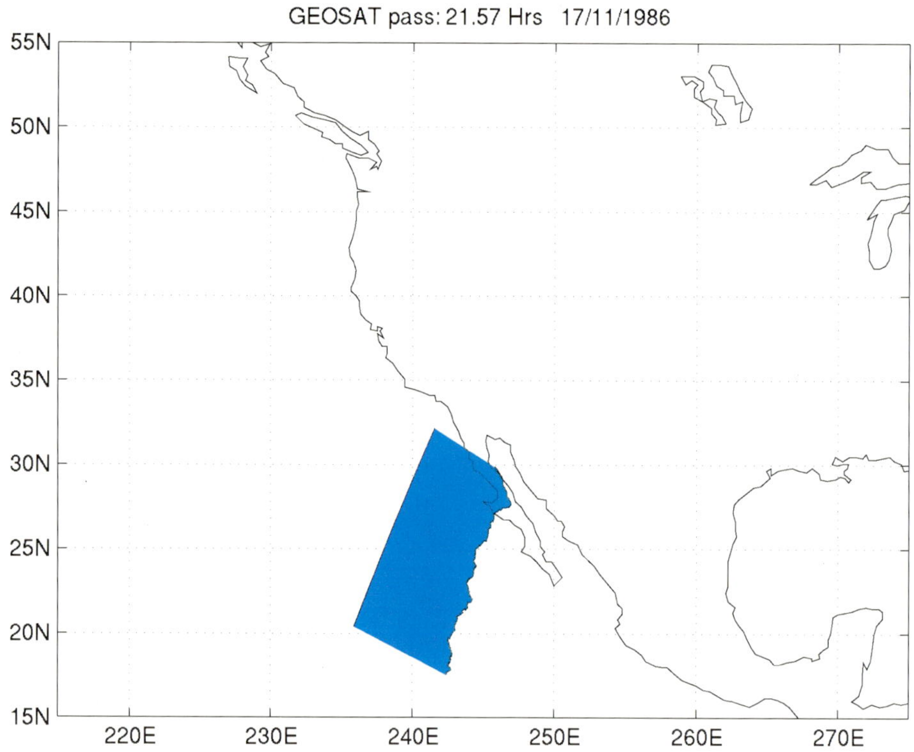

Fig. 5.5 Variation in the significant wave height along the GEOSAT pass shown in Fig. 5.1. The automatic quality control process has removed the land transition errors which existed in the original data.

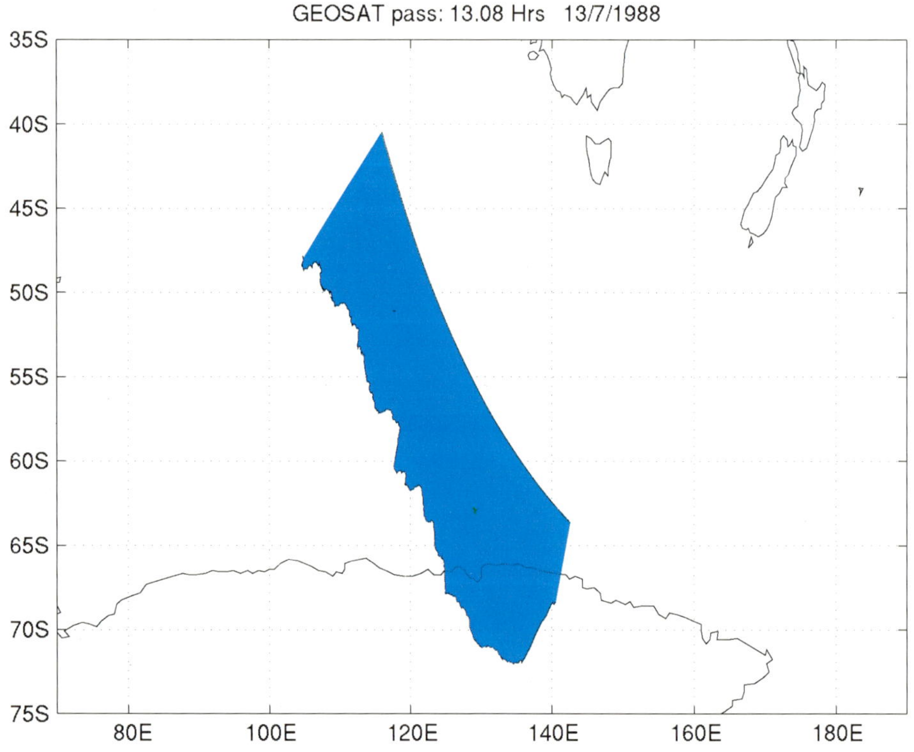

Fig. 5.6 Variation in the significant wave height along the GEOSAT pass shown in Fig. 5.2. The automatic quality control process has removed the data errors caused by the passage of the satellite over the Antarctic sea ice.

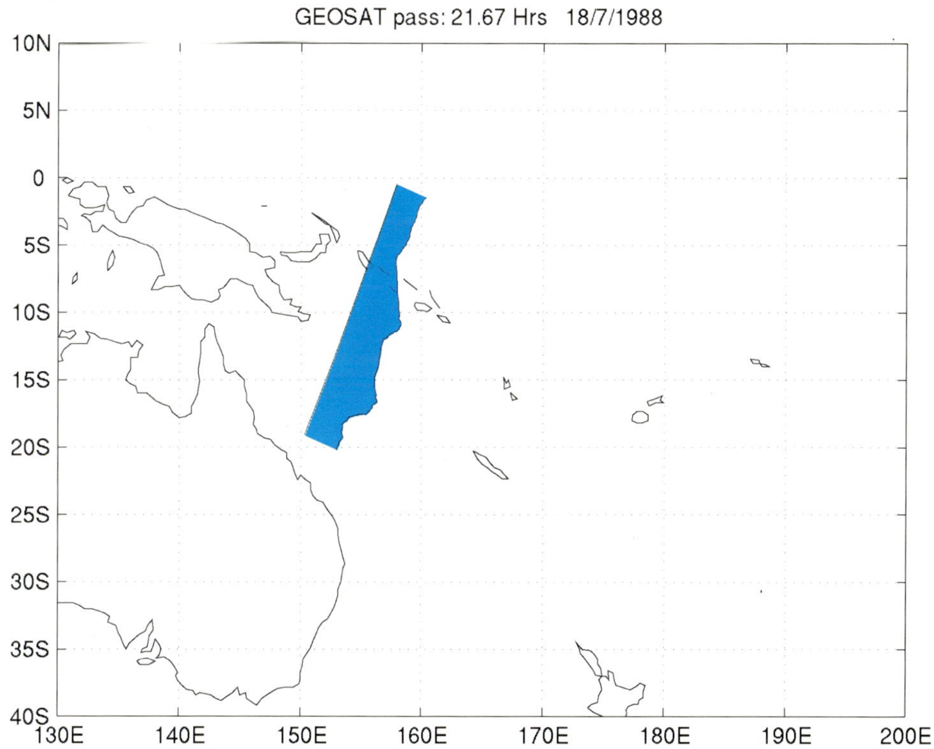

Fig. 5.7 Variation in the significant wave height along the GEOSAT pass shown in Fig. 5.3. The automatic quality control process has removed the data "spikes" caused by the isolated small islands and reefs in the Coral Sea.

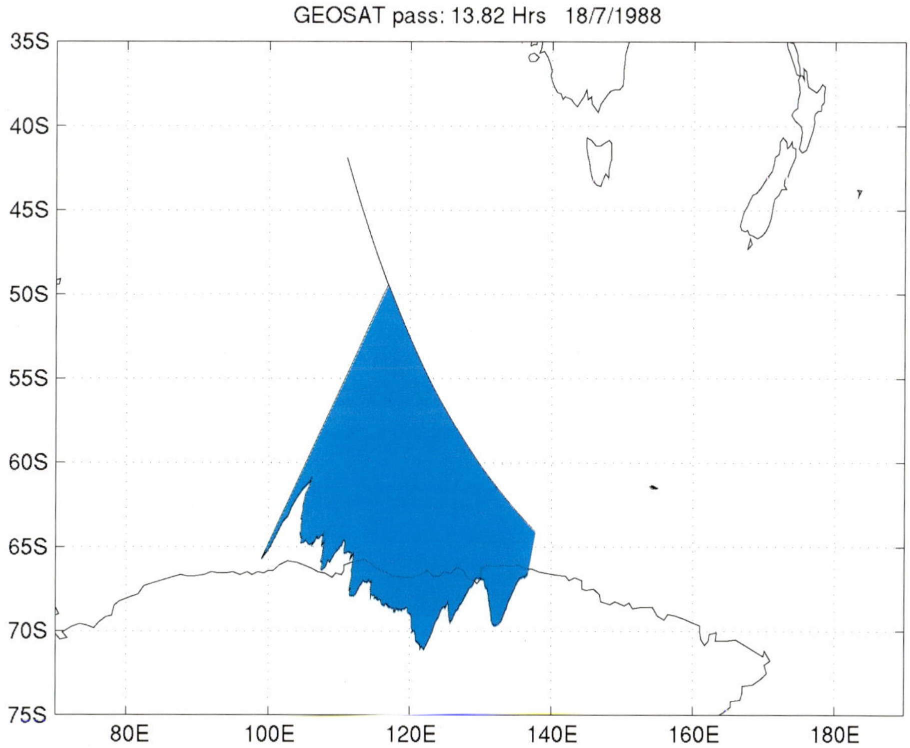

Fig. 5.8 Variation in the wind speed along the GEOSAT pass shown in Fig. 5.4. The automatic quality control process has removed the erroneously large wind speeds due to the unacceptably large satellite attitude angle.

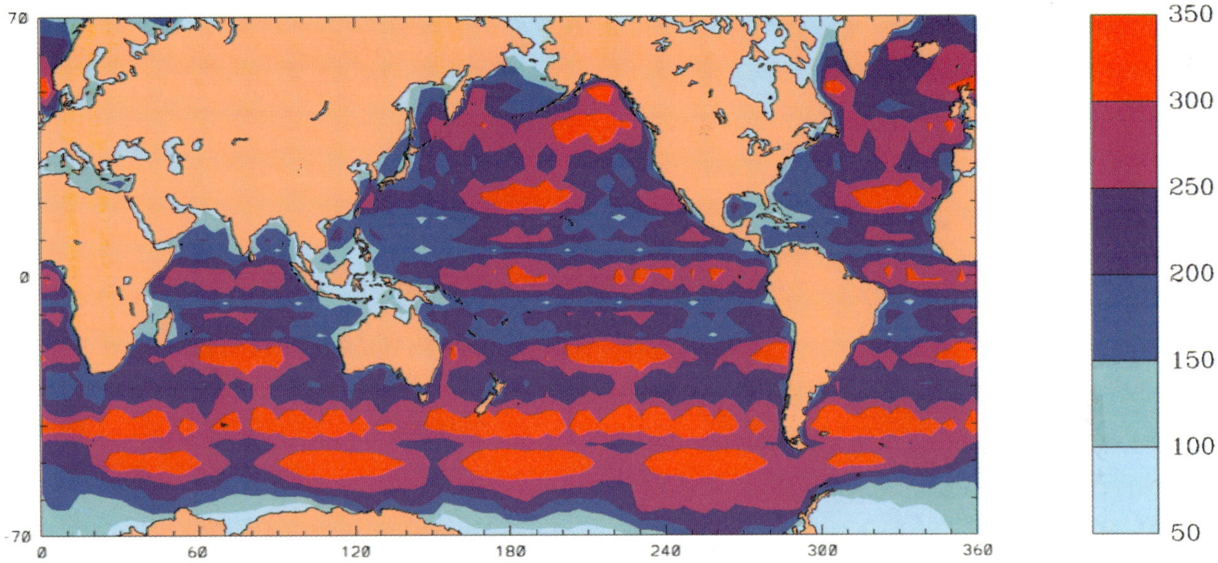

Fig. 5.9 Global coverage of GEOSAT altimeter data for the full duration of the mission. The total number of ground track passes through each 4° × 4° grid square, for the full period of the GEOSAT mission, is colour contoured according to the scale to the right of the figure.

Table 5.1 Average global density of passes per 4° × 4° grid square as a function of month

Month	Data density per grid square
January	20.4
February	23.4
March	27.7
April	26.6
May	21.1
June	19.7
July	20.1
August	19.5
September	19.3
October	17.1
November	24.0
December	31.5

tinent in the Southern Hemisphere summer than in winter. A similar trend is not clear in the Arctic regions due to the much larger amount of land in these regions. The global variability between months shown in Table 5.1 is largely due to satellite performance during the mission. In particular, the progressive degradation in data return towards the end of the mission has an influence on the monthly totals. The month of December has a noticeably higher data density than all other periods. This occurs since there were four months of December with high data return during the mission, compared to three for all other months.

The data density results are presented here to provide potential users with an indication of the reliability of the statistics for particular locations and periods. Clearly, regions and periods with more available data will produce more reliable statistical estimates. There are typically approximately 25 passes through each grid square for each month summed over the full mission. This is a relatively small number of observations from which to determine statistical properties of the wind or wave conditions. For this reason, only the most robust of statistical properties, the mean, has been calculated for each month. As will be shown in Chapter 6, comparisons with *in situ* data indicate that the seasonal variations in mean monthly conditions appear stable and unbiased.

5.4 Mean Monthly Statistics

Mean month variations in the significant wave height and the wind speed were calculated in the following manner.

1. Each GDR (Geographical Data Record) for the full GEOSAT mission was scanned.

2. The calculated values of significant wave height and wind speed were stored based on the recorded date, latitude and longitude of the observation.

3. Data were stored in one of 12 data files, depending on the month of the observation. The appropriate 4° × 4° grid square was allocated depending on the latitude and longitude of the observation.

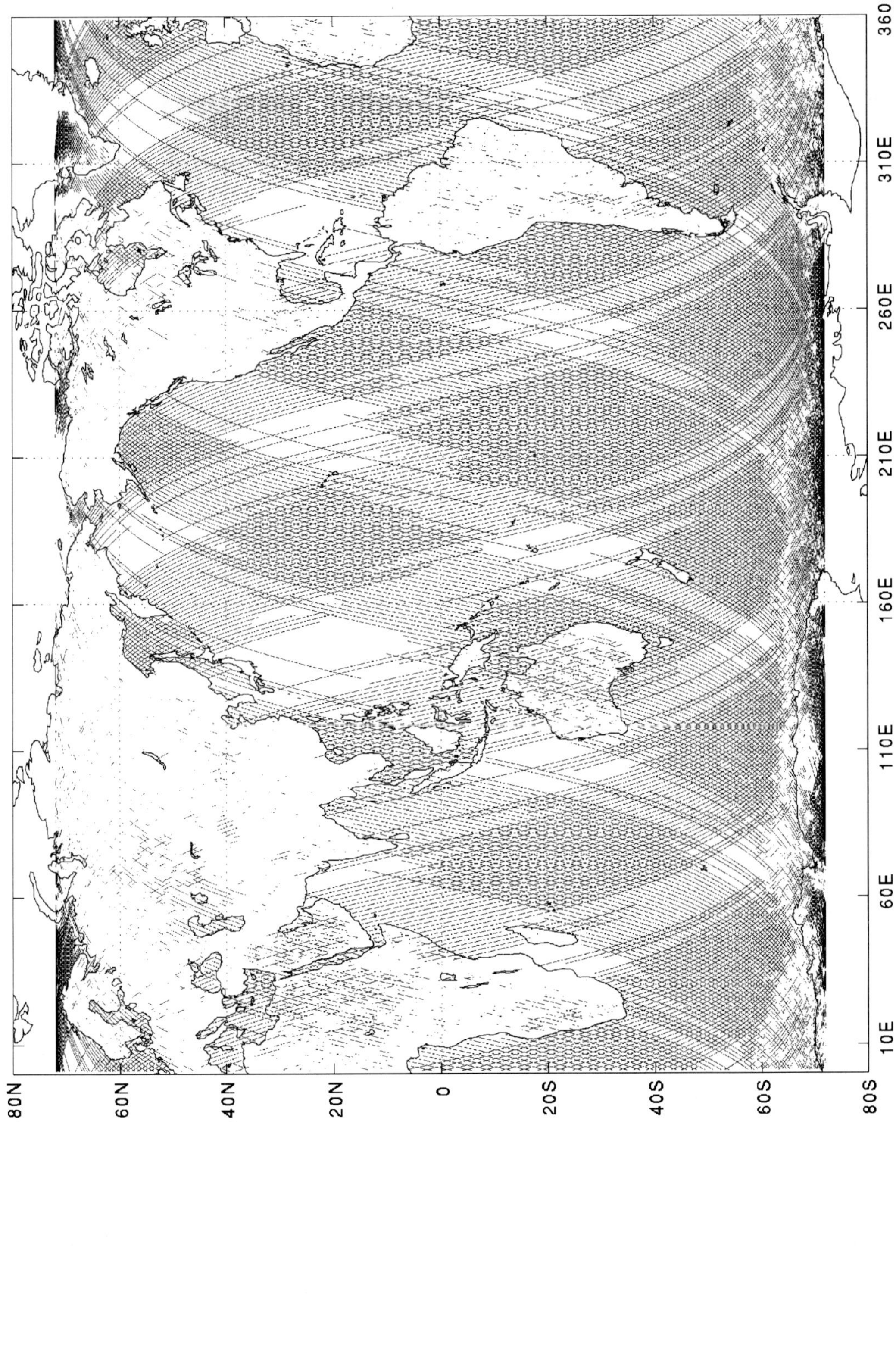

Fig. 5.10 Global ground track pattern for repeat cycle 37 of the GEOSAT satellite: 12 July 1988–28 July 1988.

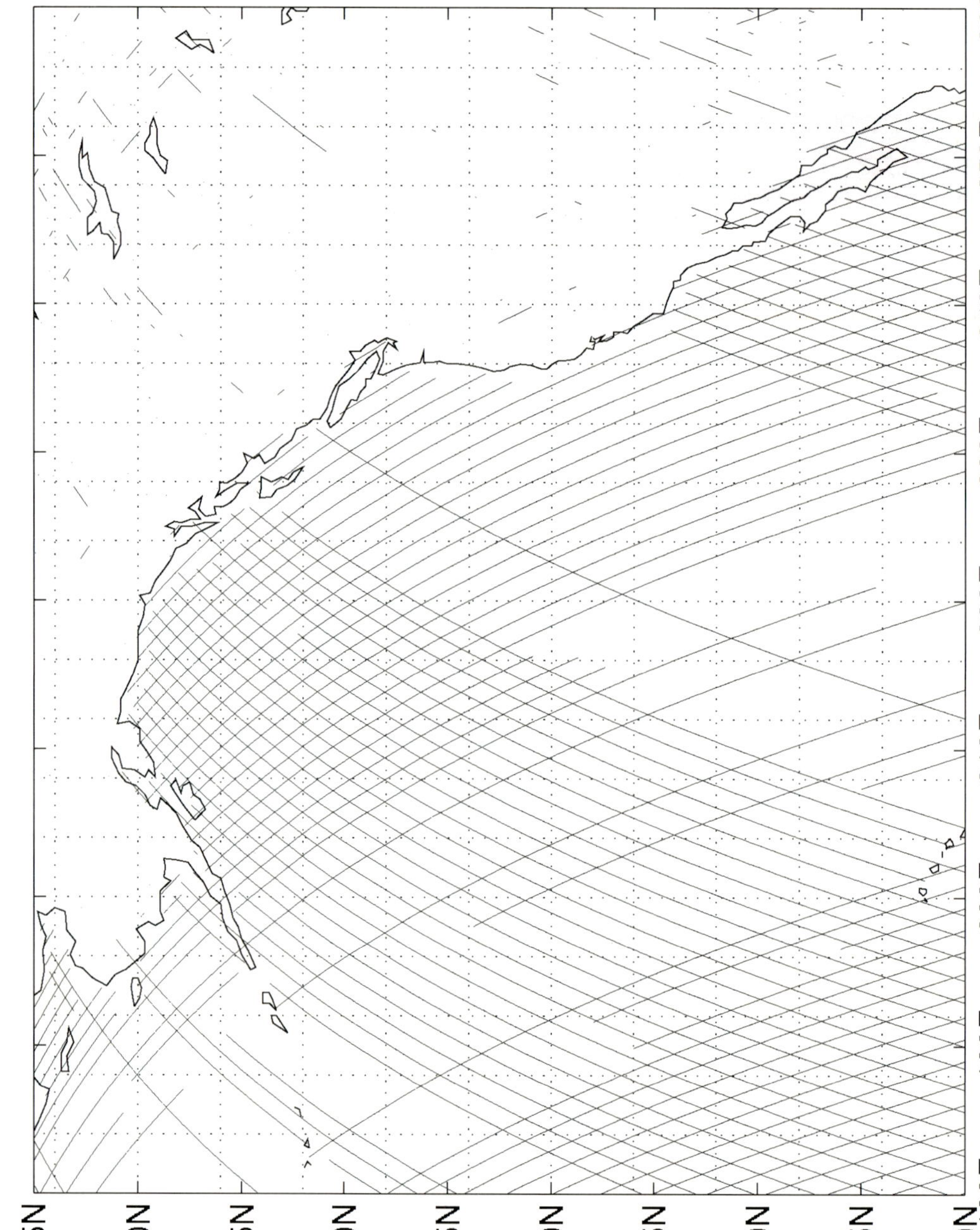

Fig. 5.11 An expanded region of the ground track pattern for repeat cycle 37 (see Fig. 5.10) taken over the north-eastern Pacific Ocean. The $4° \times 4°$ data grid is superimposed on the ground track pattern.

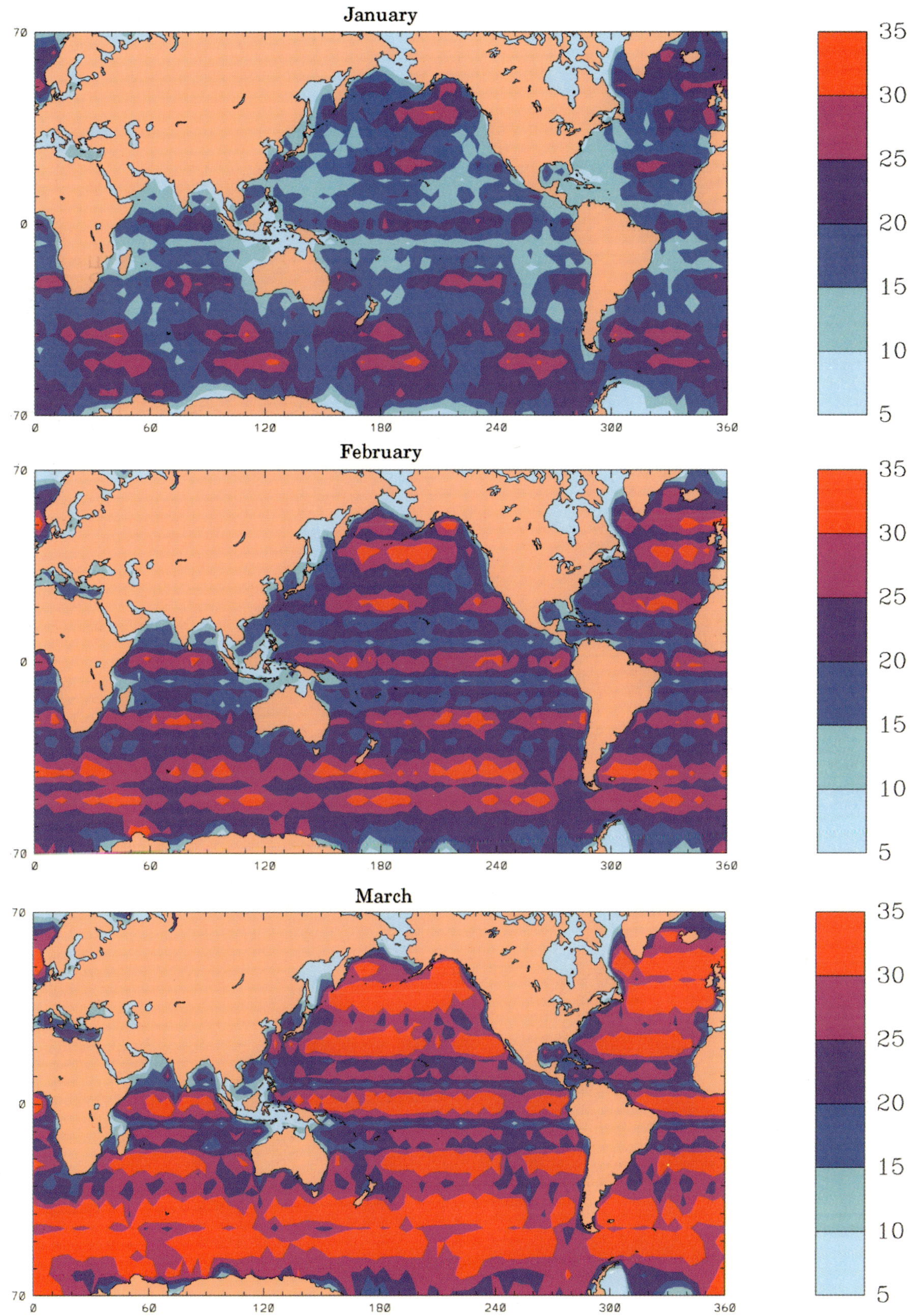

Fig. 5.12 Density of ground track passes for the months of January (top), February (middle) and March (bottom). Results represent the cumulative total of the full GEOSAT mission.

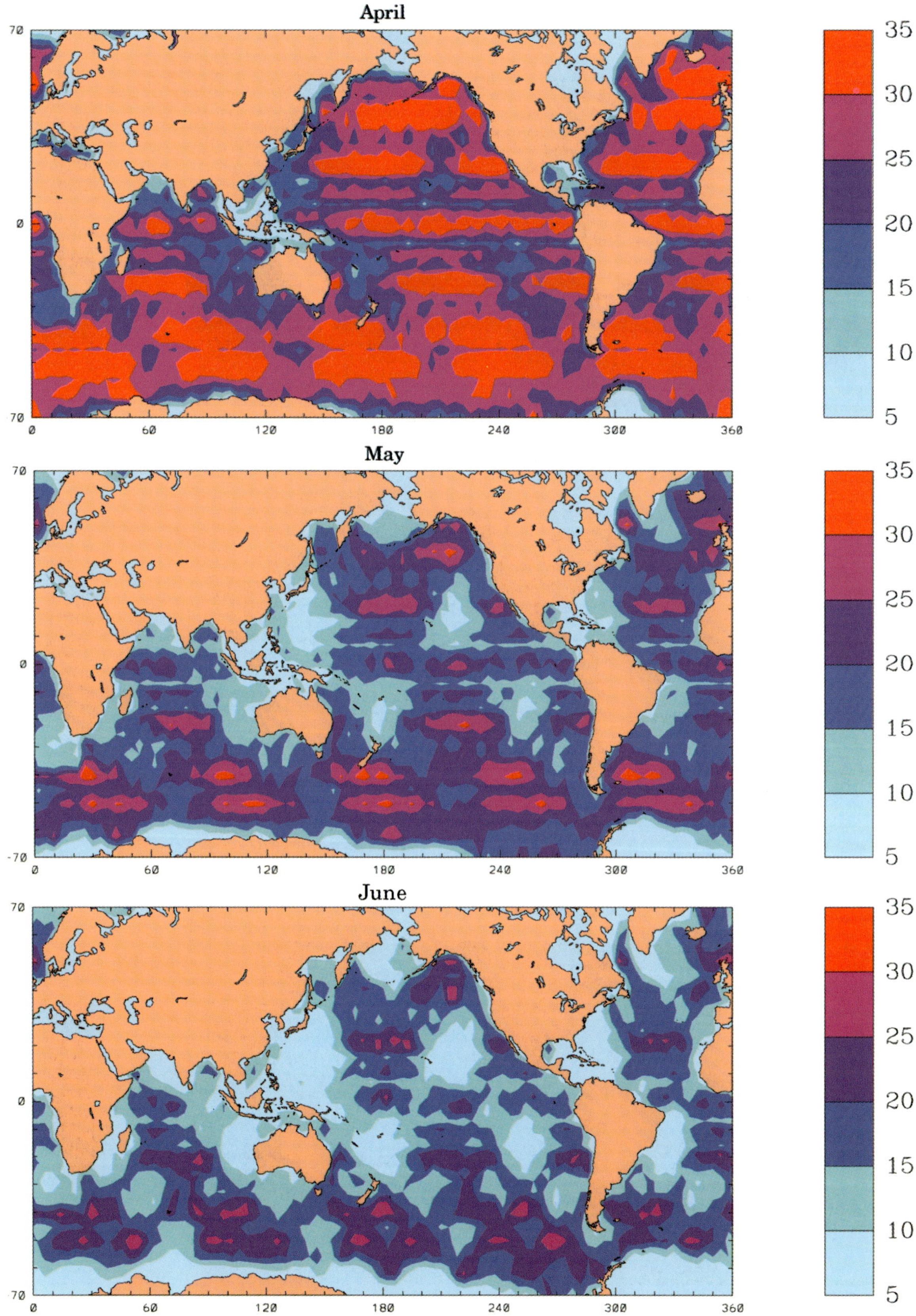

Fig. 5.13 Density of ground track passes for the months of April (top), May (middle) and June (bottom). Results represent the cumulative total of the full GEOSAT mission.

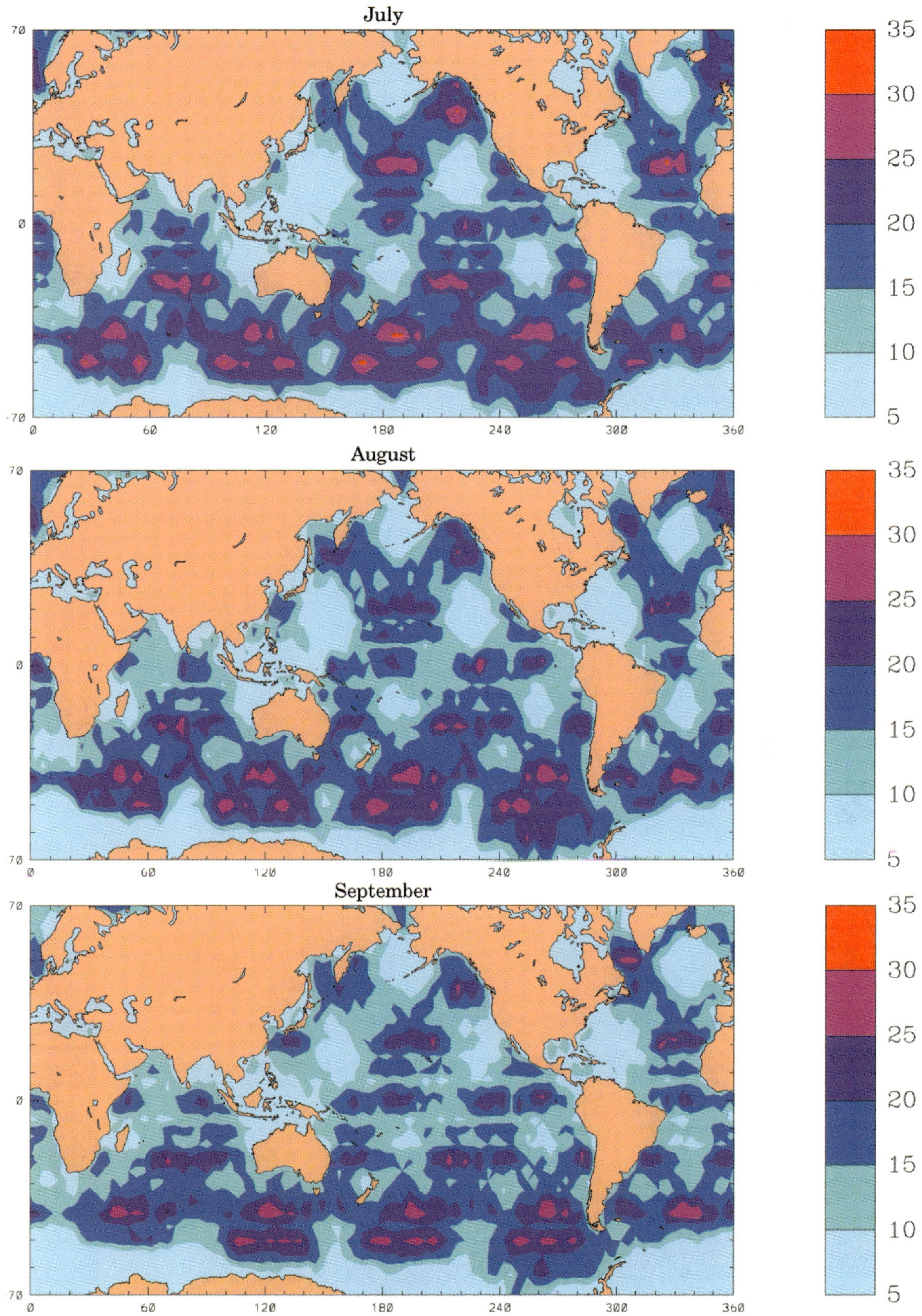

Fig. 5.14 Density of ground track passes for the months of July (top), August (middle) and September (bottom). Results represent the cumulative total of the full GEOSAT mission.

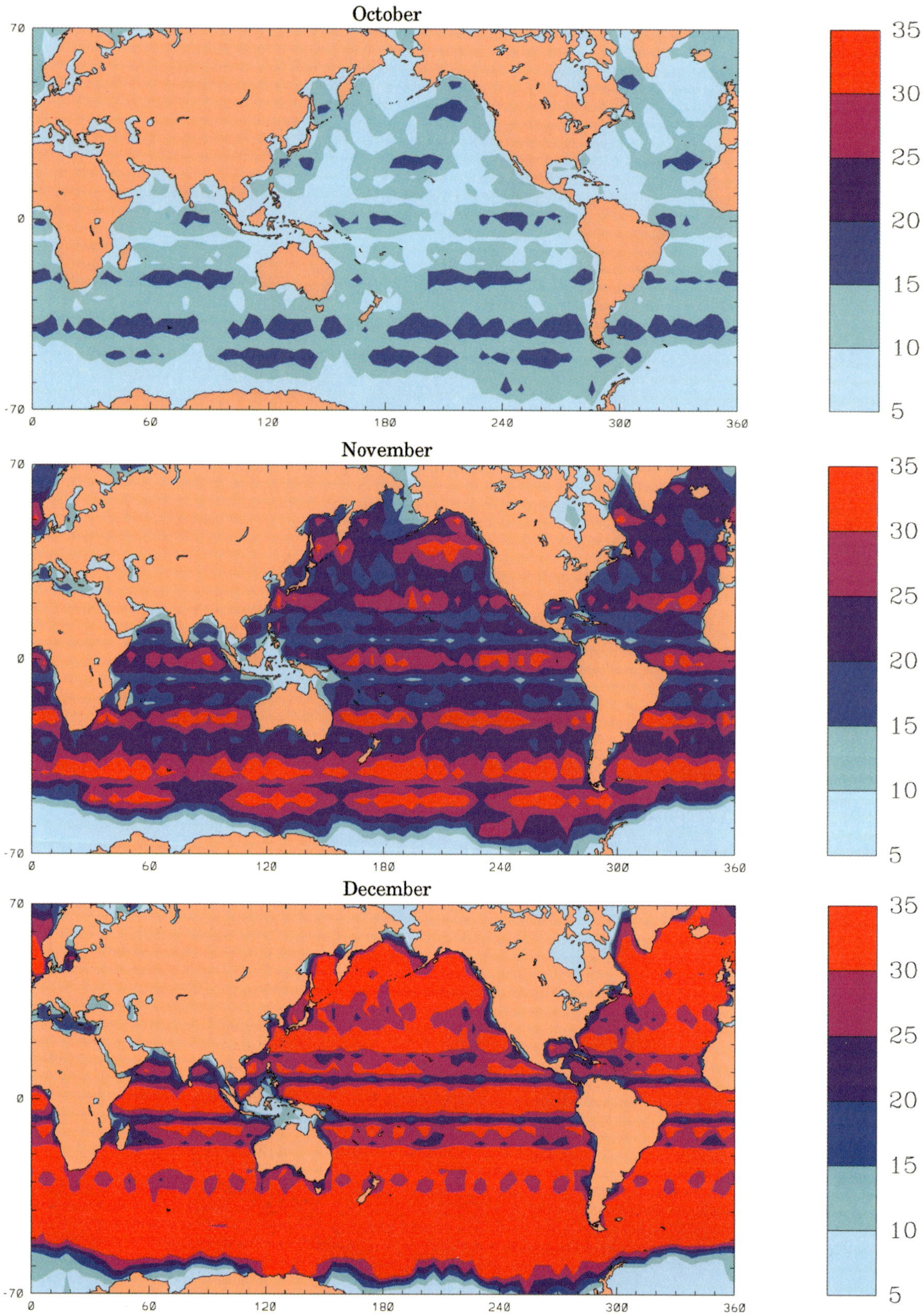

Fig. 5.15 Density of ground track passes for the months of October (top), November (middle) and December (bottom). Results represent the cumulative total of the full GEOSAT mission.

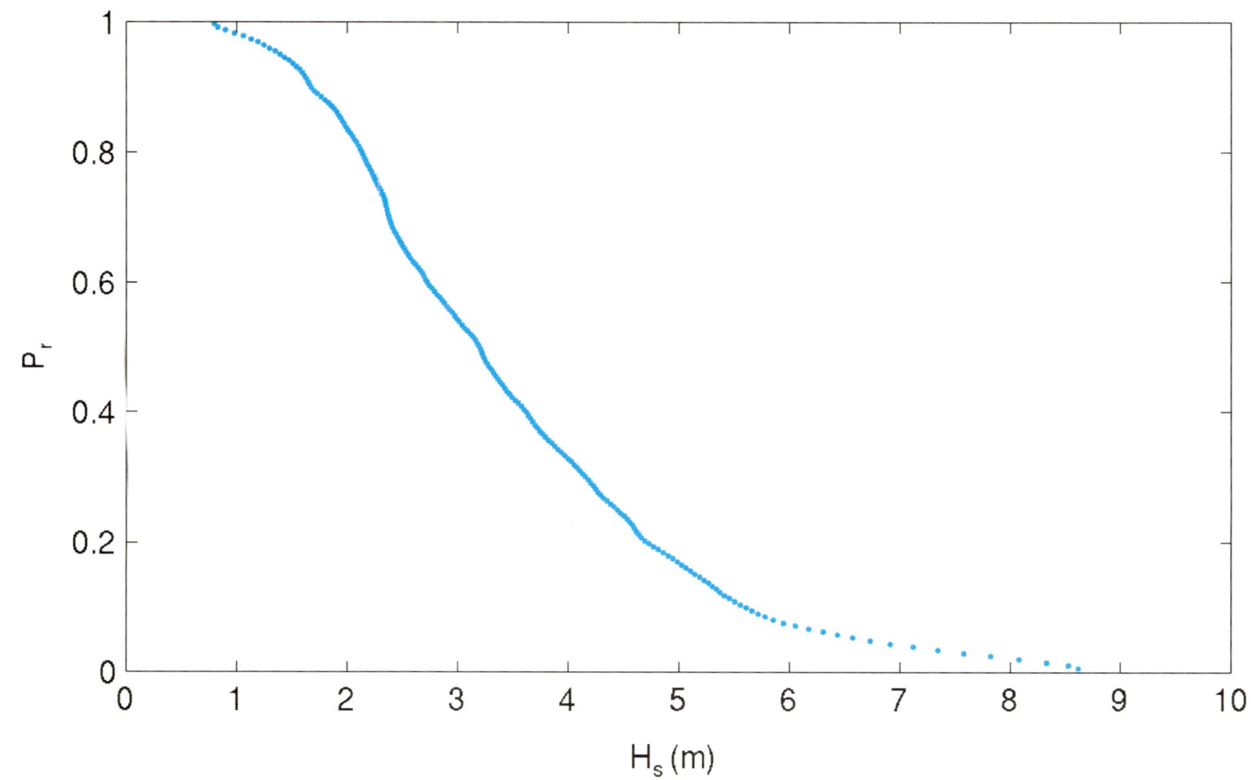

Fig. 5.16 An example of the probability distribution function (i.e. probability of exceedence, P_r as a function of the significant wave height, H_S). The example shown is from the north-eastern Atlantic Ocean (54°N, 346°E).

4. For each grid square and for each of the monthly files, the mean values of significant wave height and wind speed were determined.
5. The result is a set of twelve matrices (one for each month), for each of wind speed and significant wave height. Each matrix or table consists of 90×36 points, one for each $4^\circ \times 4^\circ$ grid square.

Contouring of the values in these matrices yields the results shown in Chapter 7.

5.5 Probability of Exceedence

The probability of exceedence, P_r, as its name suggests, represents the probability of a particular given value of wind speed or wave height being exceeded at a particular site. For instance, the probability or percentage of time that the significant wave height at a particular location would exceed 4 m. With typically only 25 observations for each grid square per month, such statistics cannot be determined on a monthly basis. The overall yearly values, where there are in the order of 300 observations, however, provide a sufficiently large data set to compile such exceedence probabilities.

The multiple observations associated with the passage of the satellite through each $4^\circ \times 4^\circ$ grid square will not, generally, be *statistically* independent. That is, they cannot really be considered as independent observations which offer new data. They are, in effect, related to each other.

On the spatial scale of the grid square, the same meteorological event is likely to have generated all the wind and wave data observed during the satellite passage across the grid square. Hence, each satellite pass through a grid square is assumed to provide only one independent observation of the wind or wave conditions at that point and time.

Two different methods of determining the values associated with each of the passes were implemented: (a) the mean value of each pass across the grid square and (b) the maximum value of the pass across the grid square. The most appropriate technique is difficult to assess. In the limit, as the size of the sampling square approaches zero, the two techniques become identical. The finite size of the grid square will, however, yield differing results for the two methods. The potential differences are assessed in the light of *in situ* data in Chapter 6.

The independent steps required to determine the probability of exceedence can be summarized as:

1. Each GDR (Geographical Data Record) for the full GEOSAT mission was scanned.
2. Based on the latitude and longitude of the observation, the values of significant wave height and wind speed were stored. Data were stored in appropriate bins based on the $4^\circ \times 4^\circ$ grid square associated with its latitude and longitude.
3. For each satellite pass through each grid square, the mean and the maximum of the pass were determined for both wind speed and significant wave height.
4. Each of these single values associated with the satellite pass across the grid square can be considered to be statistically independent, since, on average, they occur three days apart. For each grid square the values (H_s and U_{10}) were ranked in descending order of magnitude from $m = 1$ to N, where N is the total number of passes through the grid square. The probability of exceedence, P_r, was then determined using the simple plotting position formula (Gumbel, 1958)
$$P_r = \frac{m}{N+1}. \tag{5.7}$$
A typical example of the resulting relationship between P_r and significant wave height for a location in the north-eastern Atlantic Ocean (54°N, 346°E) is shown in Fig. 5.16.
5. As shown by Fig. 5.16, the result from the ranking process is a continuous (on average 300 points) relationship between P_r and either wind speed or wave height. In order to produce charts of exceedence probability, the wind speed and significant wave height associated with probabilities of exceedence of 0.1–0.9 in steps of 0.1 were determined and stored.
6. The result is a set of nine matrices (one for each exceedence probability level), for each of wind speed and significant wave height. Each matrix or table consists of 90×36 points, one for each $4^\circ \times 4^\circ$ grid square.

The matrices obtained in this fashion can be contoured (see Chapter 7). Each of these contoured charts indicates the wind speed or wave height which one could expect to be exceeded for a particular percentage of the time. For instance, it may indicate that a significant wave height in excess of 4 m could be expected to be exceeded 20% of the time at a particular location.

Chapter 6

DATA VALIDATION

6.1 Validation Sites

Although the accuracy with which the GEOSAT altimeter can measure both significant wave height and wind speed was considered in Chapter 4, the reliability of the statistical quantities derived from these individual measurements must also be considered. In particular, the implications of the relatively short data period (three years) and the spatial extent of the grid squares ($4^\circ \times 4^\circ$) must be determined. This can most easily be achieved through comparisons with *in situ* data. For such comparisons to be meaningful, however, the *in situ* data must have been recorded for a sufficiently long period to also be reliable and must be free from the influence of coastal effects. The $4^\circ \times 4^\circ$ resolution of the grid used for the GEOSAT data analysis means that the data is of only limited value near shore. In such coastal areas, the wave climate will be heavily influenced by the proximity of land and finite depth influences. Not surprisingly, the vast majority of *in situ* data have been collected in coastal sites.

For the purposes of data validation, two open ocean data sites have been identified where *in situ* data are available. The first of these comes from the Southern Ocean site of Macquarie Island ($-54^\circ 30'$S, 159°E). Both wave buoy and wind data were obtained from this site by Steedman (1993) for the period November 1988–October 1989. This is a period largely independent of the GEOSAT data set. A wave buoy was located 10 km north-west of Macquarie Island in a water depth of approximately 130 m. As waves generally approach the site from between NNW and SW, little influence from the island could be expected, the wave climate being representative of deep water. Wind data were recorded at an Australian Bureau of Meteorology site on Macquarie Island. The site has a ground elevation of 6 m, with the anemometer vane being located 3.1 m above ground level.

Validation data was also taken from the wind and wave data published by Zambresky (1989) for the Hawaiian Islands (22°N, 198°E). Data were available in the form of mean monthly wind speeds and significant wave heights for three different buoys in the same general area.

6.2 Mean Monthly Significant Wave Heights

Figure 6.1 shows a comparison between the Macquarie Island *in situ* wave data and the GEOSAT mean monthly values. The same seasonal trends are clear with the most extreme waves occurring in the Southern Hemisphere winter. The satellite data produce a lower extreme value, peaking at 4.4 m in May whereas the buoy data yield a maximum in this same month of 5.3 m. It should be remembered that the buoy data are from only one year of observations and some annual variability must be expected. The difference between a temporal average at a single point (buoy data) and a spatial average at a number of almost instantaneous but intermittent times (satellite data) could also be expected to produce some differences. Clearly, the size of the region (sampling square) over which the satellite data are considered should not be so large as to span areas with significantly varying climatology nor should it be so small as to under-sample the data. Tournadre and Ezraty (1990) have investigated this problem and found spatial averaging scales comparable to that adopted here introduce no measurable bias.

Data obtained from ship observations and presented in the USSR Climatological Atlas (1974) are

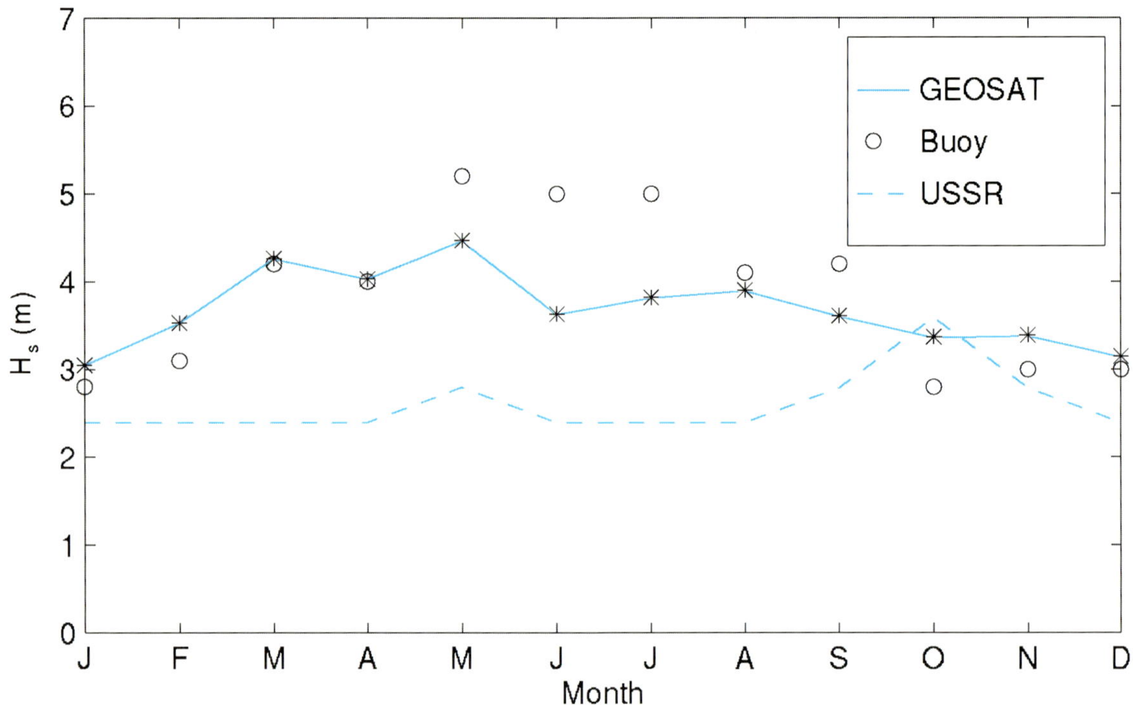

Fig. 6.1 Seasonal variation in mean monthly values of significant wave height for the Southern Ocean site of Macquarie Island (−54°30′S, 159°E). The buoy data of Steedman (1993) are compared with the GEOSAT-derived results for the grid square centred at −54°S, 158°E. In addition, data obtained from limited ship observations and published in the USSR Climatological Atlas (1974) are also presented.

also presented in Fig. 6.1. Whereas the GEOSAT derived results and *in situ* buoy data are in good agreement, the ship observations are poorly correlated with the buoy results. The ship observations show no summer–winter variation in wave conditions as intuitively expected and confirmed by both the *in situ* buoy data and the GEOSAT satellite observations. This poor agreement is a clear indication of the unreliable nature of ship observations. This is firstly caused by the subjective nature of visual observations of wave height and secondly by the intermittent nature of such observations. Voyages in the Southern Ocean, particularly during the stormy winter months, are infrequent and this paucity of data is clearly reflected in the poor quality of the resulting ship observations shown in Fig. 6.1.

A measure of the statistical stability of the data can be obtained by comparison of neighbouring 4° × 4° grid squares at the same latitude. In regions far from land, such as the Southern Ocean, little spatial variability should occur over this scale. Figure 6.2 shows a comparison between the mean monthly values of significant wave height for the grid square discussed in Fig. 6.1 and its neighbours immediately to the east and west. The results are very similar, indicating that there are sufficient passes through each grid square to form reliable mean monthly values of significant wave height.

In contrast to the extreme wave conditions at high latitudes, lower latitudes are relatively calm, all year round. Figure 6.3 shows a comparison between the available *in situ* buoy data of Zambresky (1989) and the GEOSAT altimeter-derived results for the region of the Hawaiian Islands. The data of Zambresky come from three buoys in the region:

- Buoy 51001—23.4°N, 197.7°E
- Buoy 51002—17.2°N, 202.2°E
- Buoy 51003—19.2°N, 199.8°E

The GEOSAT altimeter-derived data in Fig. 6.3 comes from the grid square centred on 22°N, 198°E.

As for the Southern Ocean site, the GEOSAT altimeter-derived mean monthly values of significant wave height are in excellent agreement with the buoy data. Both the seasonal trends and the absolute magnitude of the wave heights agree well.

The results presented in Figs 6.1 and 6.3 clearly validate the derived values of mean monthly significant wave height. They show that the size of the grid sampling squares (4° × 4°) and the number of satellite passes per month through the squares are

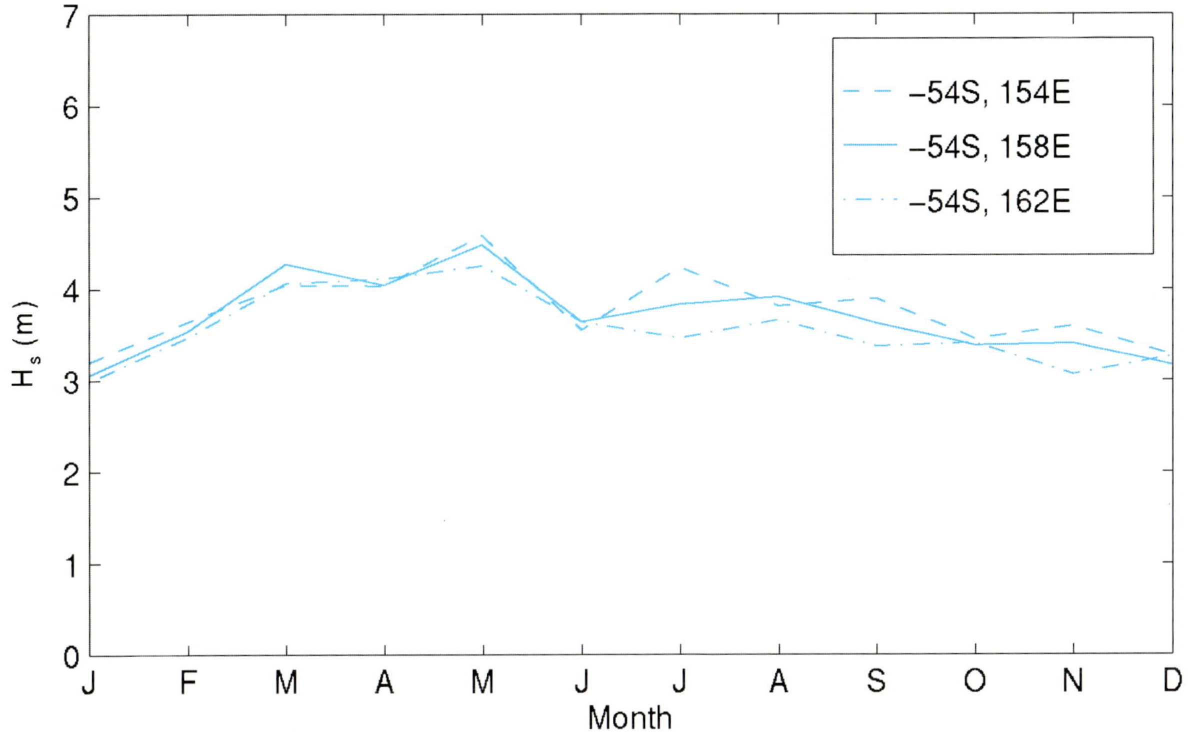

Fig. 6.2 Seasonal variation in mean monthly values of significant wave height for three neighbouring squares at the same latitude (−54°S).

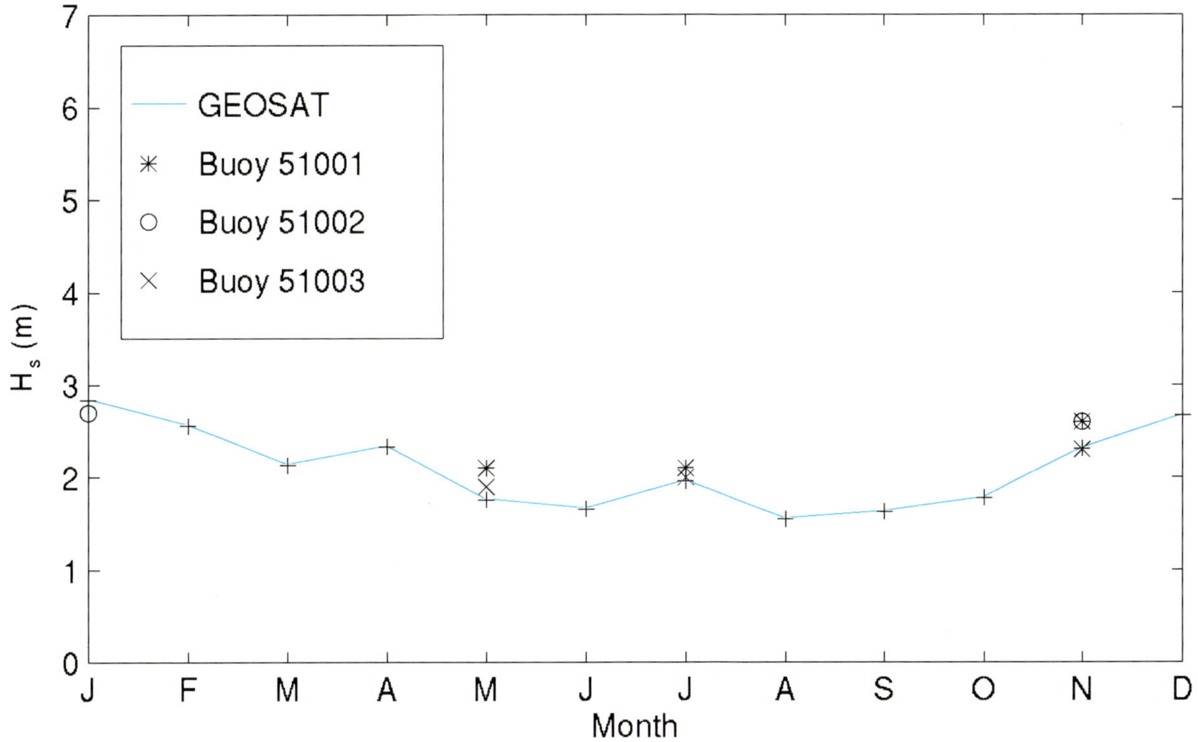

Fig. 6.3 Seasonal variation in the values of mean monthly significant wave height at Hawaii. The GEOSAT altimeter-derived data for the grid square centred on 22°N, 198°E are compared with the buoy data of Zambresky (1989).

adequate to accurately determine mean monthly significant wave heights. The two sites have very different wave climates and seasonal variations in wave conditions. That both sites agree well with the GEOSAT results also shows that no significant sea state bias exists in either the original data or the analysis procedure.

6.3 Probability of Exceedence — Significant Wave Height

Figure 6.4 shows a comparison between the probability distribution functions (probability of exceedence, P_r vs significant wave height, H_s) obtained from the GEOSAT satellite data and the *in situ* buoy data of Steedman (1993) for the Macquarie Island site. Chapter 5 described two methods of determining the exceedence probabilities from the satellite data:

- Using the mean value of each satellite pass through a grid square.

- Using the maximum value of each satellite pass through a grid square.

Both methods are shown in Fig. 6.4. In addition, results are shown not only for the grid square centred on Macquarie Island (i.e. $-54°$S, $158°$E) but also the neighbouring grid squares immediately to the east and west.

Not surprisingly, the method which used the maximum of each pass yields higher values of significant wave height for a given probability level than the *mean* method. The buoy observations are in very good agreement with both techniques, agreeing best with the *mean* method for the higher values of P_r and best with the *maximum* method for the lower values of P_r. Although the data quality control methods discussed in Chapter 5 were very reliable, the size of the data set precludes an absolute guarantee that no erroneous data survived the quality control checks. The extensive visual checking of the data revealed no obvious erroneous data. Should isolated data errors exist, they will have a very small influence on the mean monthly statistics. Since a very large number of individual observations are used in determining the mean monthly values, such infrequent errors will have an insignificant influence on the resulting means. Similarly, the *mean* method of determining P_r will be relatively insensitive to such errors. The *maximum* method, however, will be prone to significant bias as a result of such errors. An isolated data "spike", for instance, could be assigned the maximum value for a pass through the data square. Hence, the *mean* method is believed to be far more reliable and has been adopted for all subsequent analysis in this Atlas.

The probability distribution functions for the three neighbouring grid squares shown in Fig. 6.4 are very similar, indicating the derived probability distribution functions are stable. This result and the good agreement with the *in situ* buoy data indicates that the three year data period and the grid resolution are adequate to produce reliable estimates of the exceedence probability of significant wave height.

6.4 Mean Monthly Wind Speed

Figure 6.5 shows a comparison between mean monthly values of wind speed obtained from the GEOSAT altimeter data and *in situ* anemometer measurements from Macquarie Island taken by Steedman (1993). The original Australian Bureau of Meteorology data presented by Steedman (1993) were recorded at an anemometer height of 3.1 m. Since the GEOSAT altimeter wind speed algorithm recovers wind speeds equivalent to an anemometer height of 10 m, the Steedman (1993) data have been corrected to a height of 10 m, assuming a logarithmic boundary layer. The agreement between the satellite data and the anemometer measurements is very good. Both data sources show peak mean monthly wind speeds of approximately 14 ms^{-1} occurring in the May–June period. This is consistent with Fig. 6.1 which shows peak mean monthly significant wave heights occurring at this same time.

The spatial consistency of the GEOSAT altimeter estimates of wind speed are shown in Fig. 6.6, where the seasonal variation is compared for the grid square shown in Fig. 6.5 and its neighbours immediately to the east and west. The results for the three grid squares are very similar, again indicating the statistical stability of the satellite estimates of mean monthly wind speed. There is more variability than in the comparable values of mean monthly significant wave heights (Fig. 6.2). As already indicated (Section 4.5), wind speeds are less spatially coherent than significant wave heights. Consequently, the greater spatial variability of the wind speed estimates is to be expected.

Figure 6.7 shows a similar comparison between the GEOSAT mean monthly wind speeds and the buoy data of Zambresky (1989) for the Hawaiian region. Again the agreement is very good. Although there is slightly more variability in the GEOSAT altimeter estimates of mean month wind speeds as compared to mean monthly wave heights the results presented in Figs 6.5–6.7 clearly show that the GEOSAT data is statistically stable and agrees well

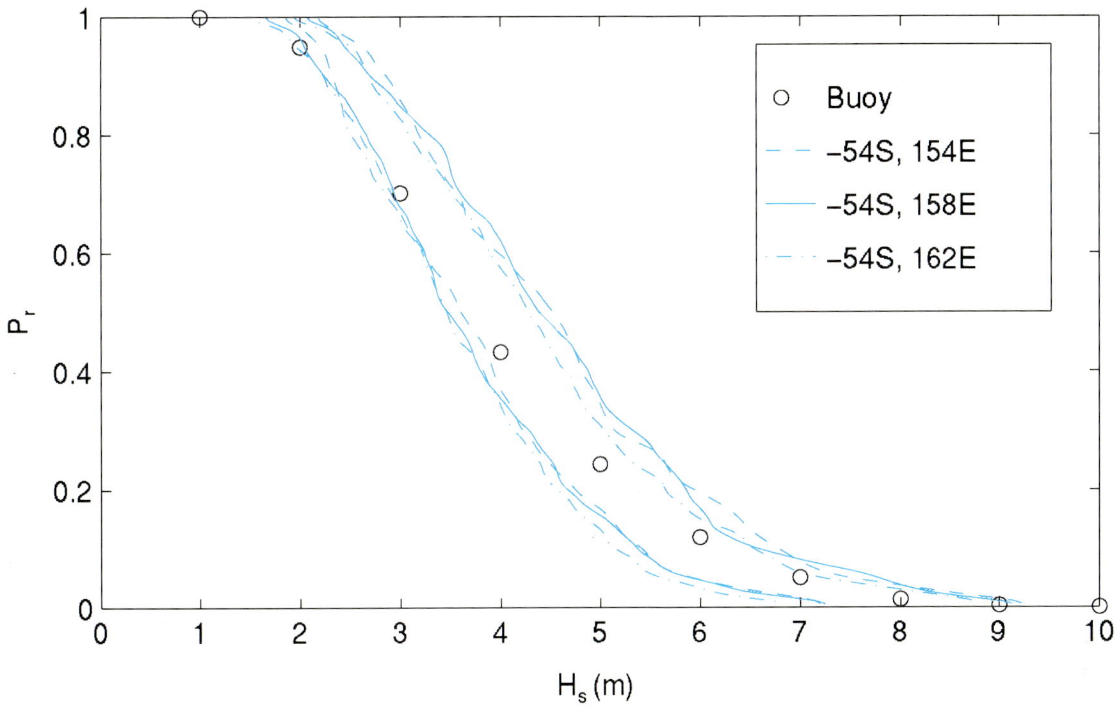

Fig. 6.4 A comparison between the probability distribution functions obtained from the GEOSAT data (lines) with that derived from the buoy data of Steedman (1993) (circles) for the Southern Ocean site of Macquarie Island (−54°30′S, 159°E). The GEOSAT results are presented for three sampling squares in the vicinity of Macquarie Island. The three curves to the left were obtained using the *mean* values of individual passes through each sampling square, the curves to the right from the *maximum* values of individual passes through each sampling square.

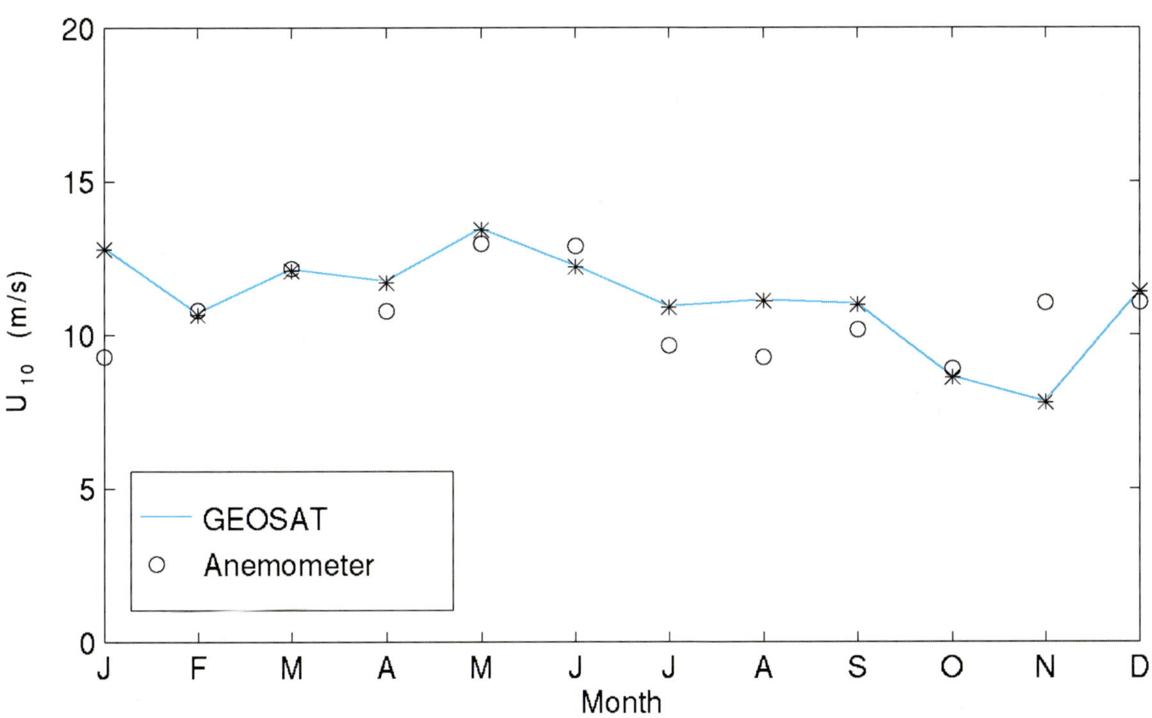

Fig. 6.5 Seasonal variation in mean monthly values of wind speed, U_{10}, for the Southern Ocean site of Macquarie Island (−54°30′S, 159°E). The anemometer data of Steedman (1993) are compared with the GEOSAT-derived results for the grid square centred at −54°S, 158°E.

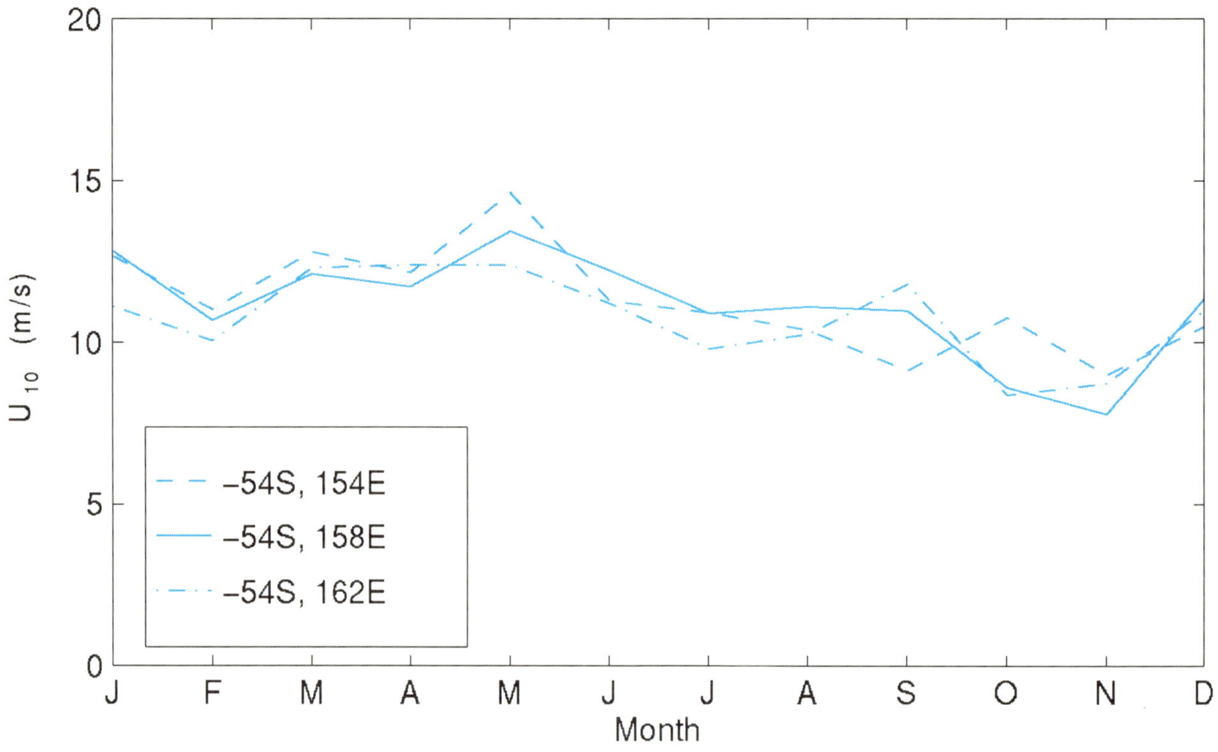

Fig. 6.6 Seasonal variation in mean monthly values of wind speed for three neighbouring squares at the same latitude (−54°S).

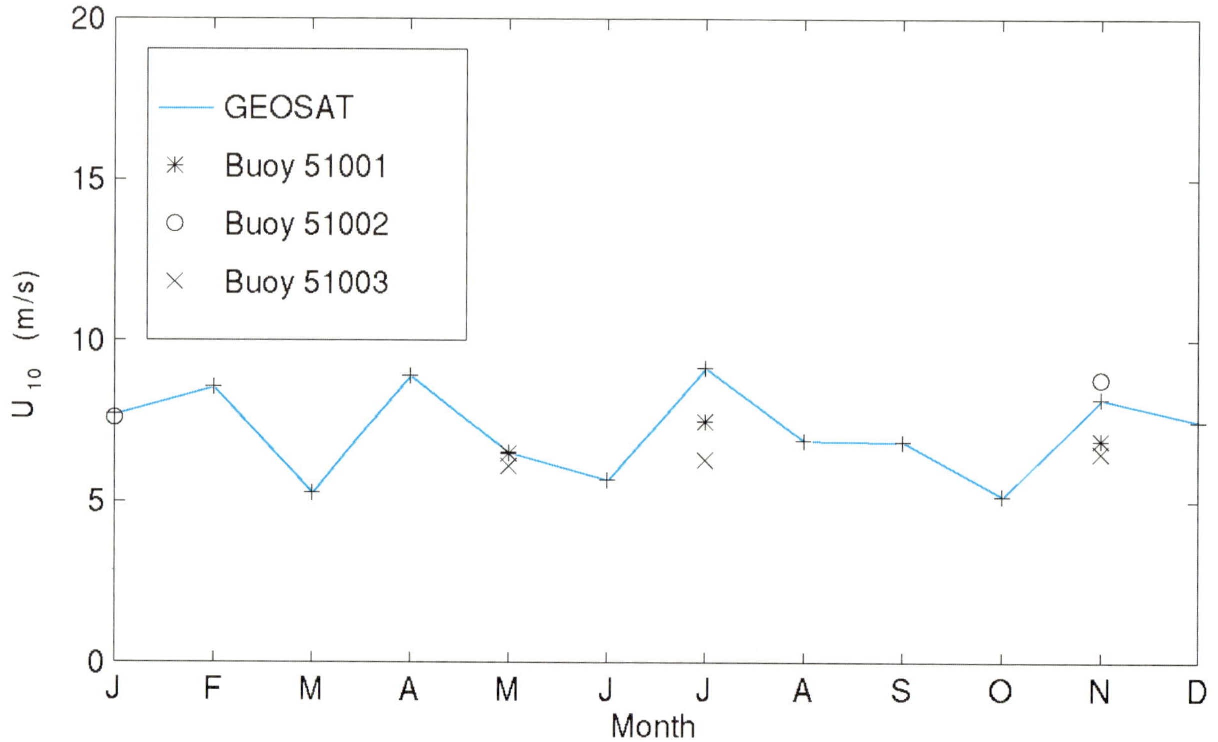

Fig. 6.7 Seasonal variation in the values of mean monthly wind speed at Hawaii. The GEOSAT altimeter-derived data for the grid square centred on 22°N, 198°E are compared with the buoy data of Zambresky (1989).

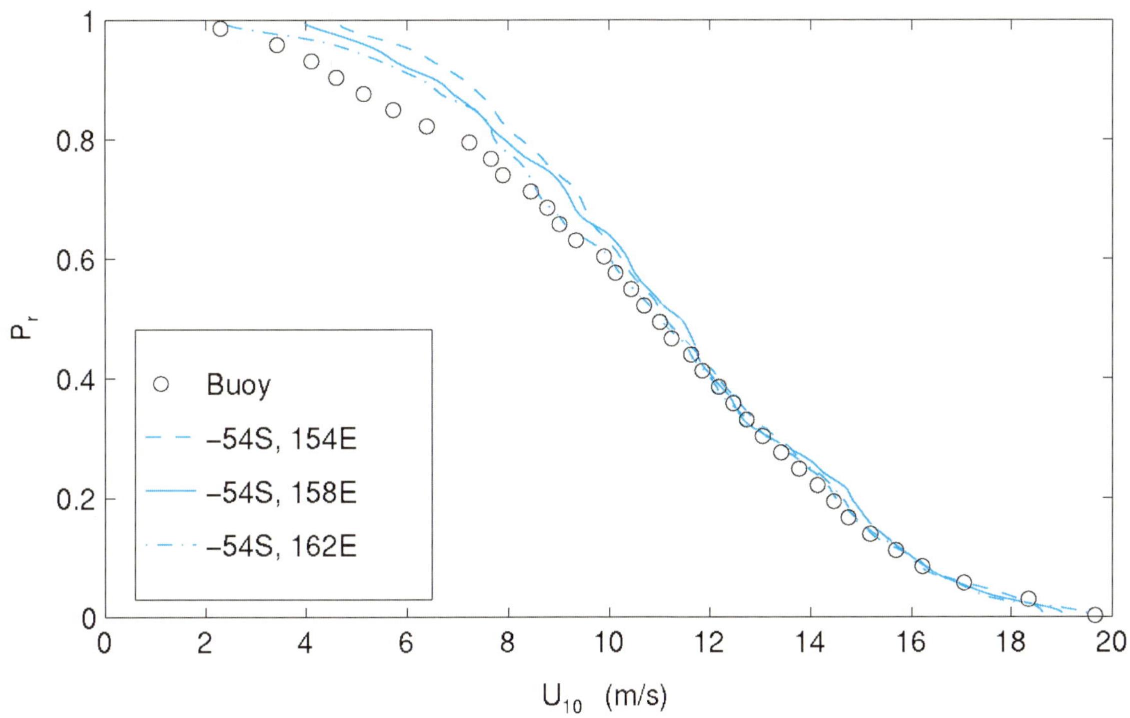

Fig. 6.8 A comparison between the probability distribution function obtained from the GEOSAT data (lines) with that derived from the anemometer data of Steedman (1993) (circles) for the Southern Ocean site of Macquarie Island (−54°30′S, 159°E). The GEOSAT results are presented for three sampling squares in the vicinity of Macquarie Island.

with *in situ* measurements.

6.5 Probability of Exceedence — Wind Speed

Figure 6.8 shows a comparison between the probability distribution functions obtained from the GEOSAT altimeter data and the *in situ* anemometer data from Steedman (1993) for the Southern Ocean site of Macquarie Island.

The comparison between the satellite and *in situ* data is again good. The satellite data appear to over-estimate the wind speeds at low values, yielding higher wind speeds for the same probability of exceedence. At higher wind speeds, the probability distributions are in excellent agreement. The higher values, derived from the GEOSAT satellite data at low wind speeds, are not surprising. The GEOSAT data used to derive the probability distribution functions consist of the averages of each pass through the appropriate 4° × 4° grid square. This averaging process would tend to have a "smearing" effect, producing higher values at low wind speeds and lower values at high wind speeds than could be expected at a single point. The fact that the satellite and anemometer data are in good agreement at high wind speeds indicates that for these high wind events there is little spatial variability across the full grid square. In contrast, low wind speeds tend to be spatially variable, accounting for the differences between the two data sets.

Figure 6.8 also shows the probability distribution functions obtained from the grid squares immediately to the east and west of that centred on Macquarie Island. The three probability distribution functions are remarkably similar, with only minor differences at low wind speeds. This spatial consistency is an indication of the statistical stability of the probability of exceedence values obtained from the GEOSAT altimeter data set.

6.6 Conclusions as to the Validity of Results

The above sections have presented comparisons between GEOSAT altimeter-derived statistics and *in situ* data for:

- Mean monthly significant wave height
- Exceedence probability levels for significant wave height
- Mean monthly wind speed
- Exceedence probability levels for wind speed

In addition, data have been presented for two regions with very different wind and wave climates (Southern Ocean and Hawaii). In all cases, the agreement between the GEOSAT altimeter results and the *in situ* data is excellent. The results provide a compelling validation of the accuracy of the GEOSAT data. In addition, the data are spatially consistent, indicating that the temporal extent of the GEOSAT mission (approximately three years) was sufficient to yield stable statistics.

Chapter 7

GLOBAL WIND AND WAVE STATISTICS

7.1 Background

The satellite observations of winds and waves are presented in the form of monthly means and exceedence probabilities for the whole globe and for a Southern Hemisphere polar perspective (Section 7.2), bimonthly details for each major ocean basin (Section 7.3), latitudinal and longitudinal sections and sections along selected great-circle routes (Section 7.4). In addition, detailed seasonal variations and exceedence probabilities are presented at selected points (Section 7.5). Each of these analyses provide fascinating evidence of the local effects on the Global circulation and the complex interactions between winds, fetches, waves and wave propagation. Some discussion of the major features accompanies each set of analyses.

7.2 Global Monthly Means and Exceedence Probabilities (Figs 7.1–7.5)

Mean wind directions for each season are shown in Fig. 7.1. Since the satellite data cannot provide wind direction, this was derived from a Global atmospheric circulation model and should be used as indicative of the major circulation features (Chapter 2), rather than for detailed comparison with the satellite observations. The monthly mean observations of wind, significant wave height and exceedence probabilities derived from the satellite data are shown in Figs 7.2 and 7.3, and in Figs 7.4 and 7.5, provide an additional polar perspective of the southern oceans.

The major regional circulation features discussed in Chapter 2 are clearly delineated by the contours of 70–80% exceedence for winds in Fig. 7.3(e). Regions of greater than 5 ms^{-1} show the major trade wind and westerly flow regimes and emphasize the rough and windy "roaring forties" of the southern oceans. Light winds of less than 5 ms^{-1} follow the axes of the subtropical ridges and equatorial doldrums or ITCZ.

In high latitudes of the Northern Hemisphere the winds and wave heights are generally low in summer [Fig. 7.2(e) and (f)]. Broad belts of strong westerly flow with mean speeds exceeding 10–15 ms^{-1} are observed in winter [Fig. 7.2(a)] with the North Atlantic having the most extreme conditions. The associated mean significant wave heights are greater than 3 m and have maxima over 5 m [Fig. 7.2(b)]. Waves are generated in the western ocean regions and propagate eastward over several thousand kilometers under the influence of the westerly winds. Thus, the peak wave heights are observed in the central and eastern North Pacific and North Atlantic ocean basins, to the east of the wind maxima [Figs 7.2(a) and (b)].

A marked band of strong westerlies indicates the "roaring forties" between 35 and 50°S. The strongest winds of over 15 ms^{-1} occur in the South Indian Ocean in July and August [Fig. 7.2(e)] and the seasonal change is small, unlike that in the Northern Hemisphere. Since the southern oceans are largely free of continents, very long fetches and broad areas of high waves are observed. Peak mean waves of over 5 m are observed in the South Indian Ocean during July and August [Fig. 7.2(f)] and wave heights generally exceed 2–3 m throughout the year (Fig. 7.2). Winter ice and scattered islands restrict the propagation of waves between South America and Antarctica and also disrupt the westerly flow, leading to a sharp reduction of wave heights from the west to the east of Drake passage [e.g. Figs 7.2(f), 7.3(f) and 7.5(b)]. Regular development of large anticyclones reduces the frequency of strong westerly winds in and poleward of the Tasman Sea [Fig. 7.3(a) and (c)].

The subtropical ridge axis is indicated by the band of minimum wind speed that remains fairly steady along 20–30°N throughout the year but migrates from 20–30°S between the austral summer and winter [Fig. 7.2(a) and (e)]. Swells generated in the westerly belt propagate equatorward into this region. The propagation from the southern oceans, and the sheltering effect of the major land masses are clear in the 50–60% exceedence heights of greater than 2 m of Fig. 7.3(d). Thus, the weak wind regimes of the subtropical ridges are not associated with a similar reduction of significant wave heights [Fig. 7.3(c) and (d)].

The near equatorial oceans experience generally weak winds and low seas. The mean winds are easterly throughout the year in most of the Pacific and Atlantic Oceans (Fig. 7.1), but change direction from easterly to westerly associated with the ebb and flow of the Asian/Australasian monsoon in the western North Pacific and Indian Ocean. The 10% exceedence analyses in Fig. 7.3(b) clearly indicate the protected regions of nearly flat seas east of the major continents in the tropics and throughout the myriad of tropical islands north of Australia. Because of the short time series for the satellite data and narrow observing swath of individual orbits, very severe systems, such as tropical cyclones, are not resolved in these analyses.

7.3 Details of Each Ocean Basin (Figs 7.6–7.20)

High resolution views of each of the major ocean basins are provided in Figs 7.6–7.8 for the Indian Ocean, in Figs 7.9–7.14 for the North and South Pacific, and in Figs 7.15–7.20 for the North and South Atlantic. These figures show more clearly the seasonal changes together with some interesting regional effects.

7.3.1 The Indian Ocean

The Indian Ocean (Figs 7.6–7.8) is dominated by the Asian/Australasian monsoon. In January [Figs 7.1 and 7.6(a)] the northeasterly winter monsoons of Asia result in winds exceeding 7.5 ms^{-1} in the South China Sea, the southwestern Bay of Bengal and off the west coast of Africa. The Australasian summer monsoon is established, with strong northwesterly flow throughout Indonesia and northern Australia. Because of the shielding effect of Asia and the Indonesian Islands, these strong winds result in relatively low seas, which can, however, be sharp and choppy in the shallow seas [Fig. 7.6(b)]. Generally weaker flow and lower seas occur over the whole region in March [Fig. 7.6(a) and (b)]. In May and June the massive Asian summer monsoon becomes established [Figs 7.1 and 7.6(a), (b)], leading to a strong southeasterly flow in the South China Sea, Bay of Bengal and western Arabian Sea, together with a broad expanse of easterly trade winds in the Southern Hemisphere. Weakening and reversal of flow follows in September and November as the cycle returns to the Australasian summer monsoon [Fig. 7.6(c)].

Of particular note is the band of very high winds off Somalia and Oman, the sustained trade wind jet south of the equator, the belt of westerly flow in the southern ocean, the weak winds off western Australia, and the Australasian monsoon region north of Australia [Fig. 7.6(c)]. The "Somalia Jet" (Findlater, 1969) results from southeasterly flow into the Indian summer monsoon being channelled along the West African highlands and also being focused by the sharp gradients in ocean temperatures that develop offshore. The jet develops rapidly in June and disappears in September [Fig. 7.7(a)(i)]. It results in sustained winds of over 15 ms^{-1} and the highest mean waves experienced anywhere in the tropics [Fig. 7.7(b)(i)]. A sustained region of easterly trade flow south of the equator is located equatorward of a semi-permanent high pressure region. This flow reaches a maximum strength and areal extent in the austral winter [Figs 7.6(c) and 7.7(a)(ii)]. Further poleward the westerly flow and rough seas are the strongest and have the greatest extent of anywhere on earth, thus leading to the term "roaring forties" [Fig. 7.7(a)(iii) and (b)(iii)]. The weak winds and low seas observed off the west Australian coast [Figs 7.8(a), 7.8(b), 7.7(a)(iv) and 7.7(b)(iv)] are a result of a broad trough that extends down the west coast and arises from heating of the easterly flow passing over the eastern ranges and across the inland deserts. Strong southeasterly trade flow develops along the southern side of Papua New Guinea during the Australasian winter monsoon in July, but the limited fetch restricts the development of high seas [Figs 7.6(c), 7.6(d), 7.7(a)(v) and 7.7(b)(v)].

7.3.2 The North Pacific Ocean

The North Pacific Ocean (Figs 7.9–7.11) has its northern half torn by high westerly winds of over 15 ms^{-1} in winter [Fig. 7.9(a)], which generate sustained seas of over 5 m to pound the western shore of North America [Fig. 7.9(b)]. These winter winds and high seas abate in the boreal spring and summer then redevelop in autumn [Figs 7.10(a)(i) and (ii), and 7.10(b)(i) and (ii)].

The tropical North Pacific is under the influence of the Asian monsoon on the western side and sustained trade wind flow throughout the year in central and eastern tropical regions. The easterly trade flow strengthens and extends to the equator in January [Figs 7.9(a) and 7.10(a)(iii)] and generates seas of up to 3 m [Fig. 7.10(b)(iii)]. The northeasterly winter monsoon is well established in the South China Sea in January [Figs 7.1 and 7.10(a)(iv)], then reverses to the westerly summer monsoon of the South China and Philippine Seas by July. The summer monsoon is separated from the easterly trade flow by a band of low winds along a monsoonal trough extending southeastward from Japan [Fig. 7.9(c)]. Generally light winds are experienced in the eastern equatorial Pacific throughout the year [Fig. 7.10(a)(v)], and waves of 2 m are partially maintained by swell propagation from high latitudes [Fig. 7.10(b)(v)].

7.3.3 The South Pacific Ocean

The South Pacific Ocean (Figs 7.12–7.14) experiences strong westerly flow and high seas throughout the year in its southern half, with a broad maximum extending from south of New Zealand to the Drake Passage in September [Figs 7.12(c) and (d)]. As with all southern ocean regions, these westerly winds change only slightly with season [Fig. 7.13(a)(i)]. In the tropics, the trade wind flow is less extensive and weaker than that observed in the North Pacific, although the overall wave climatology is similar [Figs 7.10(a)(iii), 7.13(a)(ii), 7.10(b)(iii) and 7.13(b)(ii)]. Propagation of swell from the generation region in the westerlies, together with regeneration in the easterly trade flow produces a broad area of seas in excess of 2 m that spreads throughout most of the eastern and central South Pacific during winter and spring [Fig. 7.12(d)]. The shielding influence of major islands on ocean swell propagating out of the westerly flow regime is illustrated by the time series for the Tasman Sea and northeast of New Zealand in Fig. 7.13(b)(iii) and (iv).

7.3.4 The North Atlantic Ocean

The North Atlantic Ocean (Figs 7.15–7.17) is the windiest and roughest ocean, with strong winter westerlies of over 15 ms^{-1} [Figs 7.15(a) and 7.16(a)(i)] generating a band of greater than 5 m seas between Newfoundland and the British Isles [Fig. 7.15(b)]. As with the Pacific, a strong seasonal variation is observed with much milder conditions occurring in summer. Shielding by the British Isles protects the North Sea from the worst of these conditions [Fig. 7.16(a)(ii) and (b)(ii)].

The trade flow in the tropics is sustained throughout the year, but is weaker than that in the North Pacific [Figs 7.10(a)(iii) and 7.16(a)(iii)]. This trade flow extends into the Caribbean Sea and Gulf of Mexico, but these regions are also effectively shielded from high seas by Florida and the ring of islands extending from Cuba to the Windward Isles [Fig. 7.16(a)(iv) and (b)(iv)]. A small monsoon is experienced in and west of the Gulf of Guinea, where generally light flow throughout much of the year becomes westerly in July [Figs 7.16(a)(v) and 7.15(c)].

7.3.5 The South Atlantic Ocean

The South Atlantic Ocean (Figs 7.18–7.20) experiences a similar seasonal cycle to the South Pacific, with strongest westerly flow and largest extent of high seas in winter and spring [Fig. 7.18(c) and (d)] and steady trade wind flow throughout the year. The barrier to the west provided by the high mountains of South America produces a slight reduction of winds and a marked reduction of waves in the south west throughout the year [Fig. 7.19(a)(i), (ii) and (b)(i), (ii)]. High seas are regenerated only in the far east of the basin [Fig. 7.18(b) and (d)]. The trade flow in this ocean is the weakest of all except the north Indian Ocean.

7.4 Latitude, Longitude and Great-Circle Sections (Figs 7.21–7.28)

The variation of mean winds and wave heights, together with probabilities of exceedence along selected latitudes, longitudes and great-circle tracks are shown in Figs 7.21–7.28. The latitude and great-circle sections confirm the marked increase in wave heights away from the east coasts of the major land masses in the westerly wind regimes. These wave increases are due to fetch increases and are not associated with a similar wind change [Fig. 7.23(a) and (b)].

7.5 Selected Output at Points (Figs 7.29–7.31)

Output in the form of graphs showing seasonal variations in wind speed and wave height and probability of exceedence as a function of wind speed and wave height are presented in Figs 7.30 and 7.31. Output has been presented at a total of 126 points throughout the world's oceans as shown in Fig. 7.29.

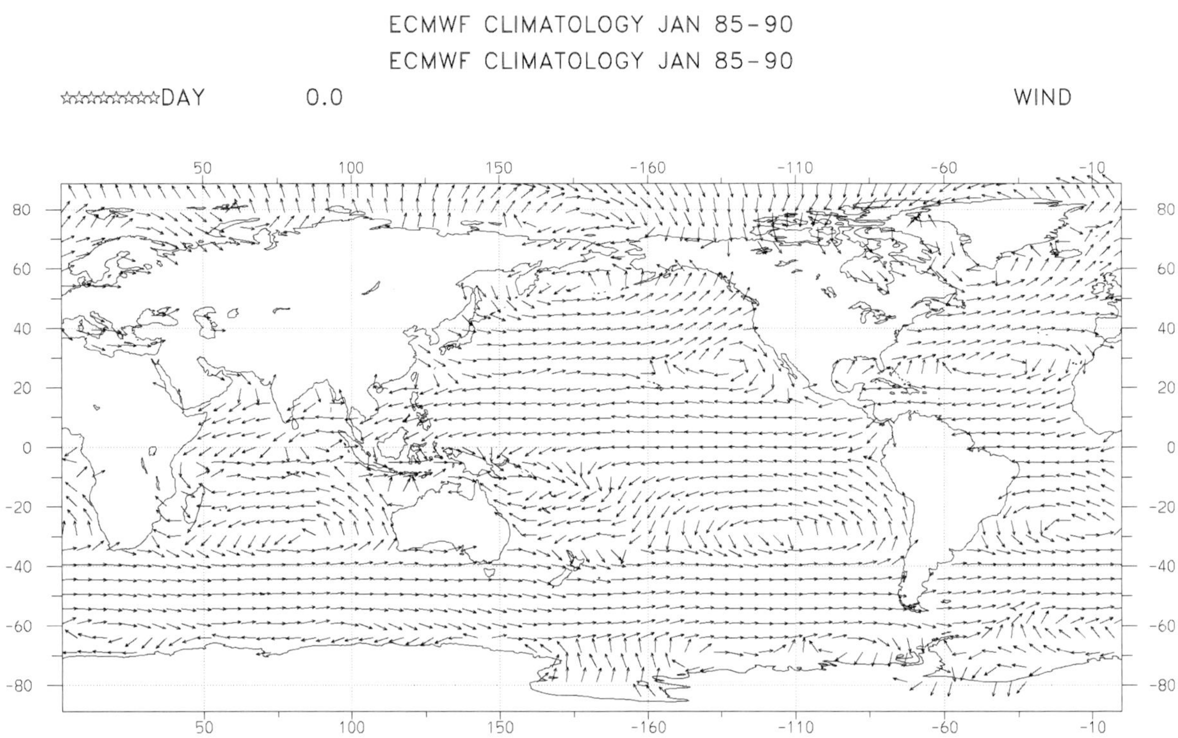

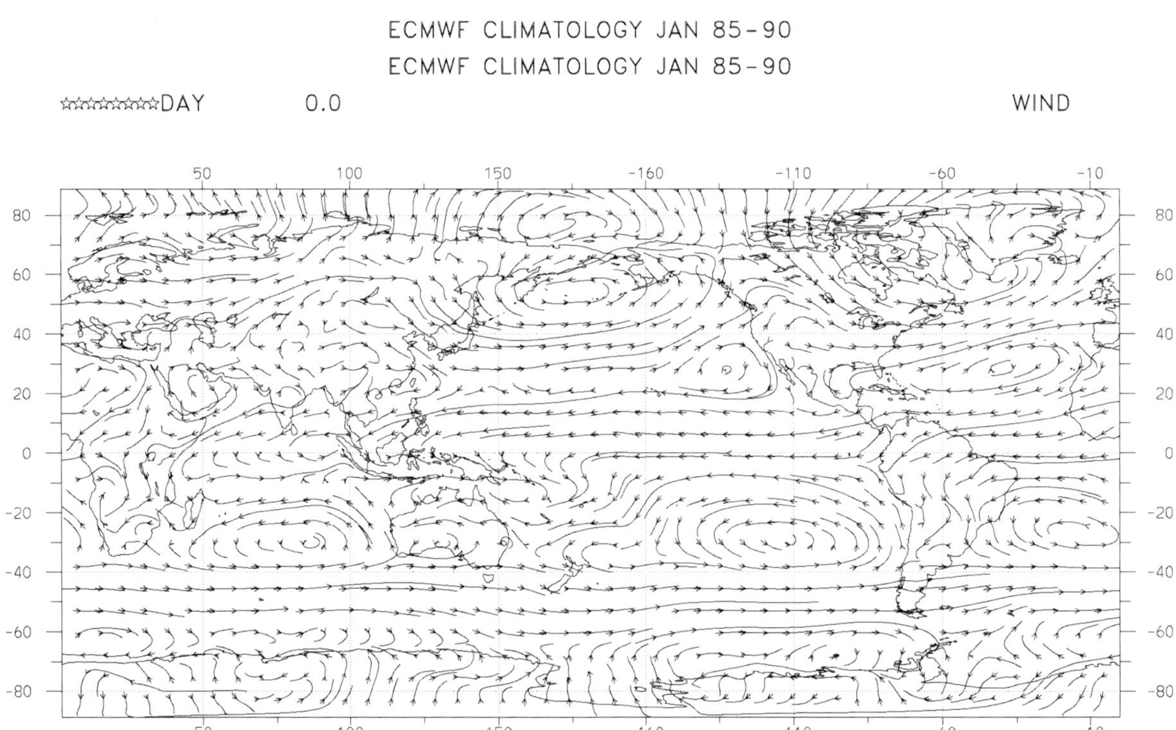

Fig. 7.1(a) Global values of Mean Wind Direction for January. Vectors showing the mean direction are shown at the top and stream lines at the bottom.

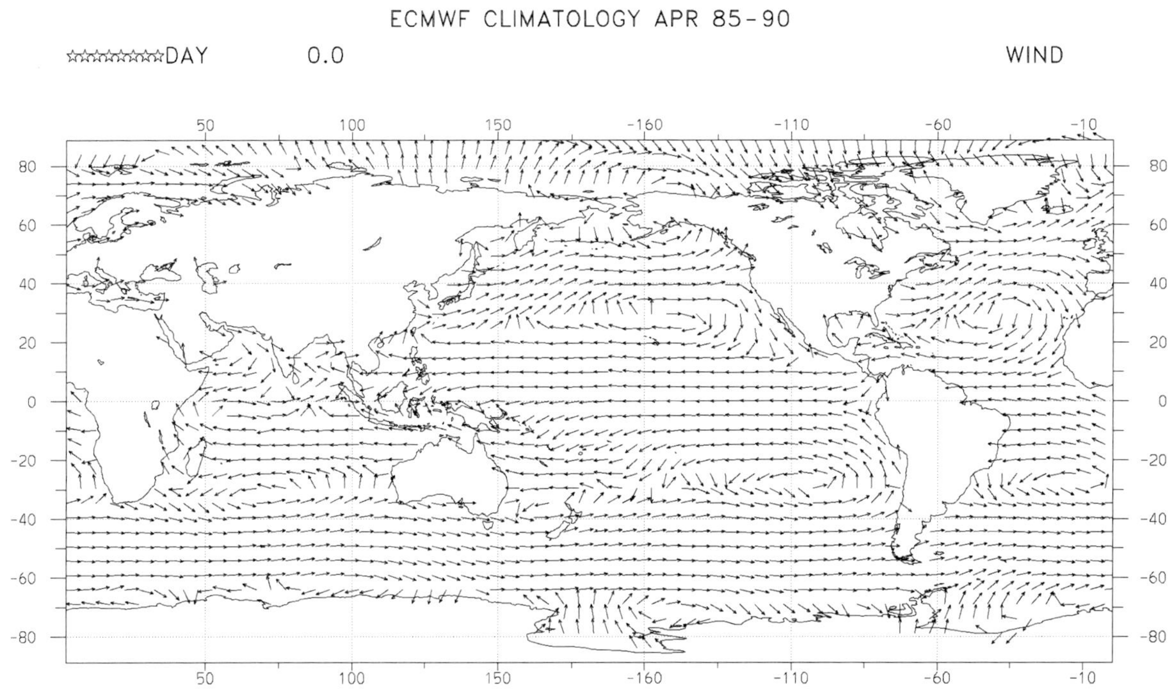

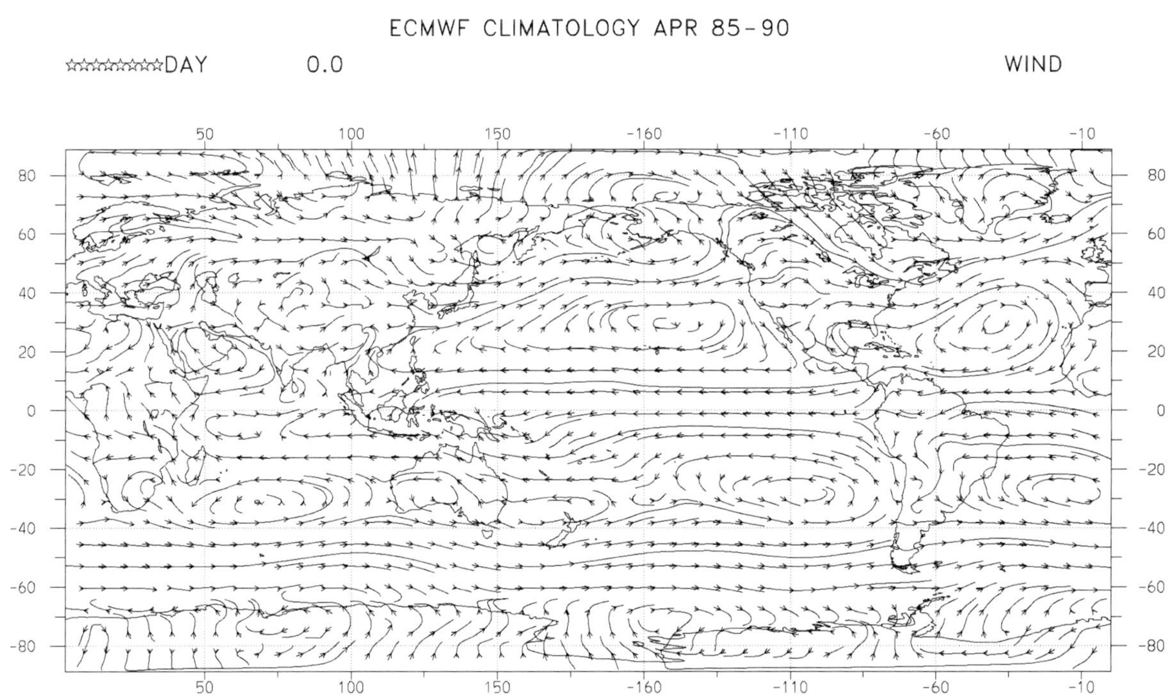

Fig. 7.1(b) Global values of Mean Wind Direction for April. Vectors showing the mean direction are shown at the top and stream lines at the bottom.

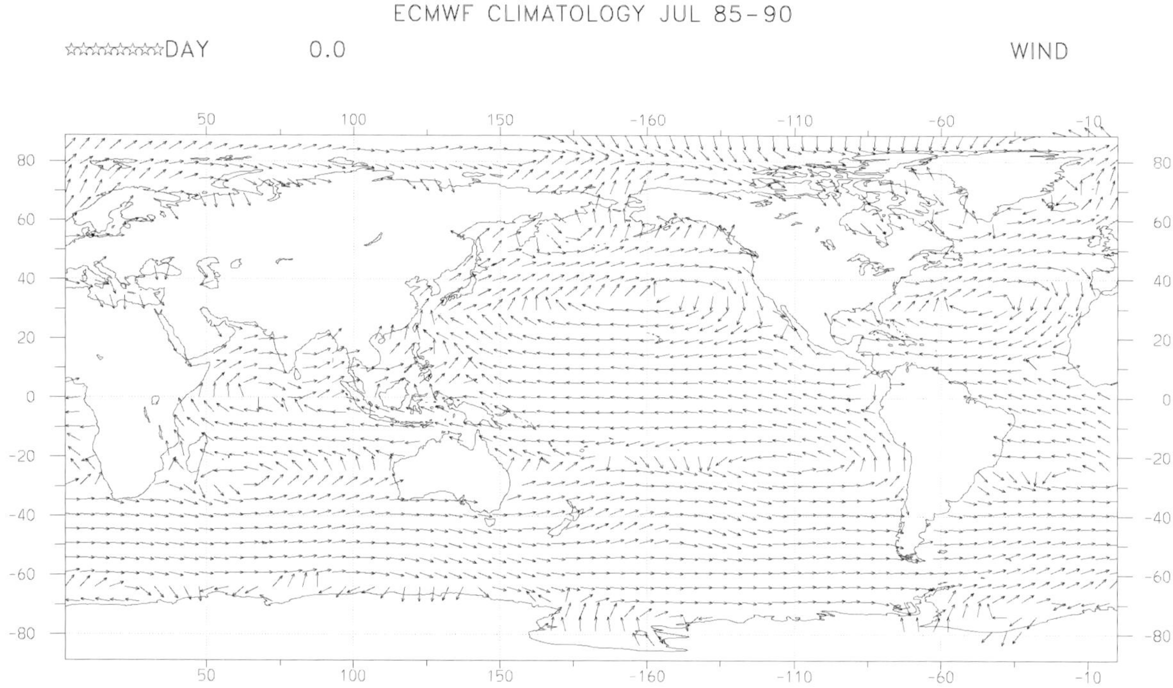

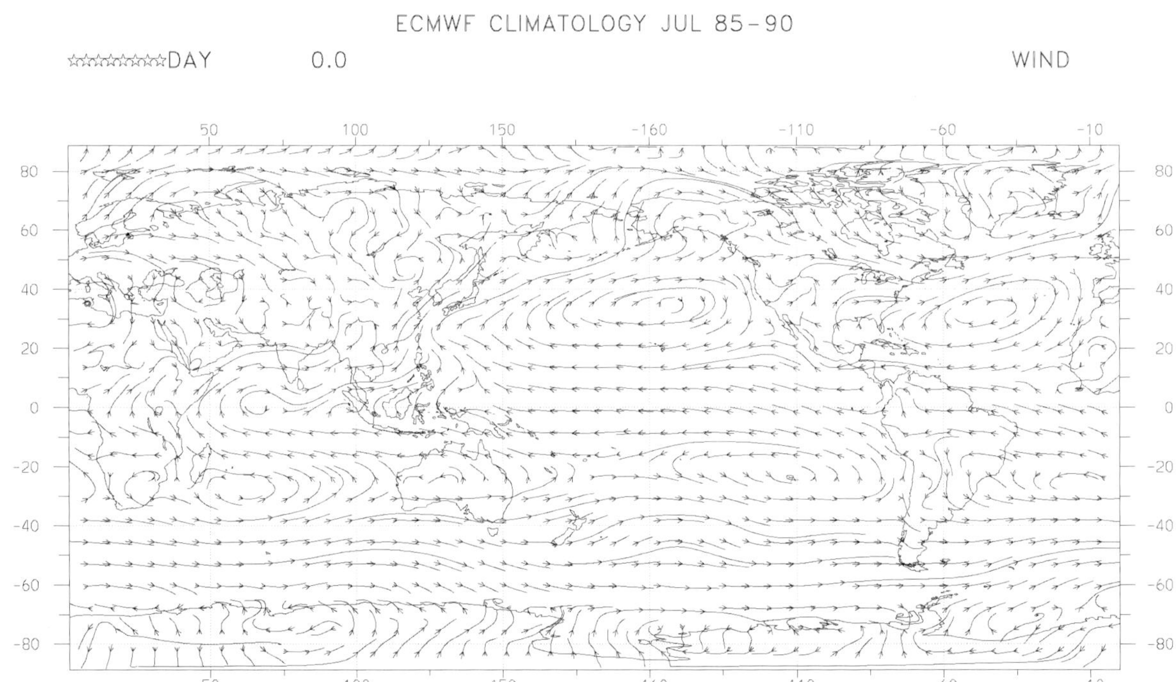

Fig. 7.1(c) Global values of Mean Wind Direction for July. Vectors showing the mean direction are shown at the top and stream lines at the bottom.

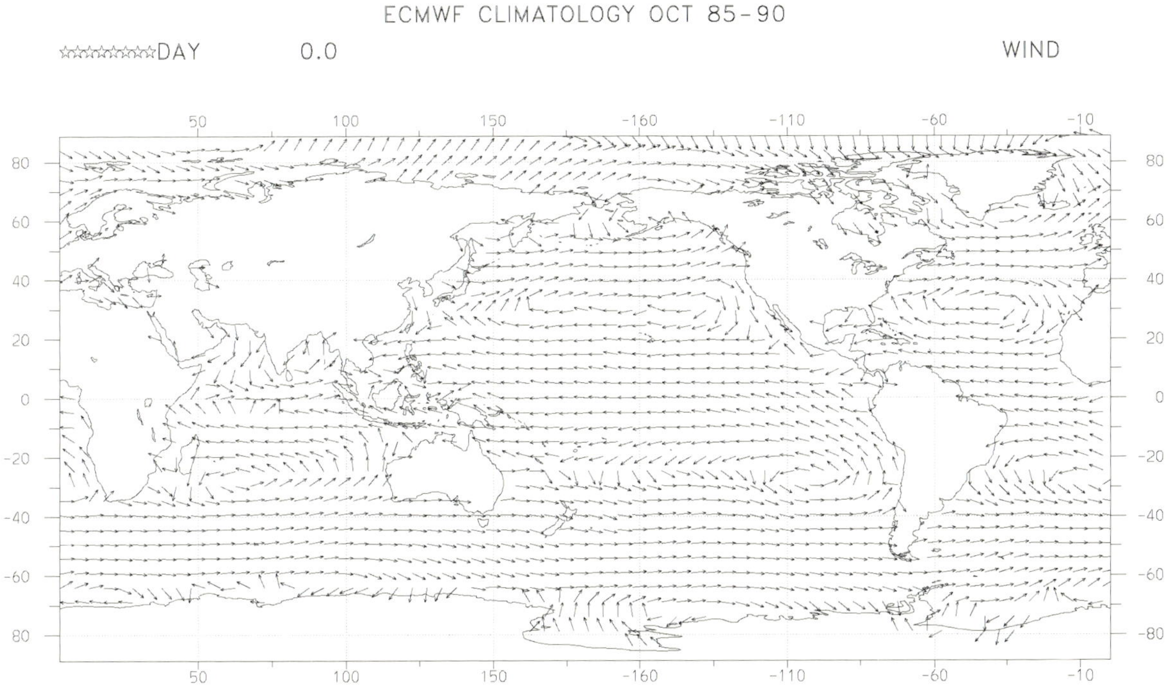

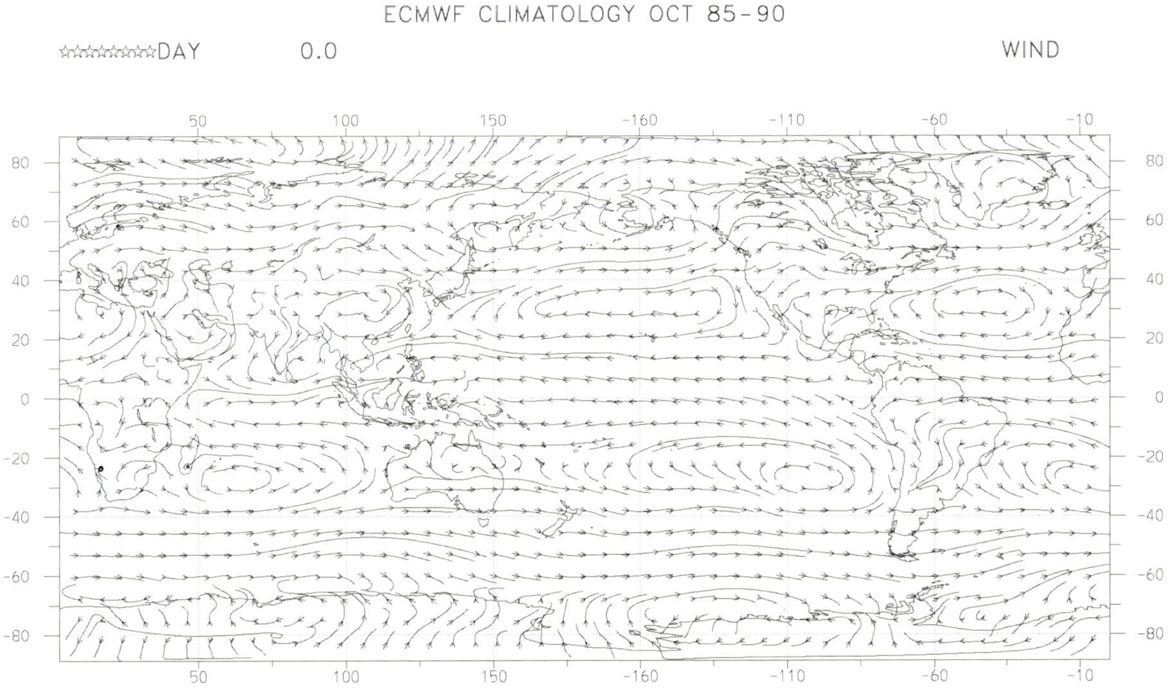

Fig. 7.1(d) Global values of Mean Wind Direction for October. Vectors showing the mean direction are shown at the top and stream lines at the bottom.

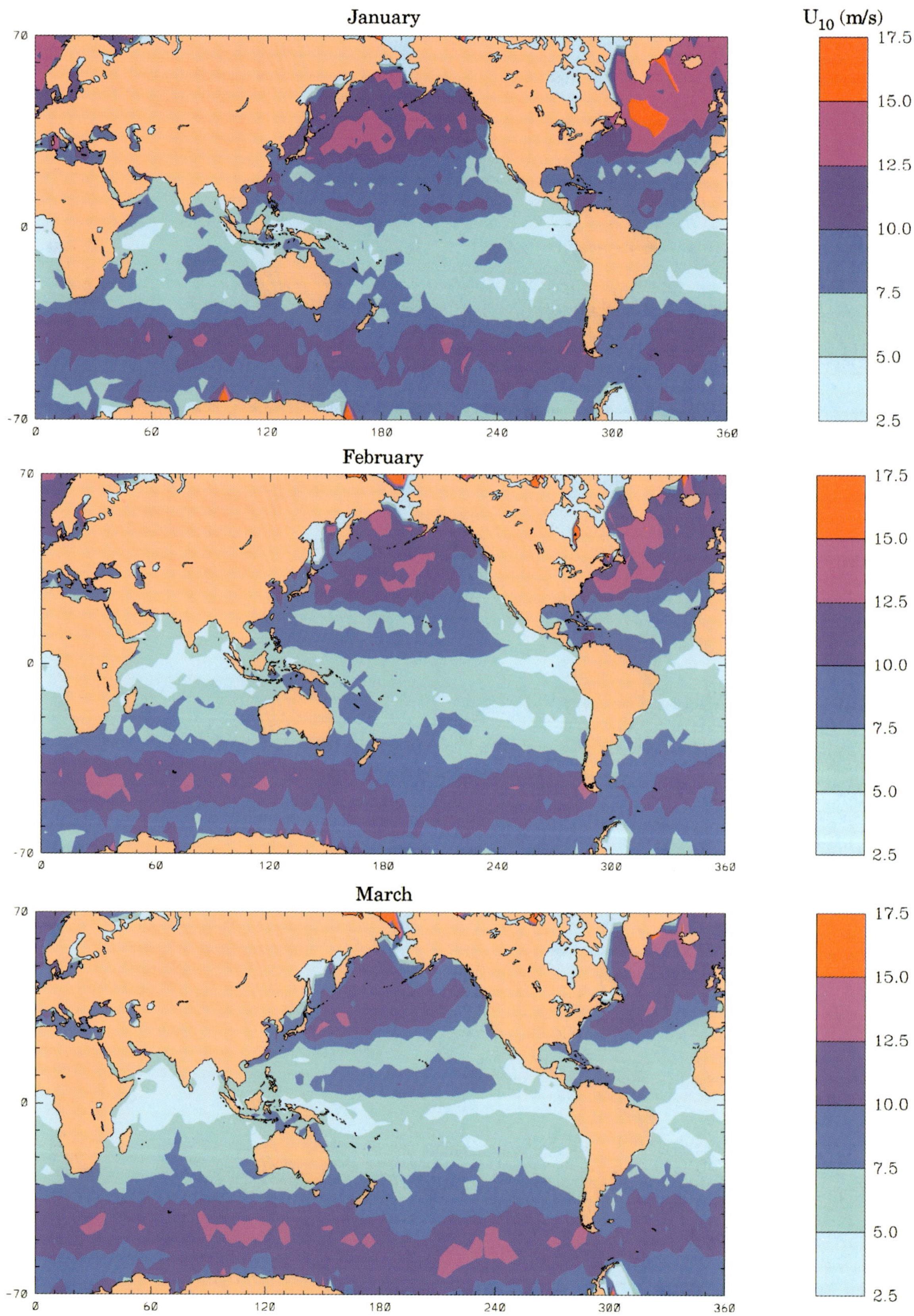

Fig. 7.2(a) Global values of Mean Monthly Wind Speed for January (top), February (middle) and March (bottom).

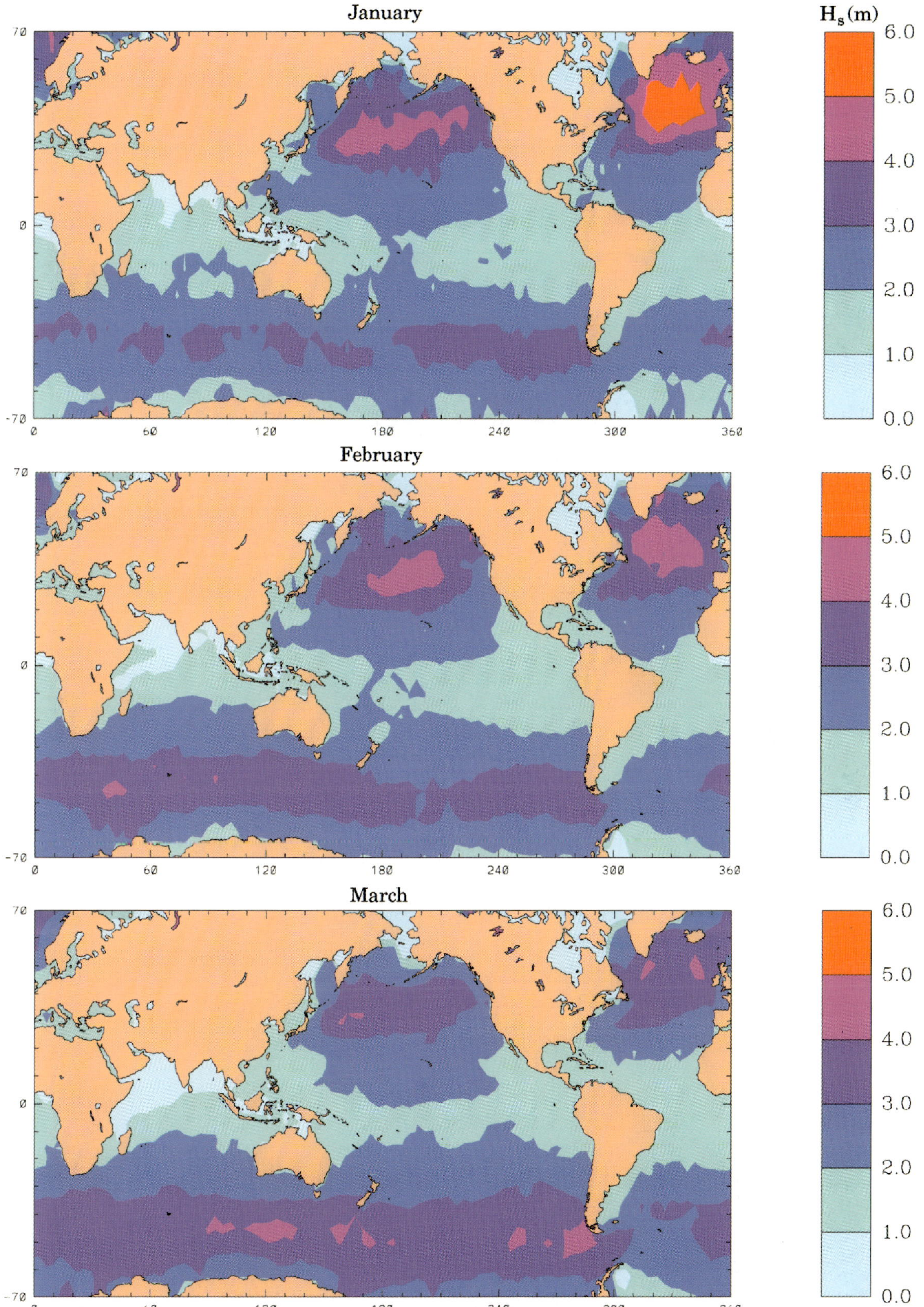

Fig. 7.2(b) Global values of Mean Monthly Significant Wave Height for January (top), February (middle) and March (bottom).

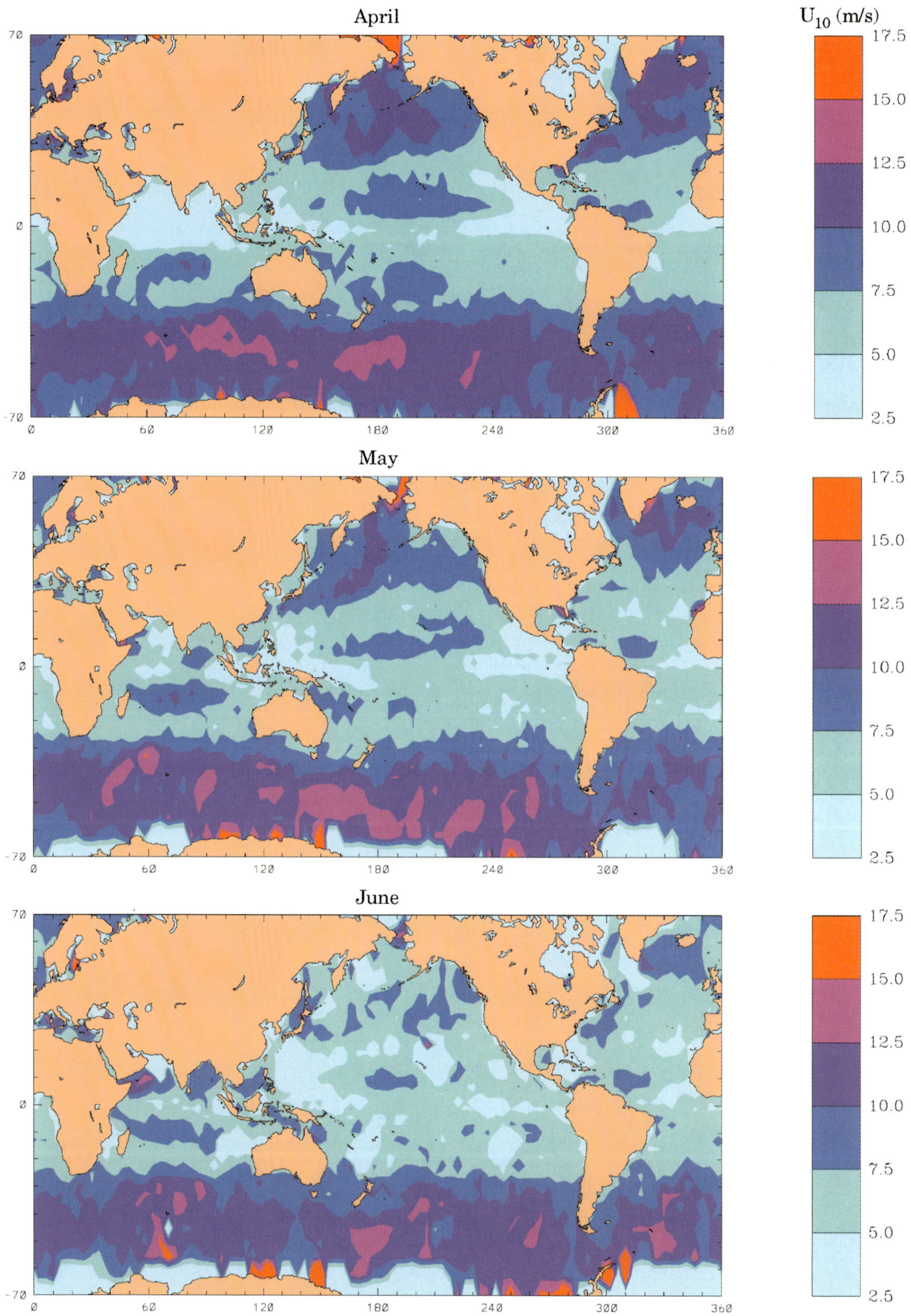

Fig. 7.2(c) Global values of Mean Monthly Wind Speed for April (top), May (middle) and June (bottom).

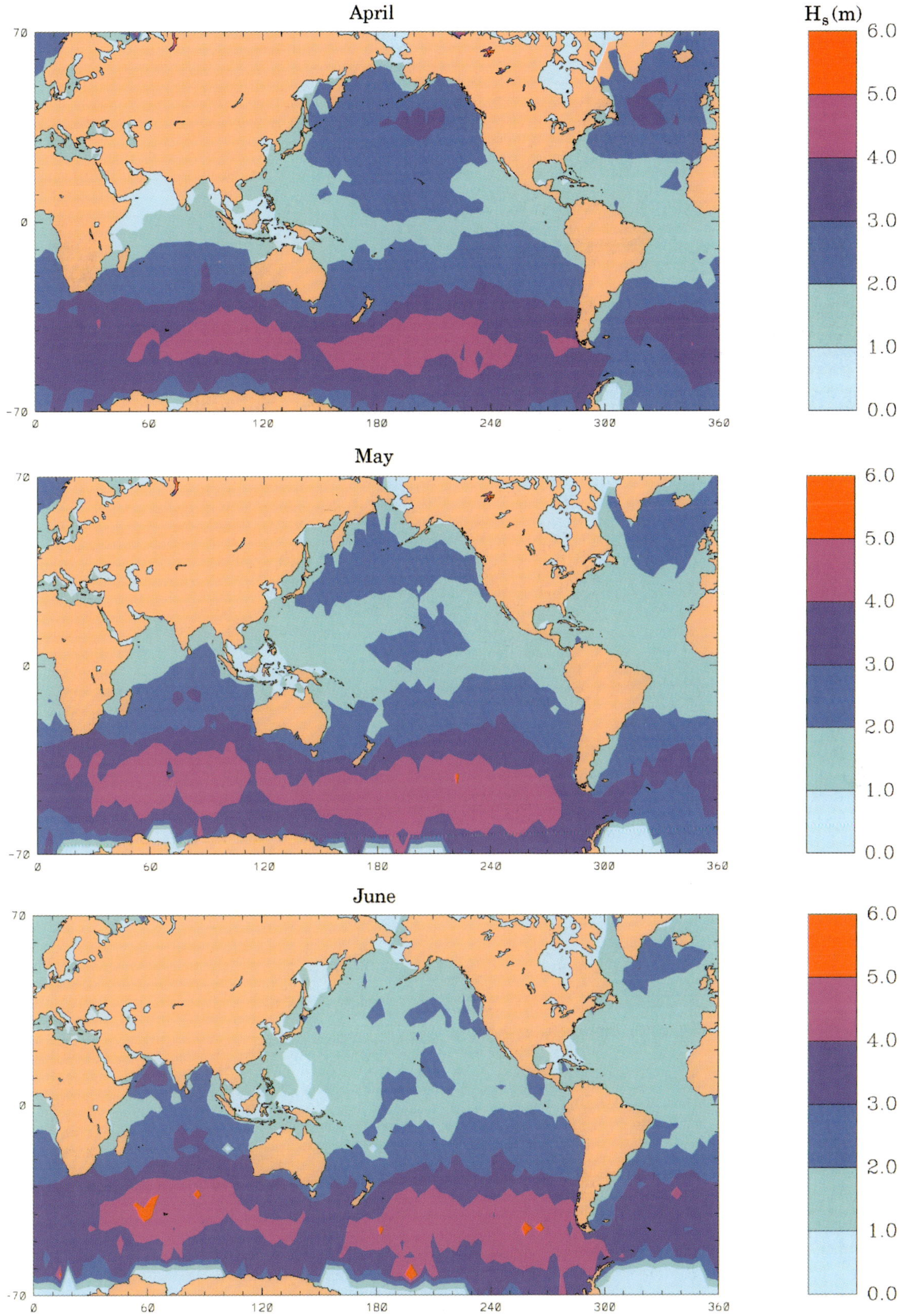

Fig. 7.2(d) Global values of Mean Monthly Significant Wave Height for April (top), May (middle) and June (bottom).

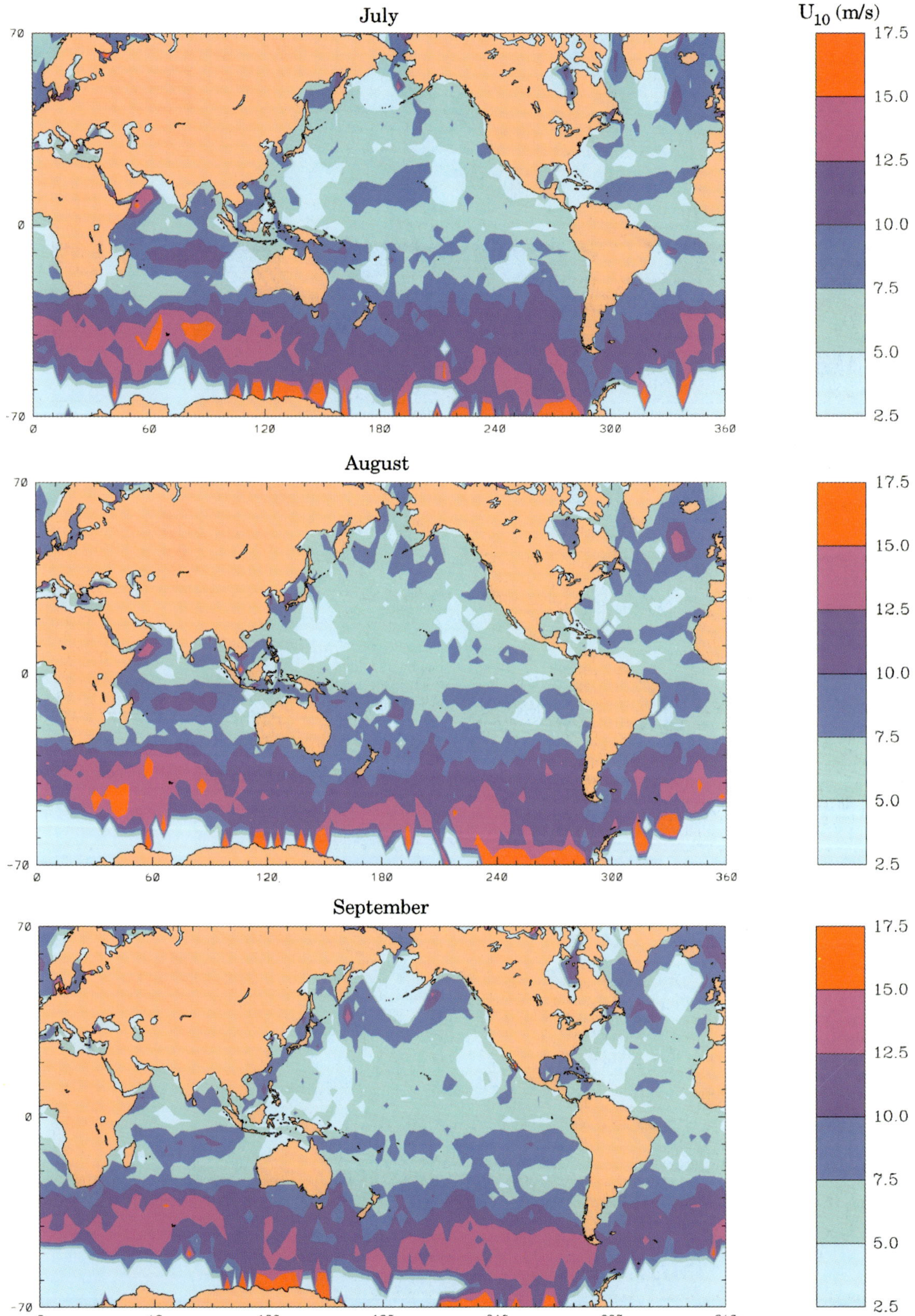

Fig. 7.2(e) Global values of Mean Monthly Wind Speed for July (top), August (middle) and September (bottom).

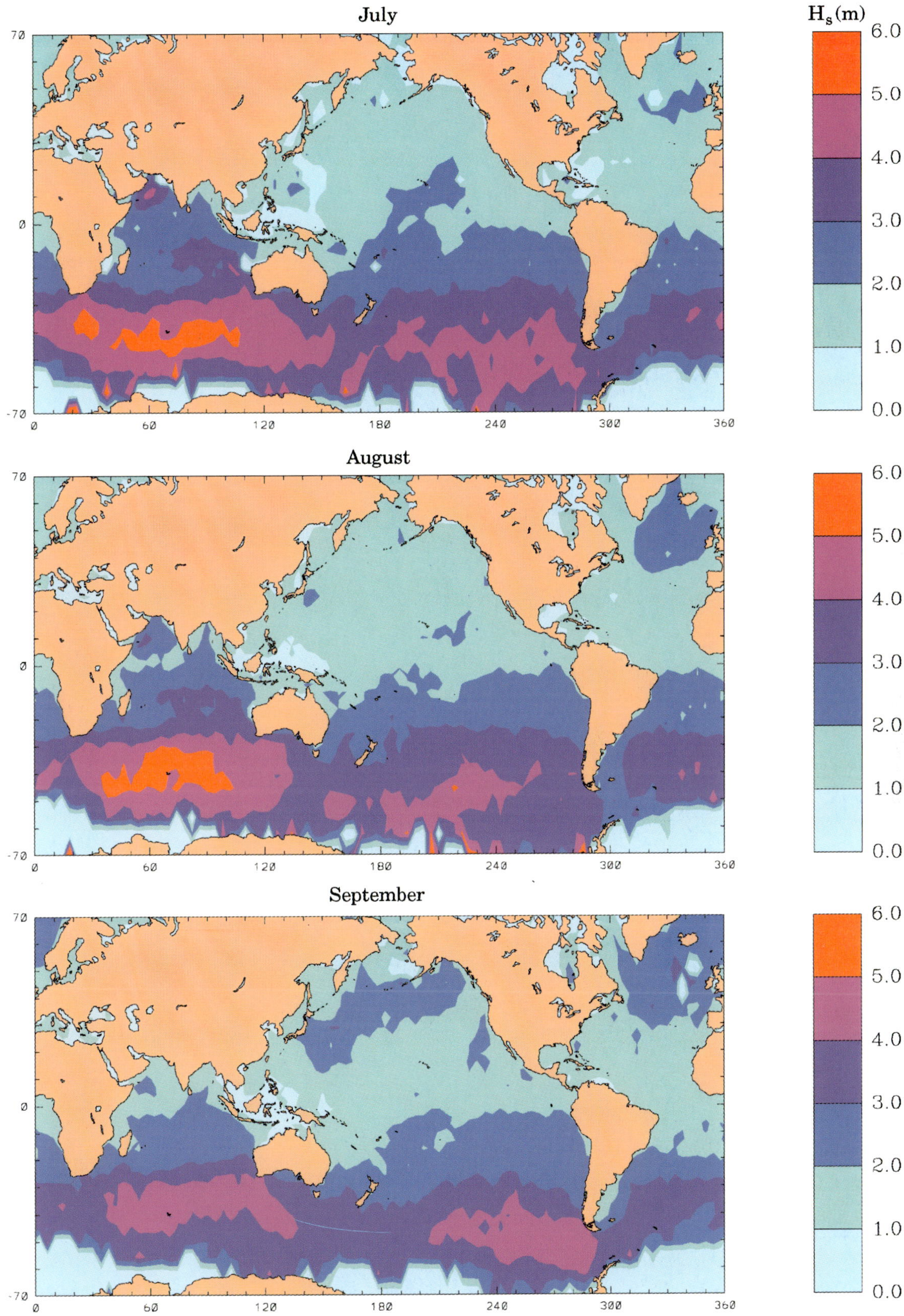

Fig. 7.2(f) Global values of Mean Monthly Significant Wave Height for July (top), August (middle) and September (bottom).

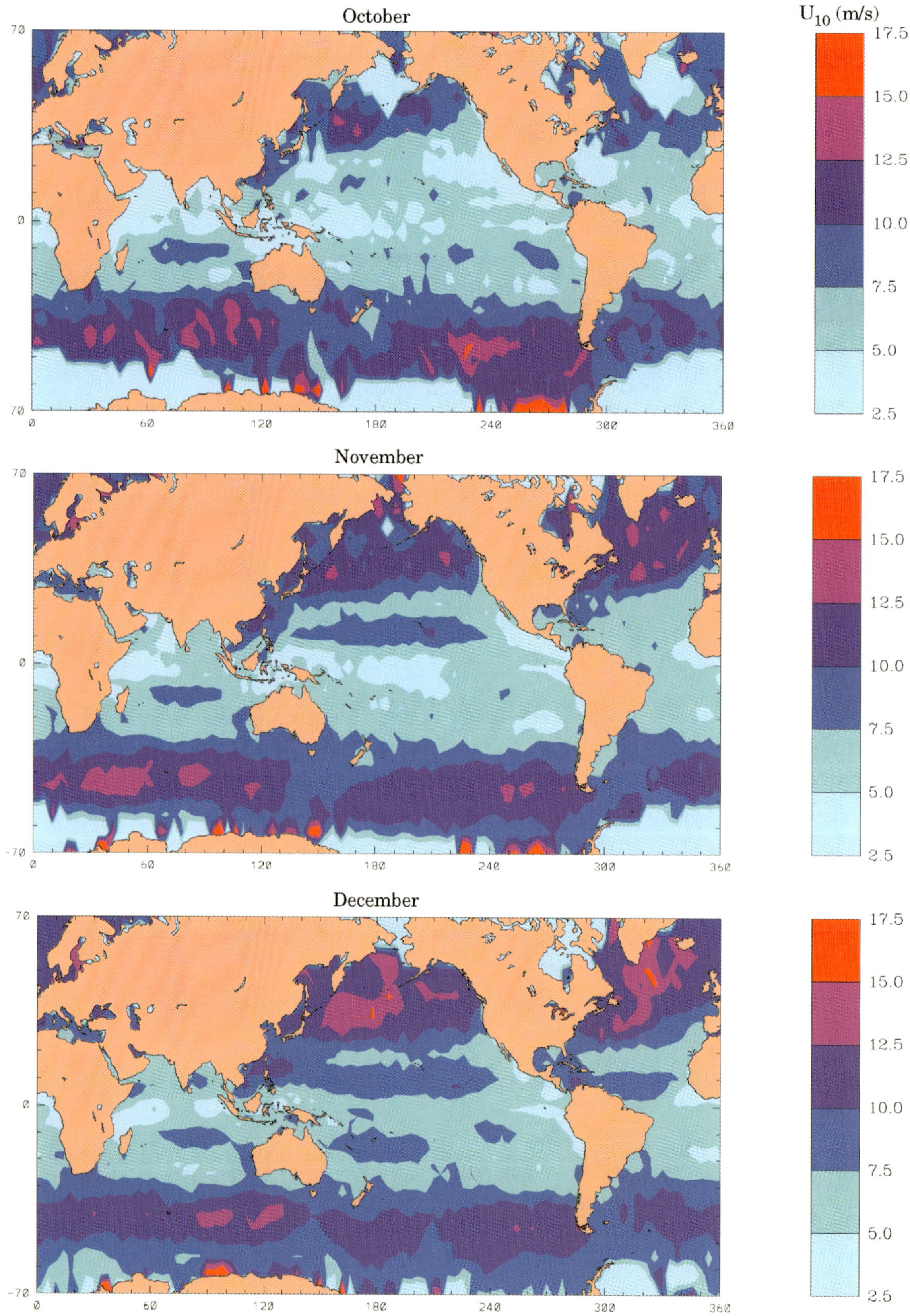

Fig. 7.2(g) Global values of Mean Monthly Wind Speed for October (top), November (middle) and December (bottom).

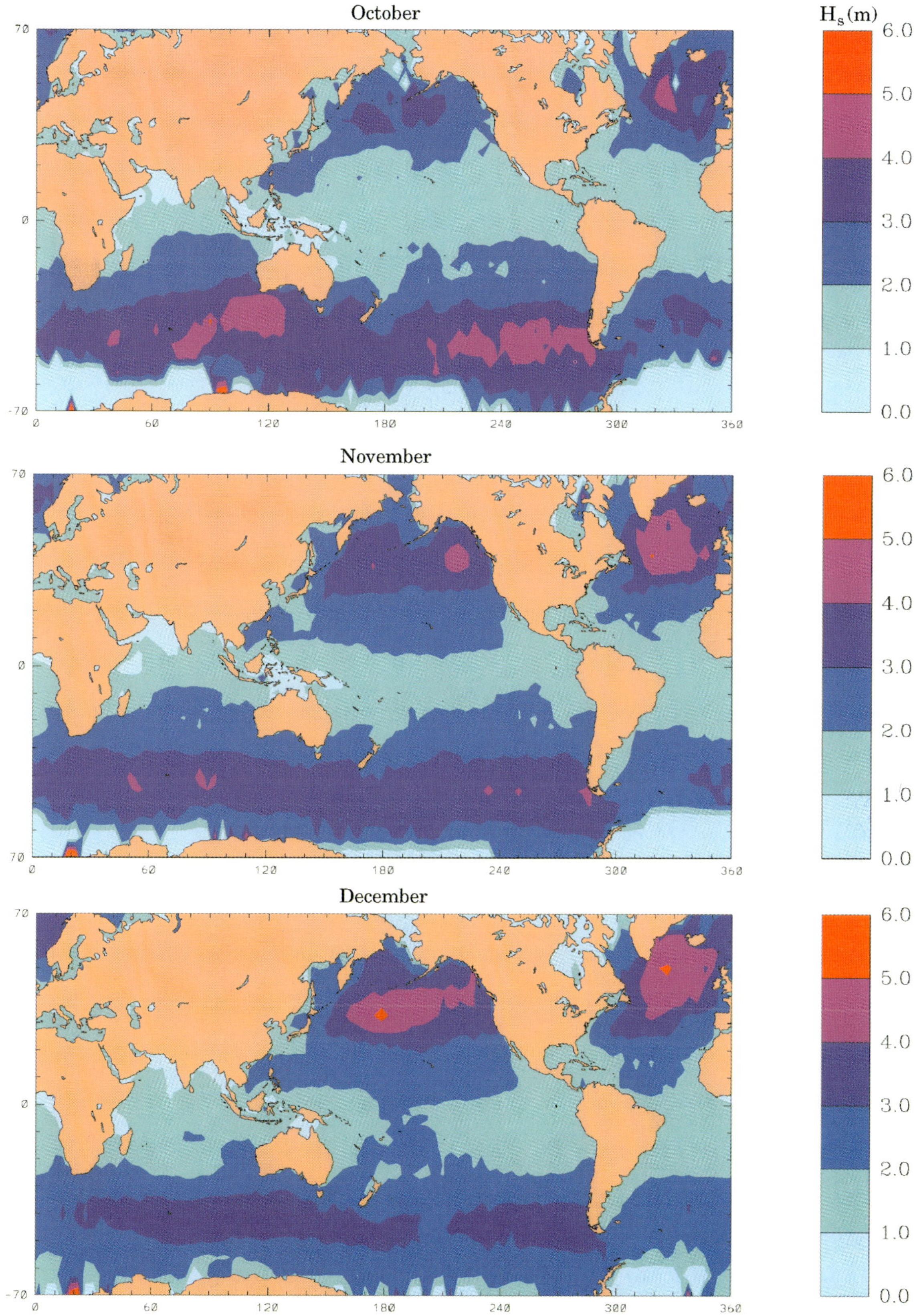

Fig. 7.2(h) Global values of Mean Monthly Significant Wave Height for October (top), November (middle) and December (bottom).

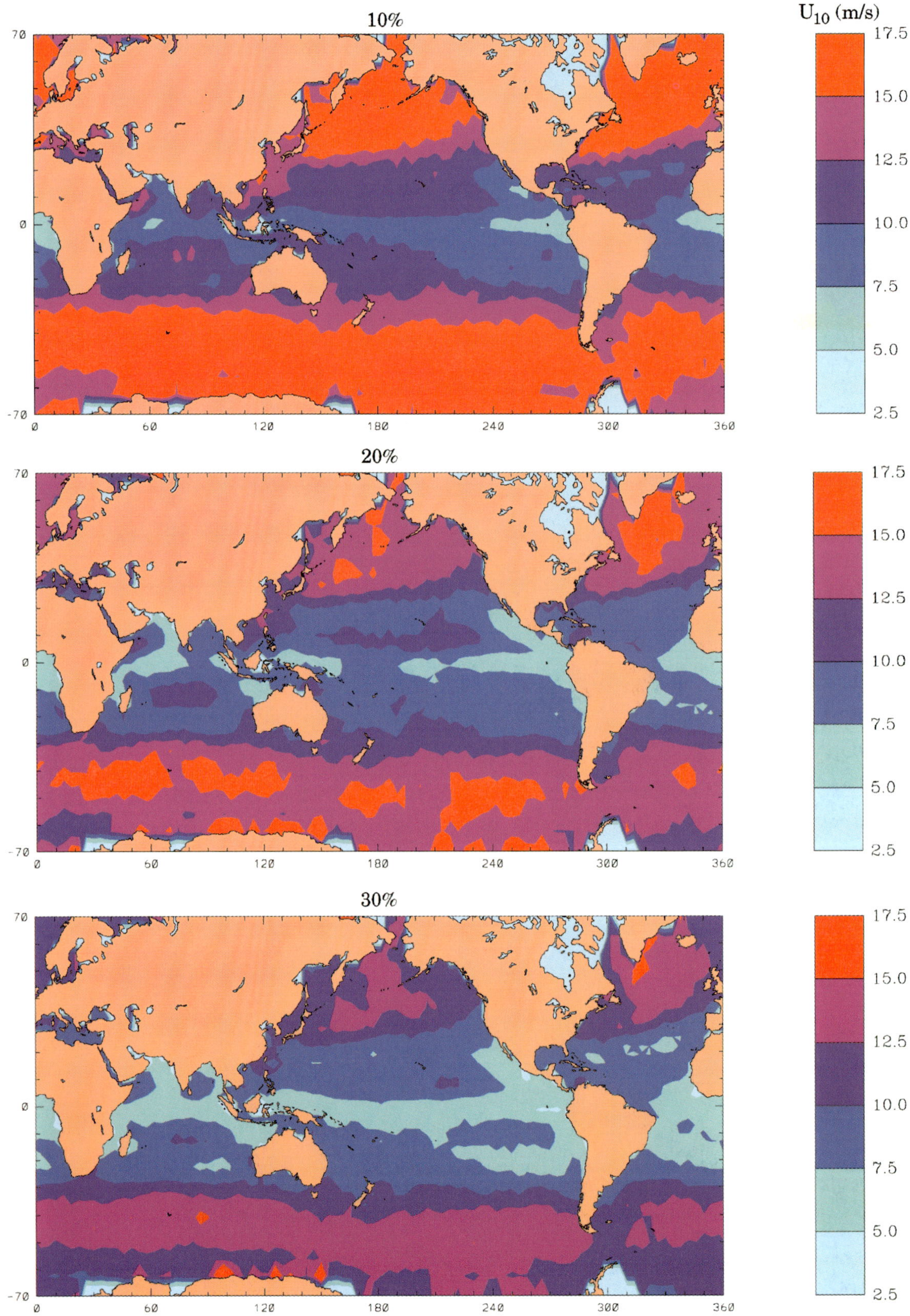

Fig. 7.3(a) Global values for the Wind Speed which will be exceeded 10% (top), 20% (middle) and 30% (bottom) of the time.

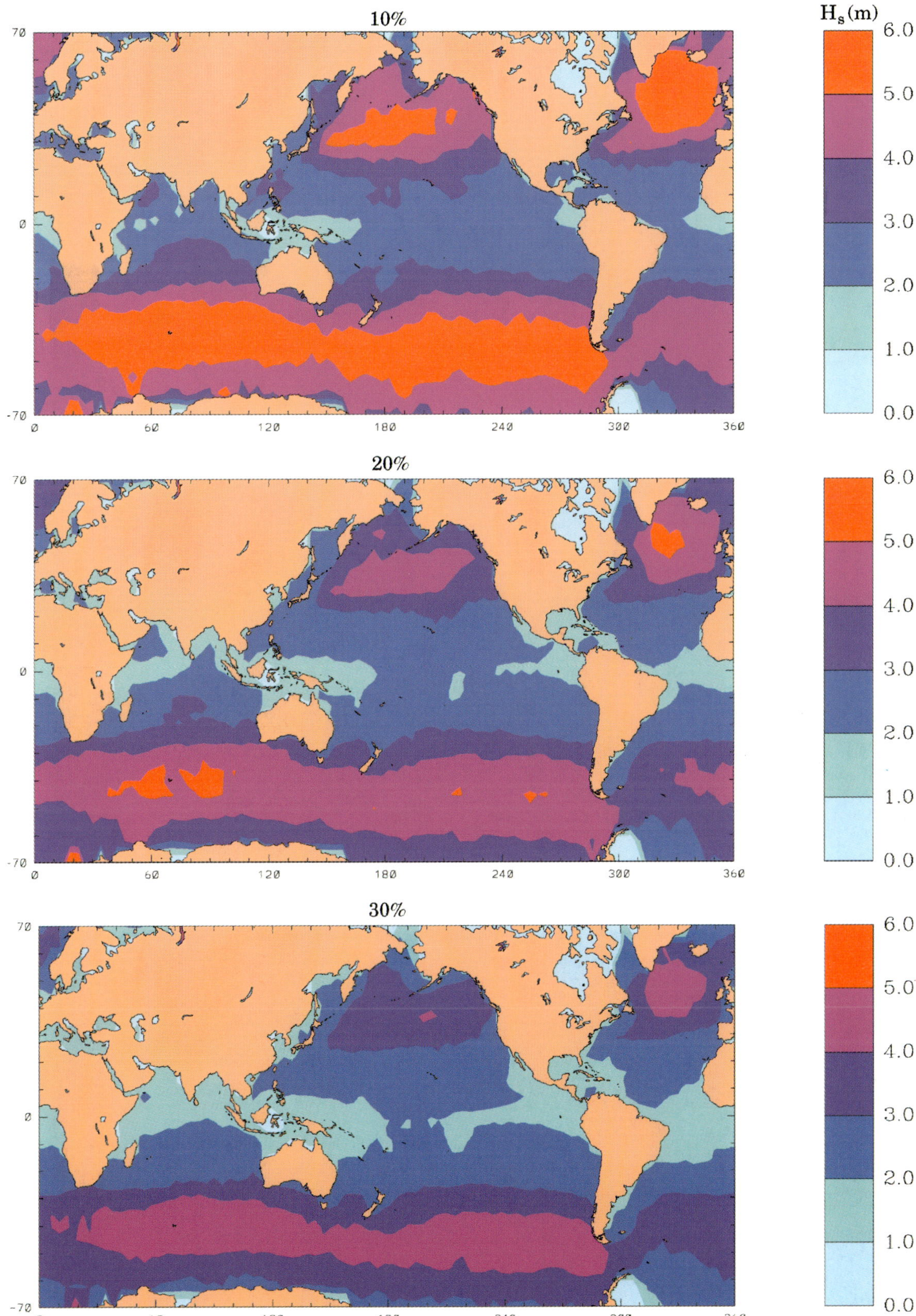

Fig. 7.3(b) Global values for the Significant Wave Height which will be exceeded 10% (top), 20% (middle) and 30% (bottom) of the time.

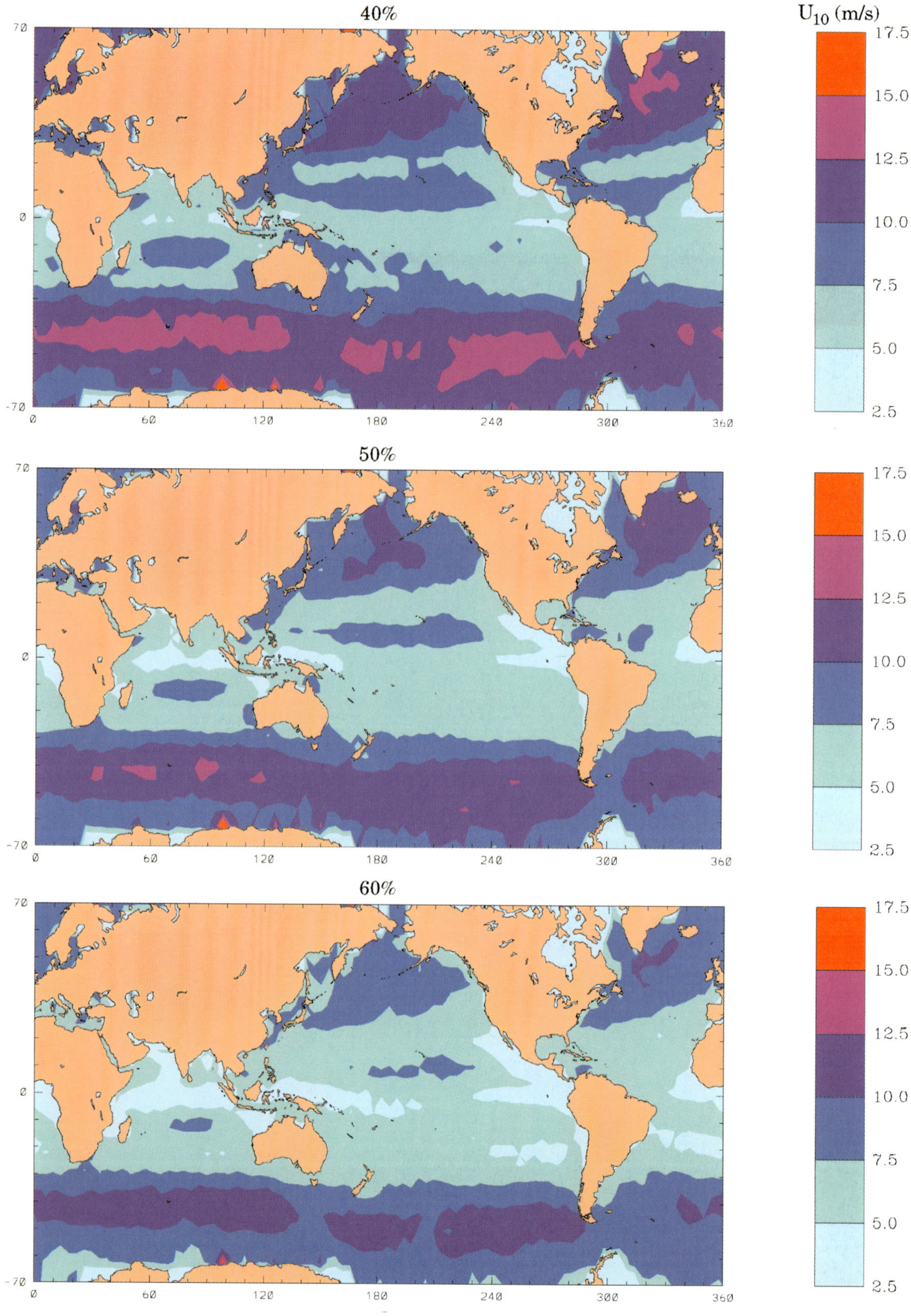

Fig. 7.3(c) Global values for the Wind Speed which will be exceeded 40% (top), 50% (middle) and 60% (bottom) of the time.

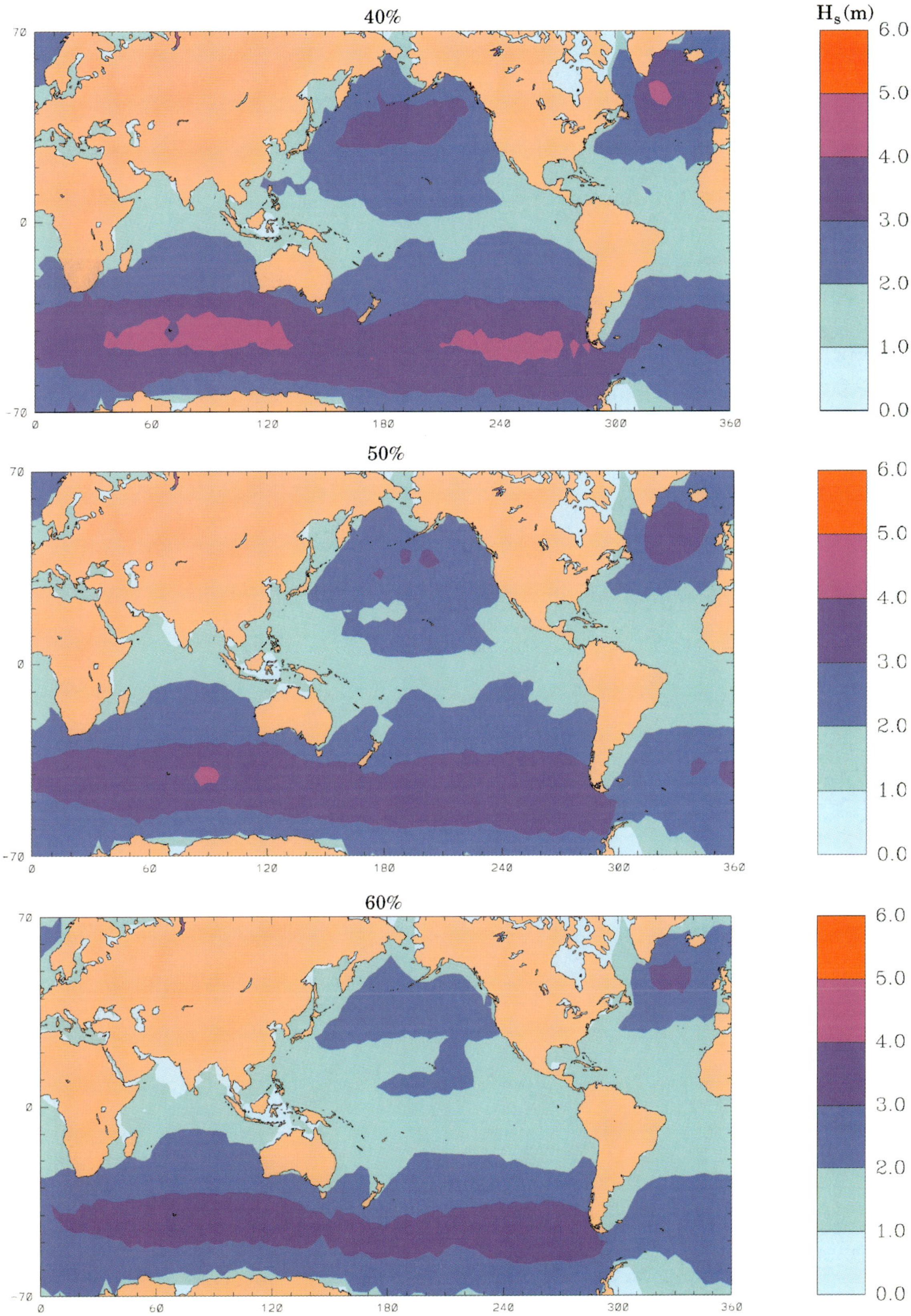

Fig. 7.3(d) Global values for the Significant Wave Height which will be exceeded 40% (top), 50% (middle) and 60% (bottom) of the time.

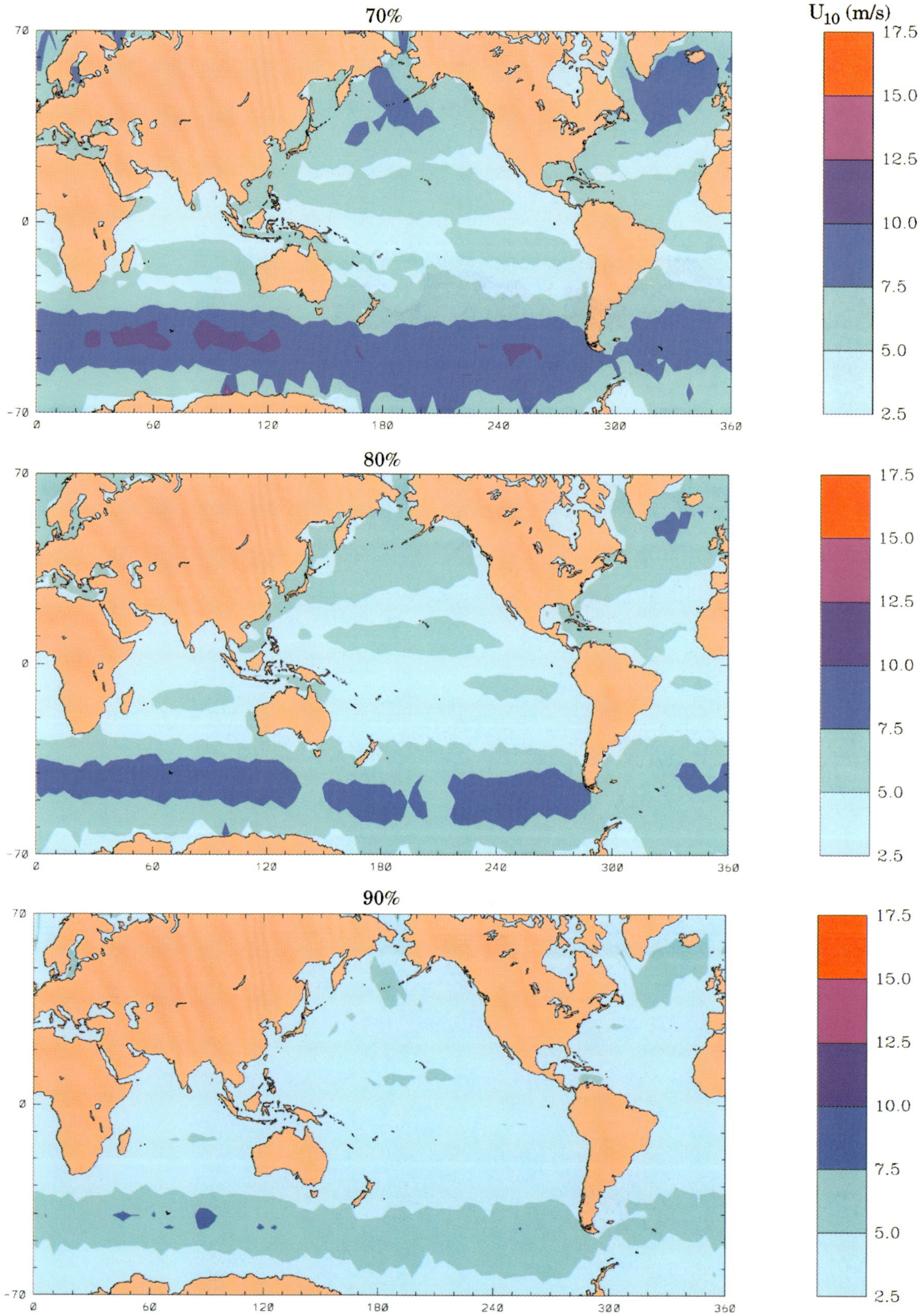

Fig. 7.3(e) Global values for the Wind Speed which will be exceeded 70% (top), 80% (middle) and 90% (bottom) of the time.

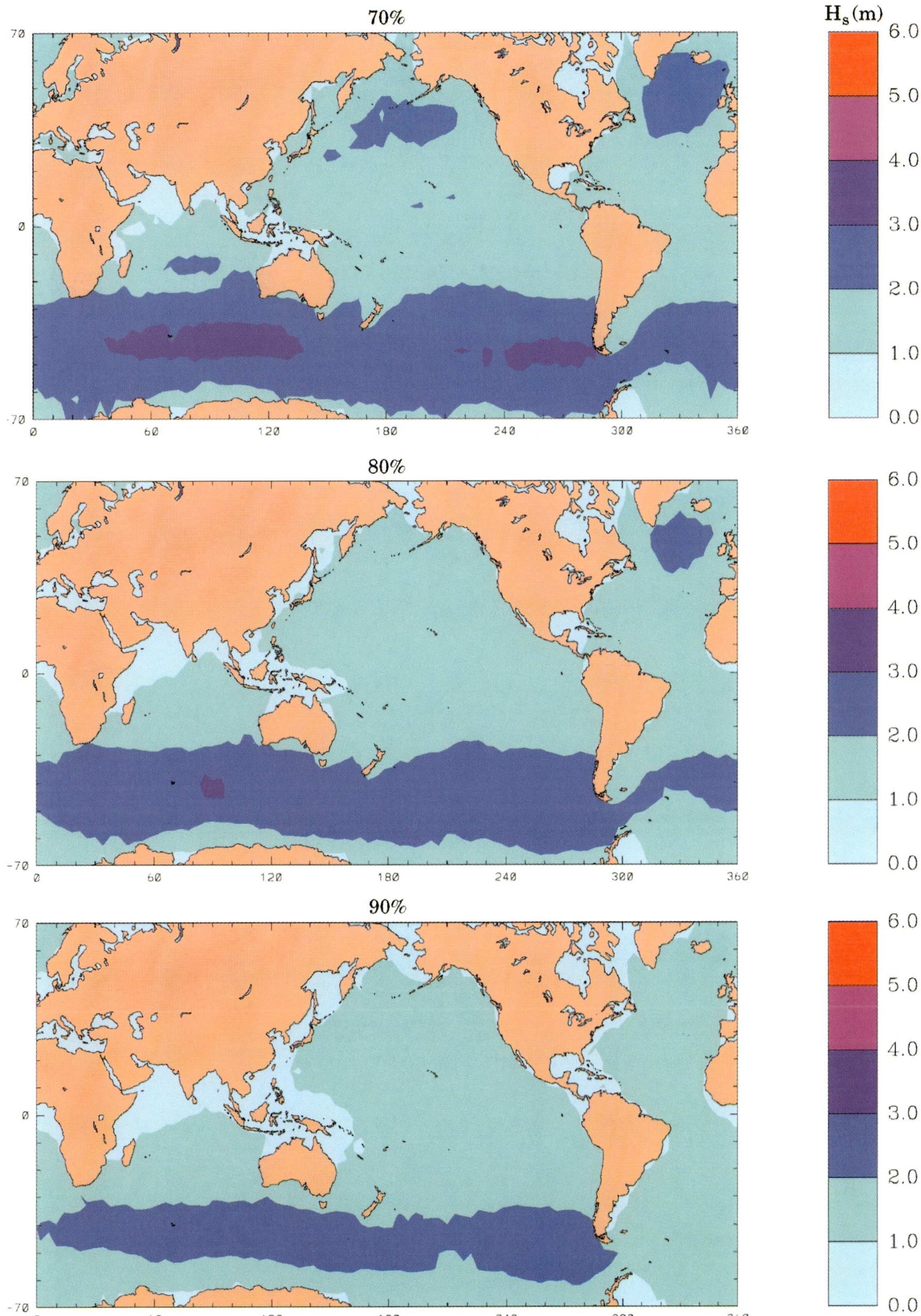

Fig. 7.3(f) Global values for the Significant Wave Height which will be exceeded 70% (top), 80% (middle) and 90% (bottom) of the time.

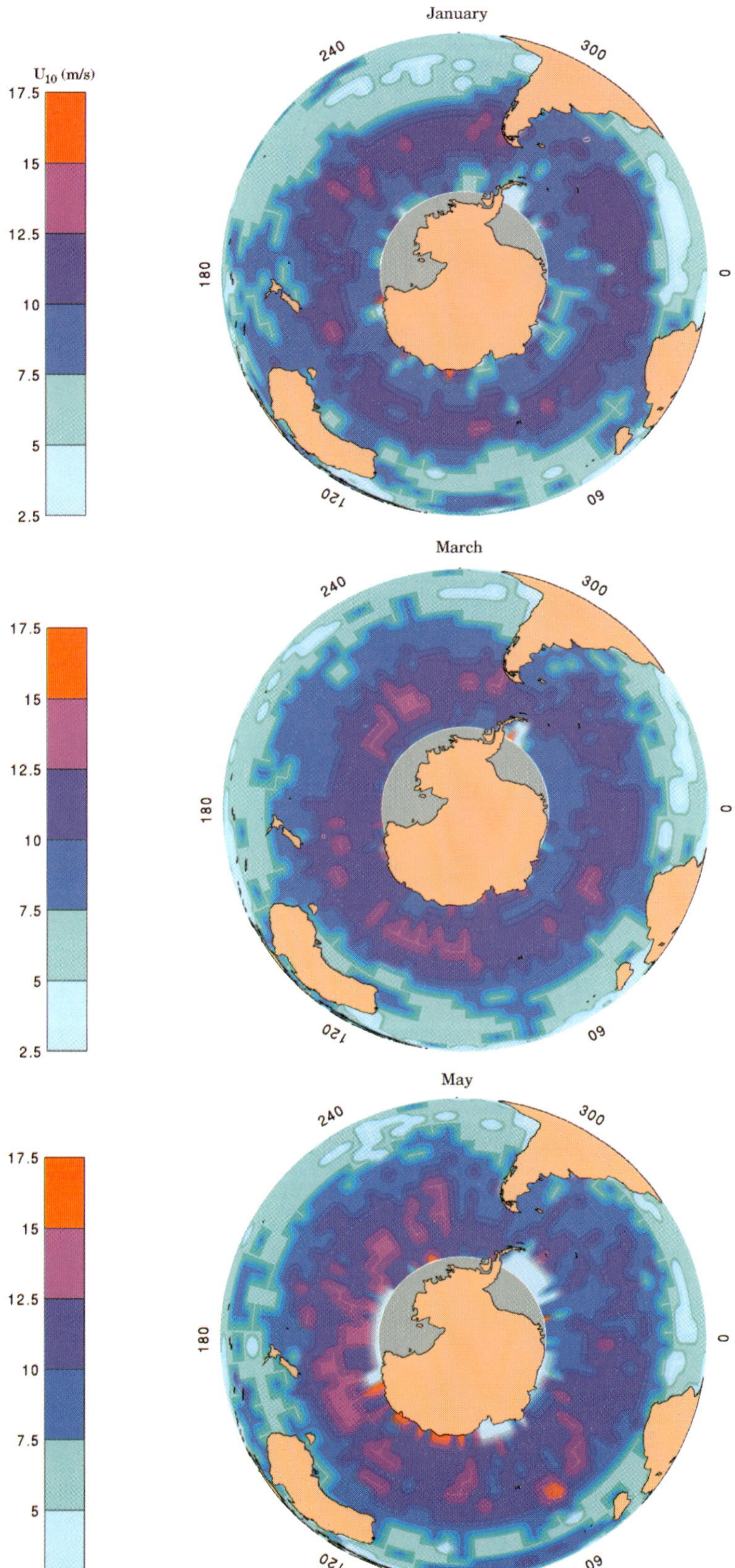

Fig. 7.4(a) Values of Mean Monthly Wind Speed for the Southern Ocean for January (top), March (middle) and May (bottom).

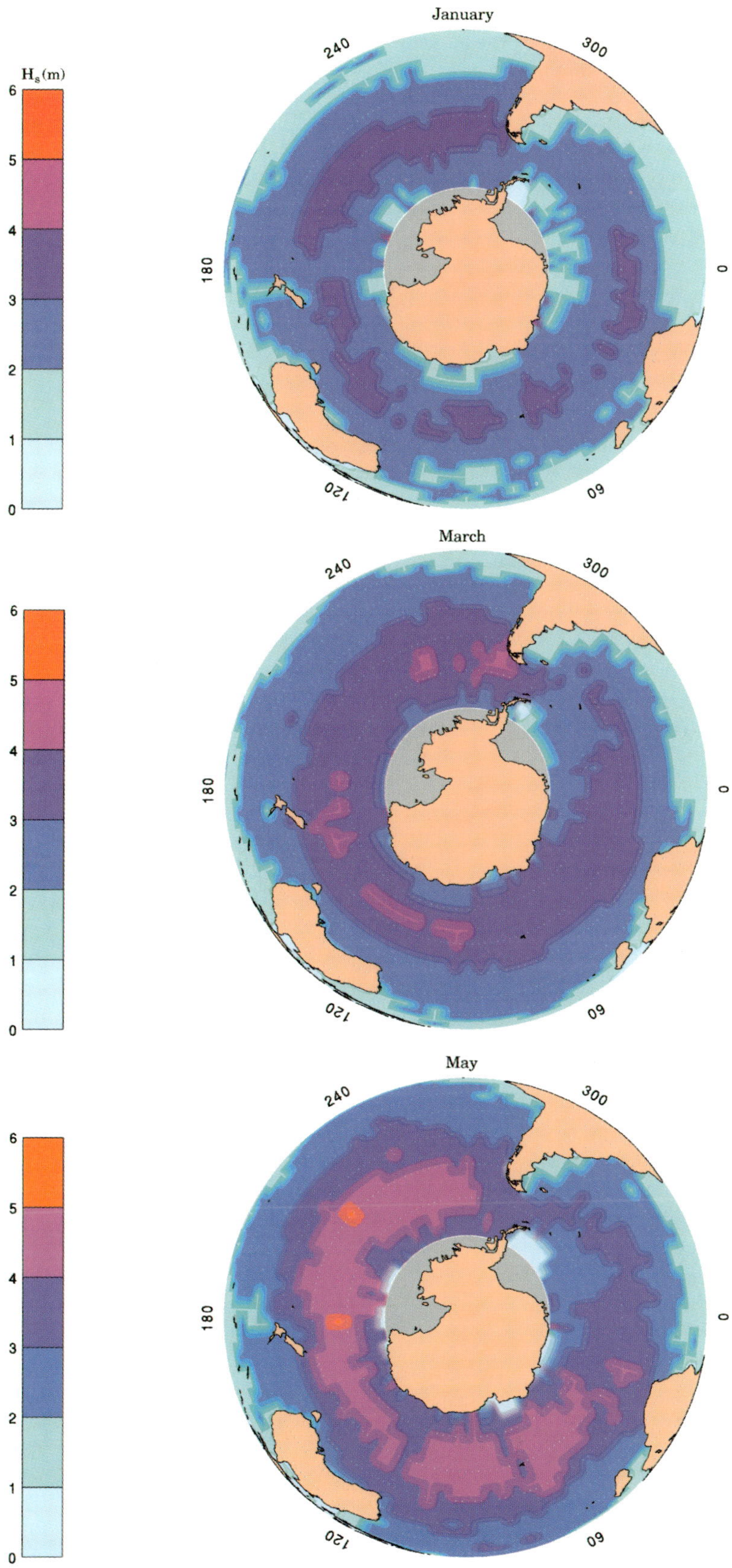

Fig. 7.4(b) Values of Mean Monthly Significant Wave Height for the Southern Ocean for January (top), March (middle) and May (bottom).

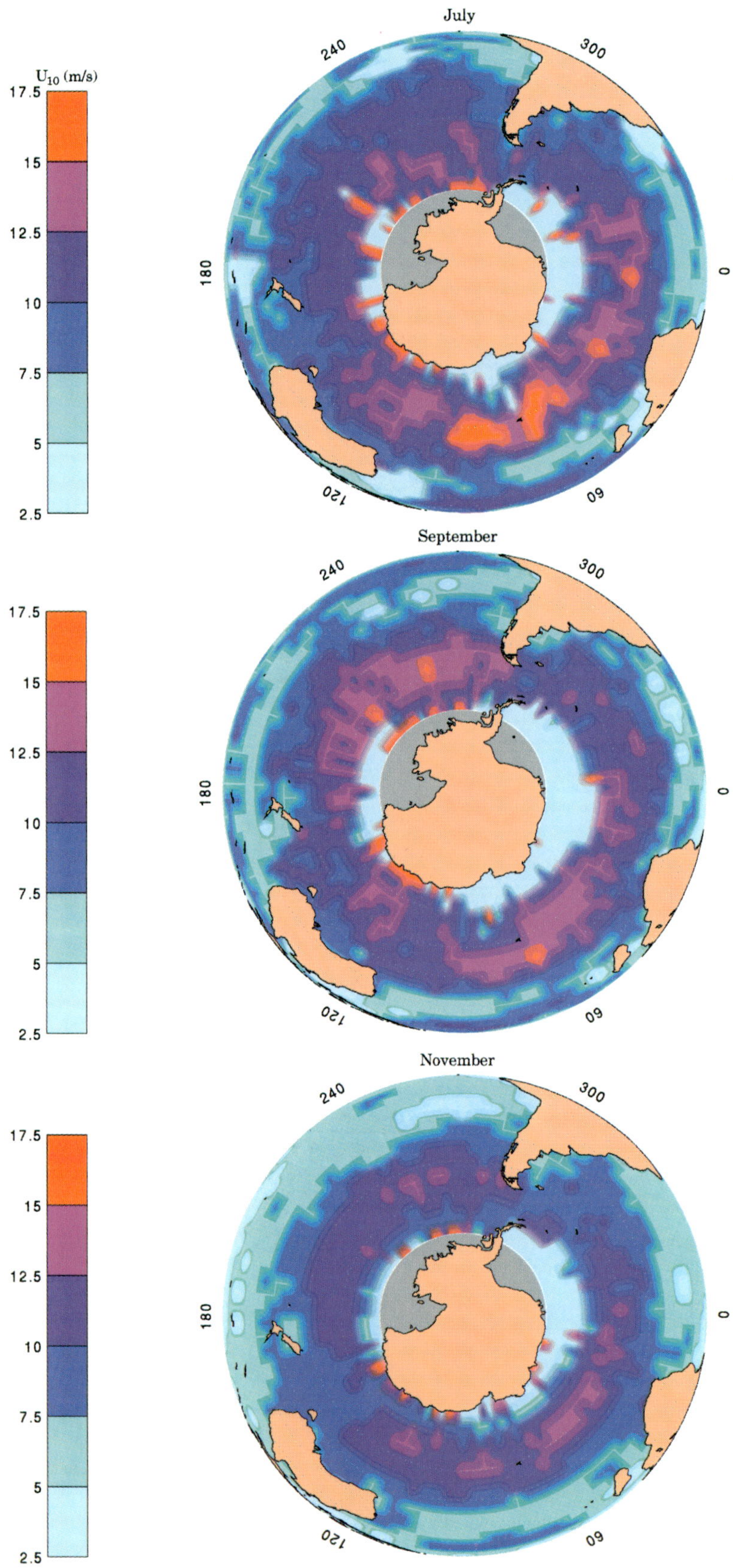

Fig. 7.4(c) Values of Mean Monthly Wind Speed for the Southern Ocean for July (top), September (middle) and November (bottom).

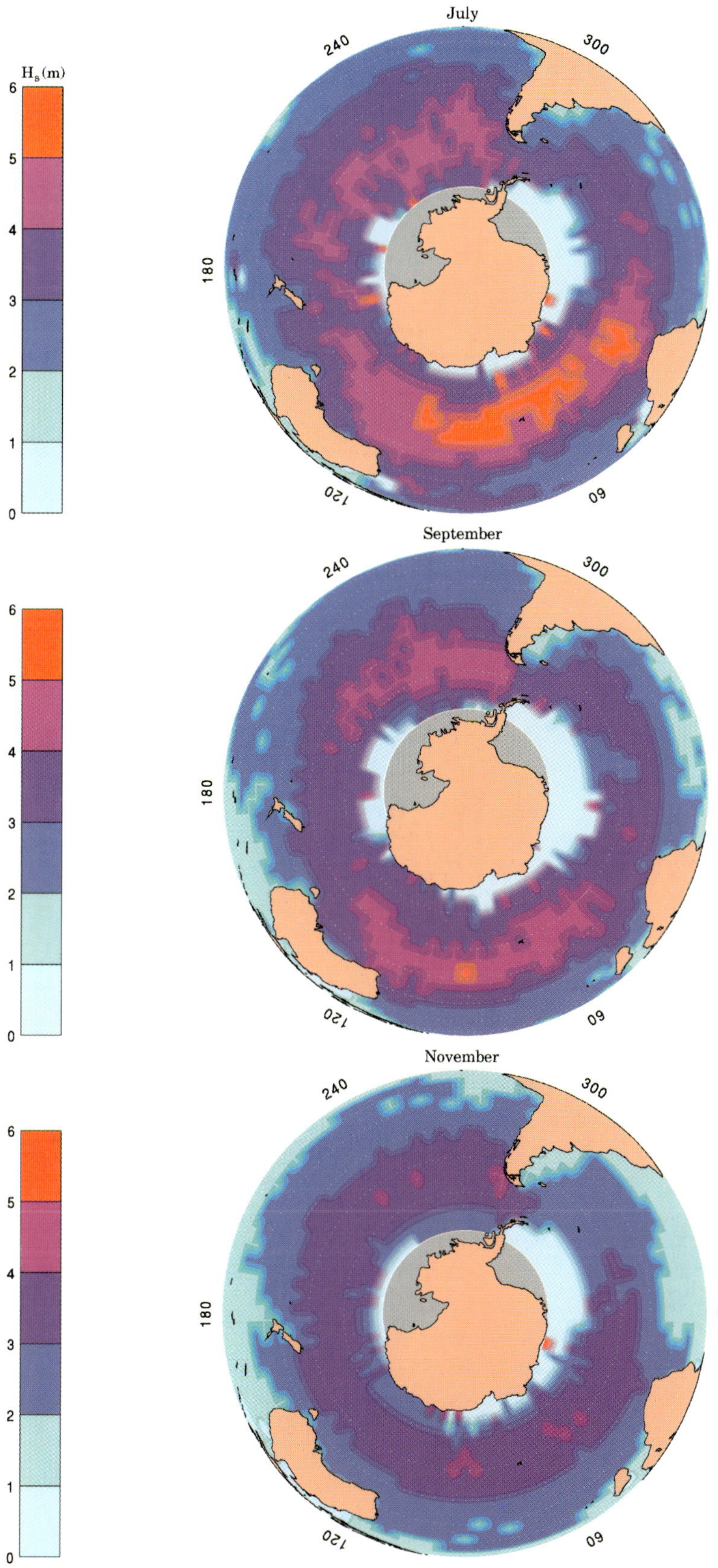

Fig. 7.4(d) Values of Mean Monthly Significant Wave Height for the Southern Ocean for July (top), September (middle) and November (bottom).

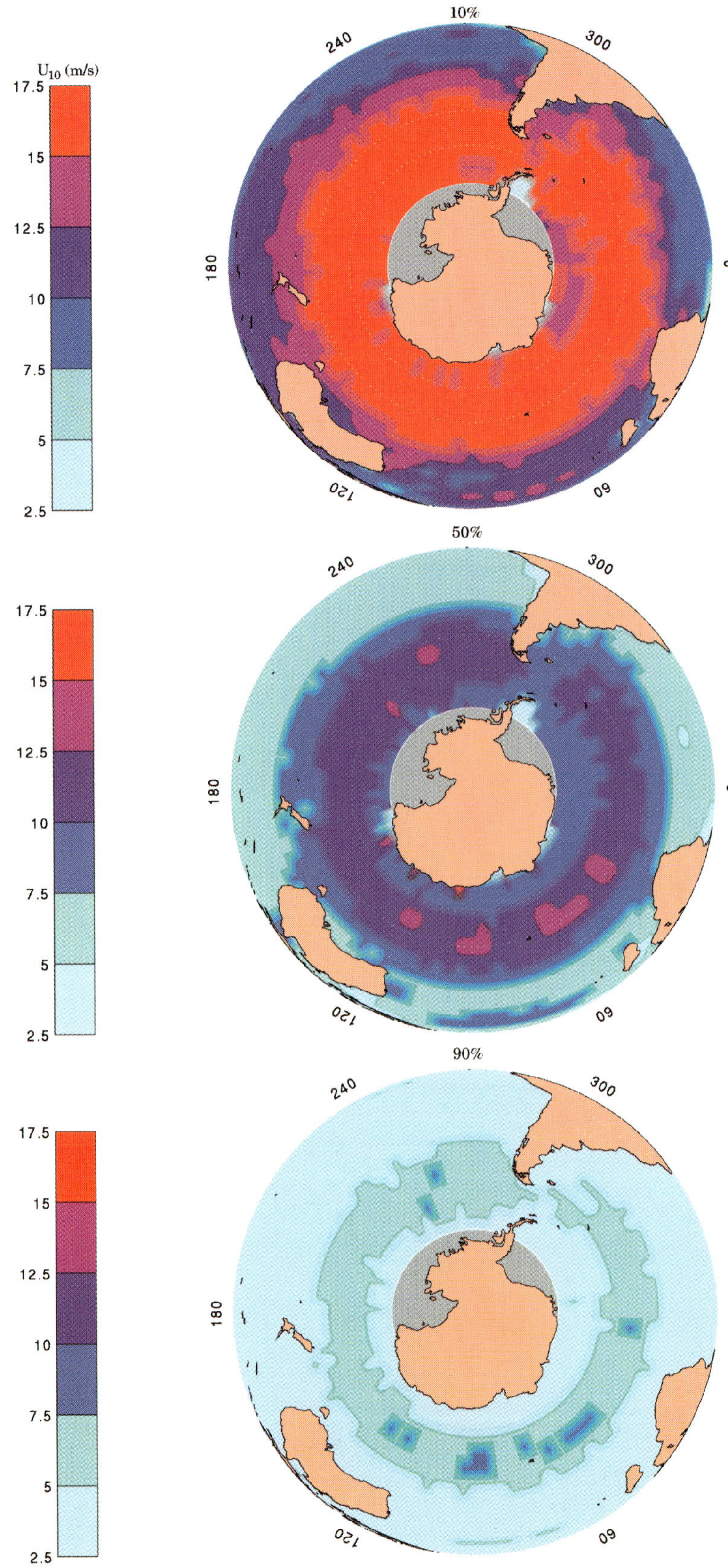

Fig. 7.5(a) Values for the Southern Ocean of the Wind Speed which will be exceeded 10% (top), 50% (middle) and 90% (bottom) of the time.

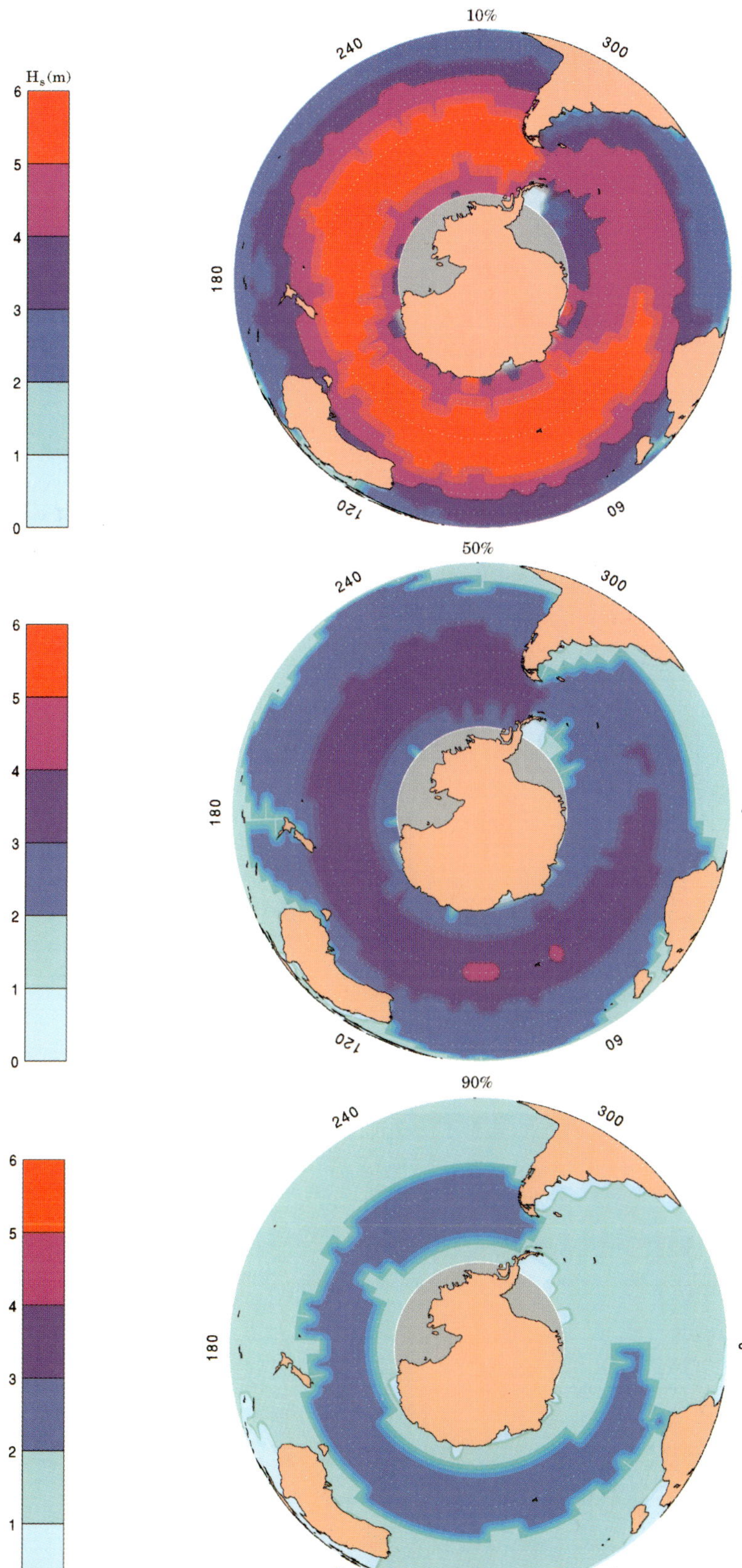

Fig. 7.5(b) Values for the Southern Ocean of the Significant Wave Height which will be exceeded 10% (top), 50% (middle) and 90% (bottom) of the time.

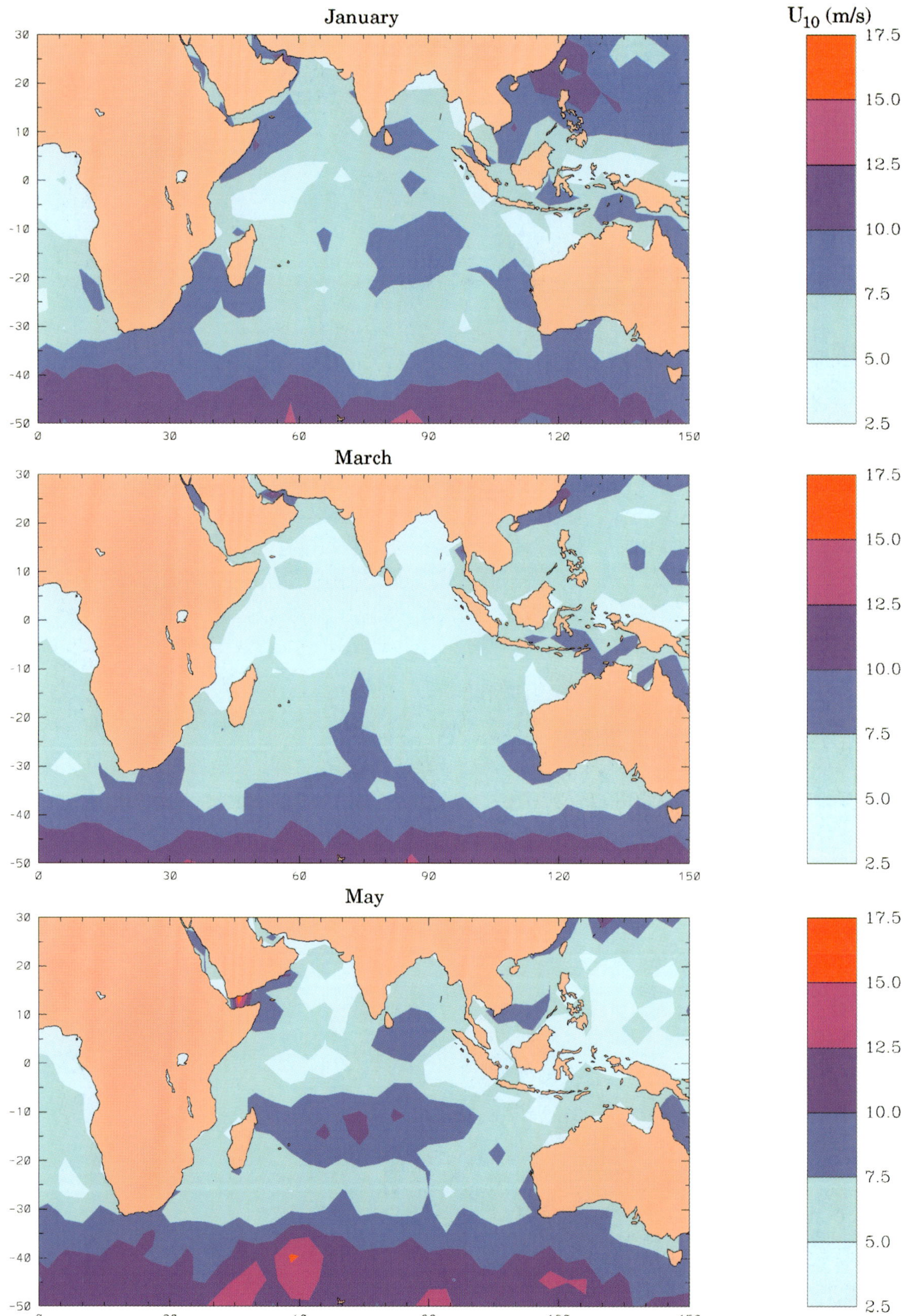

Fig. 7.6(a) Values of Mean Monthly Wind Speed for the Indian Ocean for January (top), March (middle) and May (bottom).

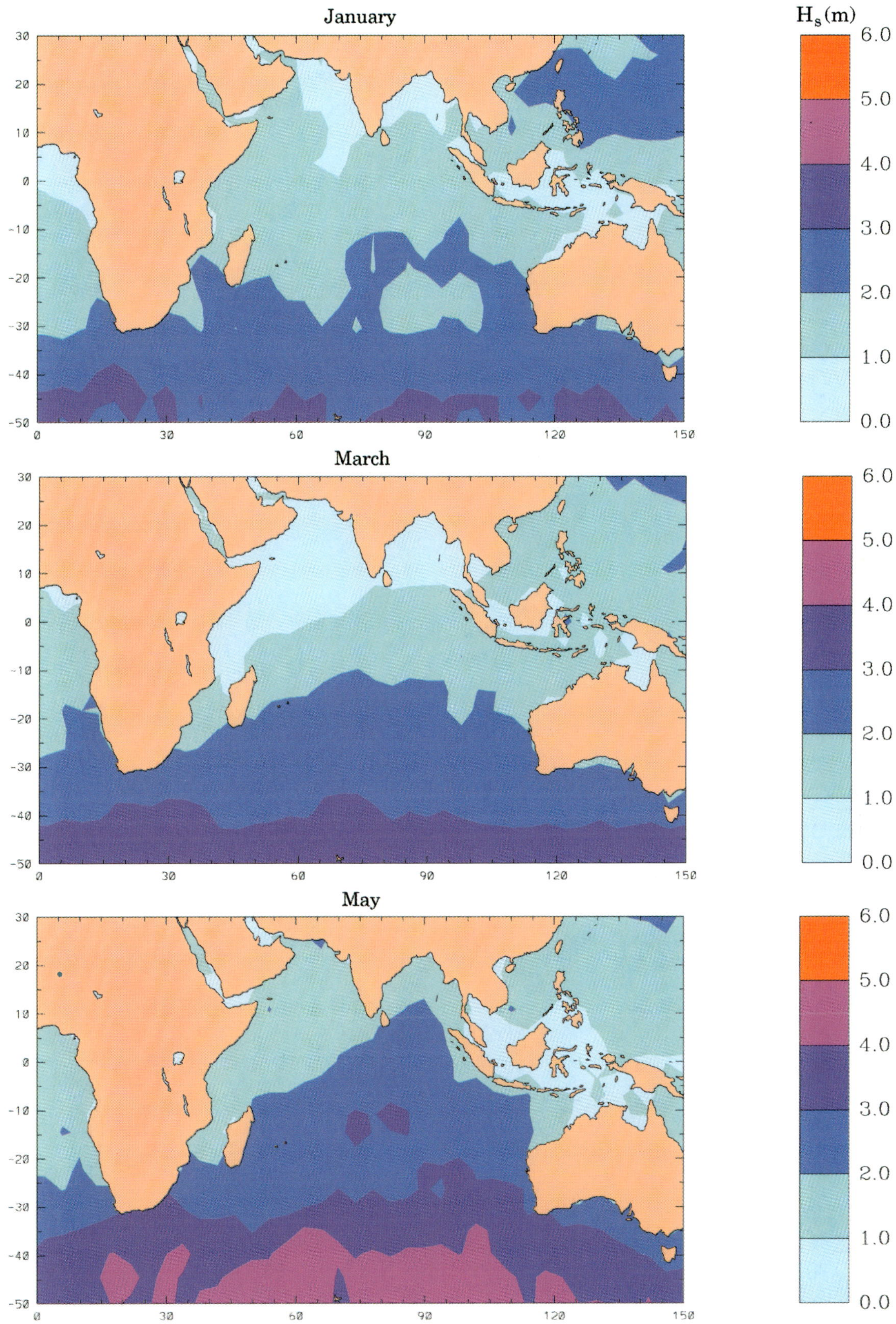

Fig. 7.6(b) Values of Mean Monthly Significant Wave Height for the Indian Ocean for January (top), March (middle) and May (bottom).

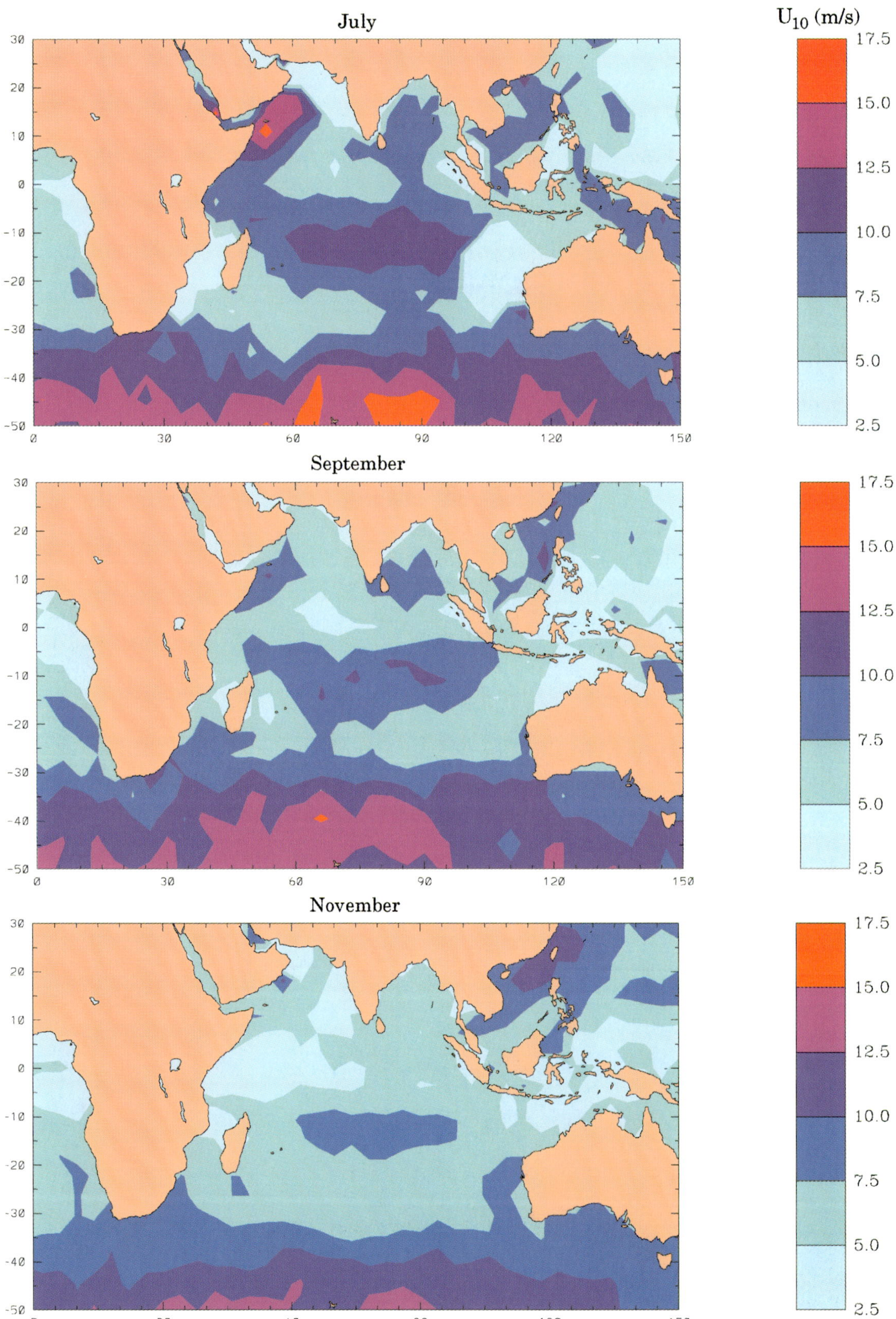

Fig. 7.6(c) Values of Mean Monthly Wind Speed for the Indian Ocean for July (top), September (middle) and November (bottom).

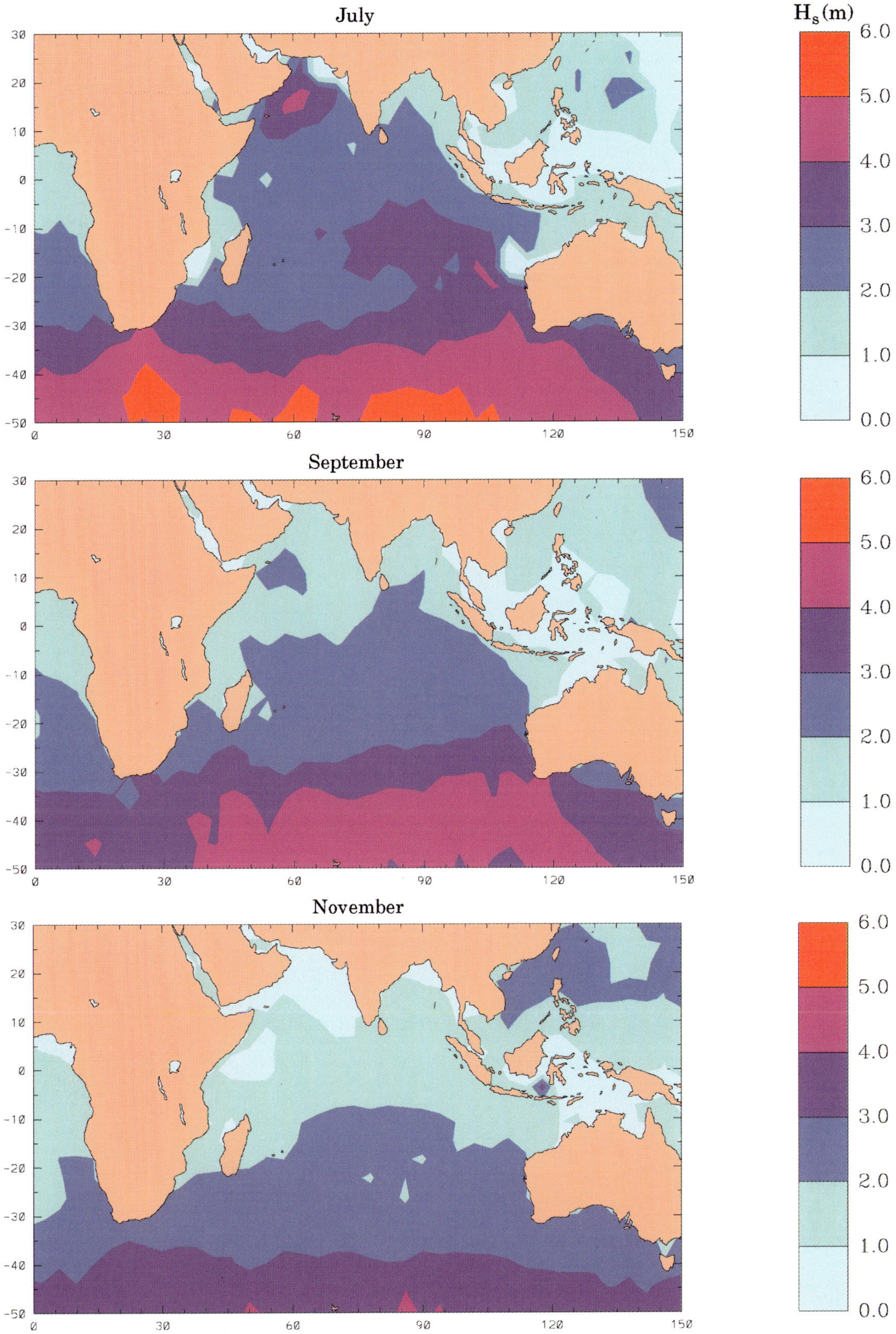

Fig. 7.6(d) Values of Mean Monthly Significant Wave Height for the Indian Ocean for July (top), September (middle) and November (bottom).

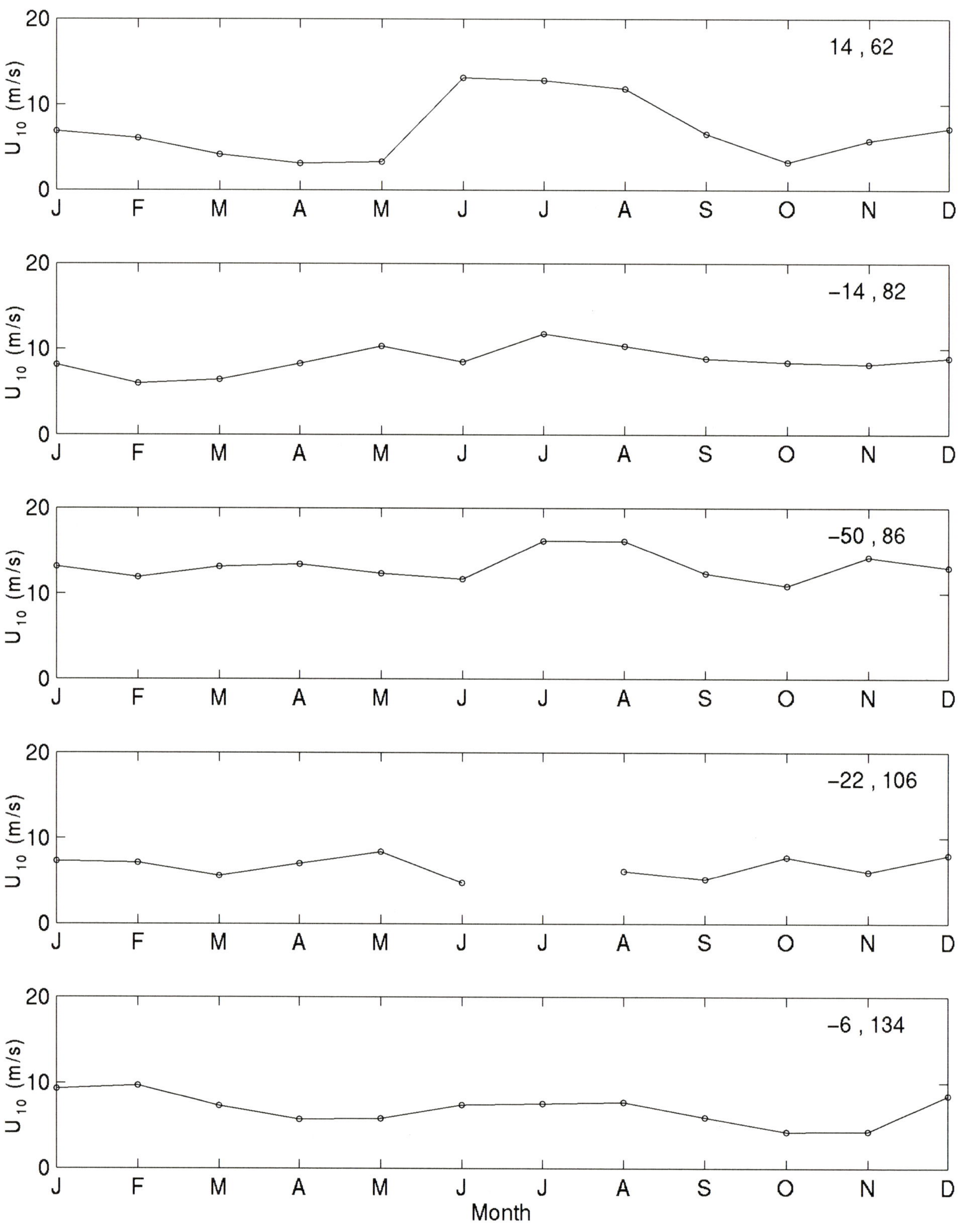

Fig. 7.7(a) Seasonal variability of wind speed at selected sites in the Indian Ocean: (i) in the vicinity of the Somali Jet (14°, 62°), (ii) in the trade wind maximum (−14°, 82°), (iii) in the westerly flow of the Southern Ocean (−50°, 86°), (iv) off the West Australian coast (−22°, 106°) and (v) in the Australasian monsoon (−6°, 134°).

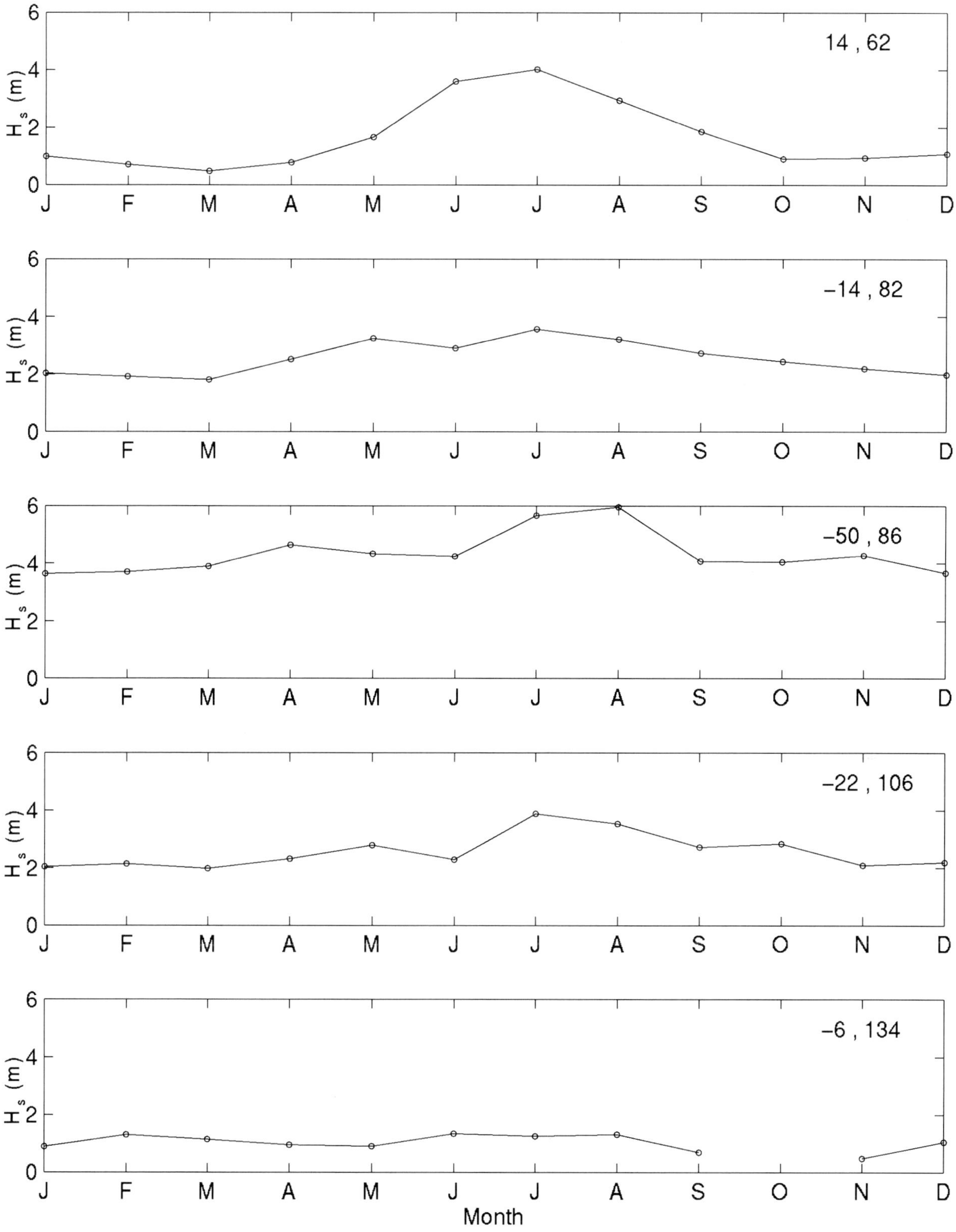

Fig. 7.7(b) Seasonal variability of significant wave height at selected sites in the Indian Ocean: (i) in the vicinity of the Somali Jet (14°, 62°), (ii) in the trade wind maximum (−14°, 82°), (iii) in the westerly flow of the Southern Ocean (−50°, 86°), (iv) off the West Australian coast (−22°, 106°) and (v) in the Australasian monsoon (−6°, 134°).

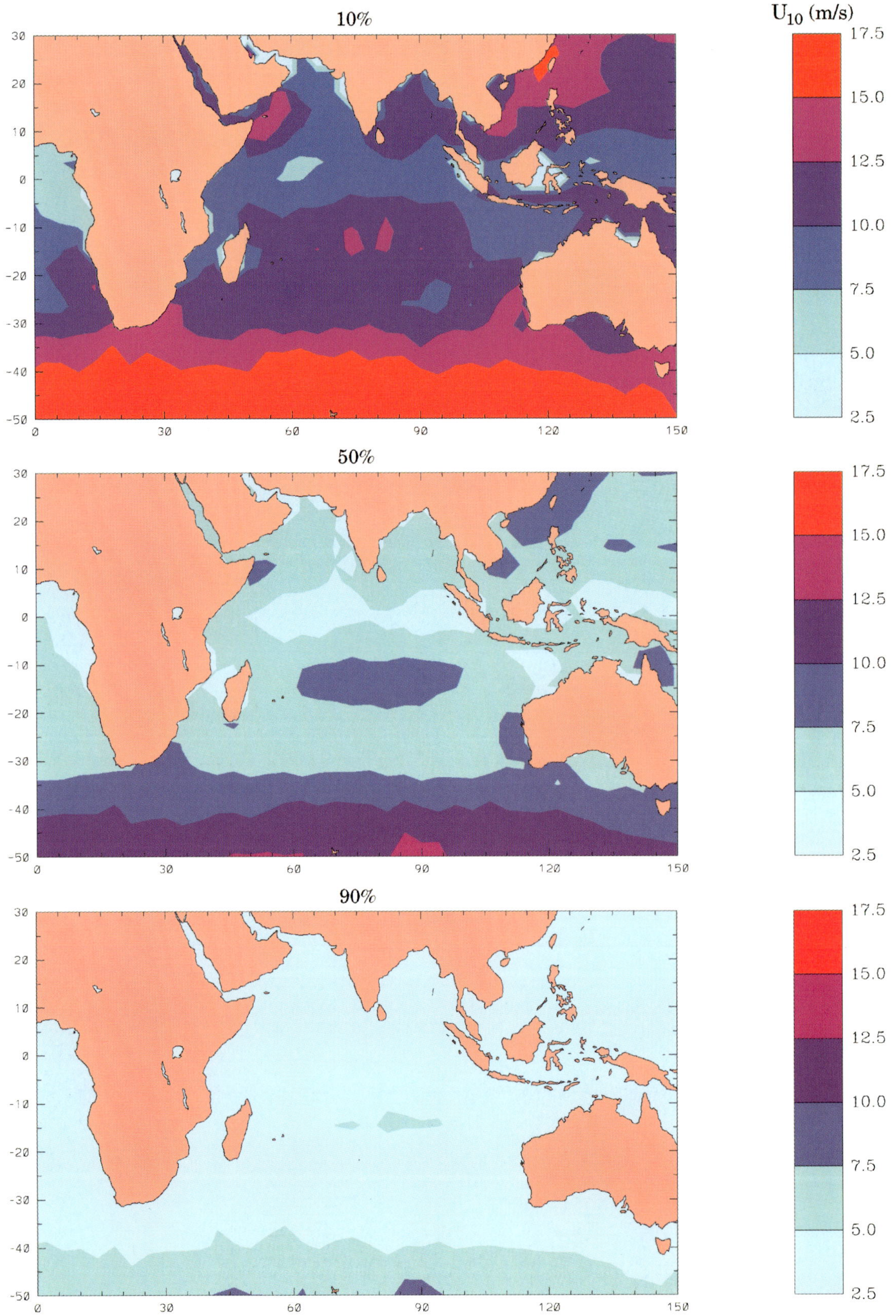

Fig. 7.8(a) Values for the Indian Ocean of the Wind Speed which will be exceeded 10% (top), 50% (middle) and 90% (bottom) of the time.

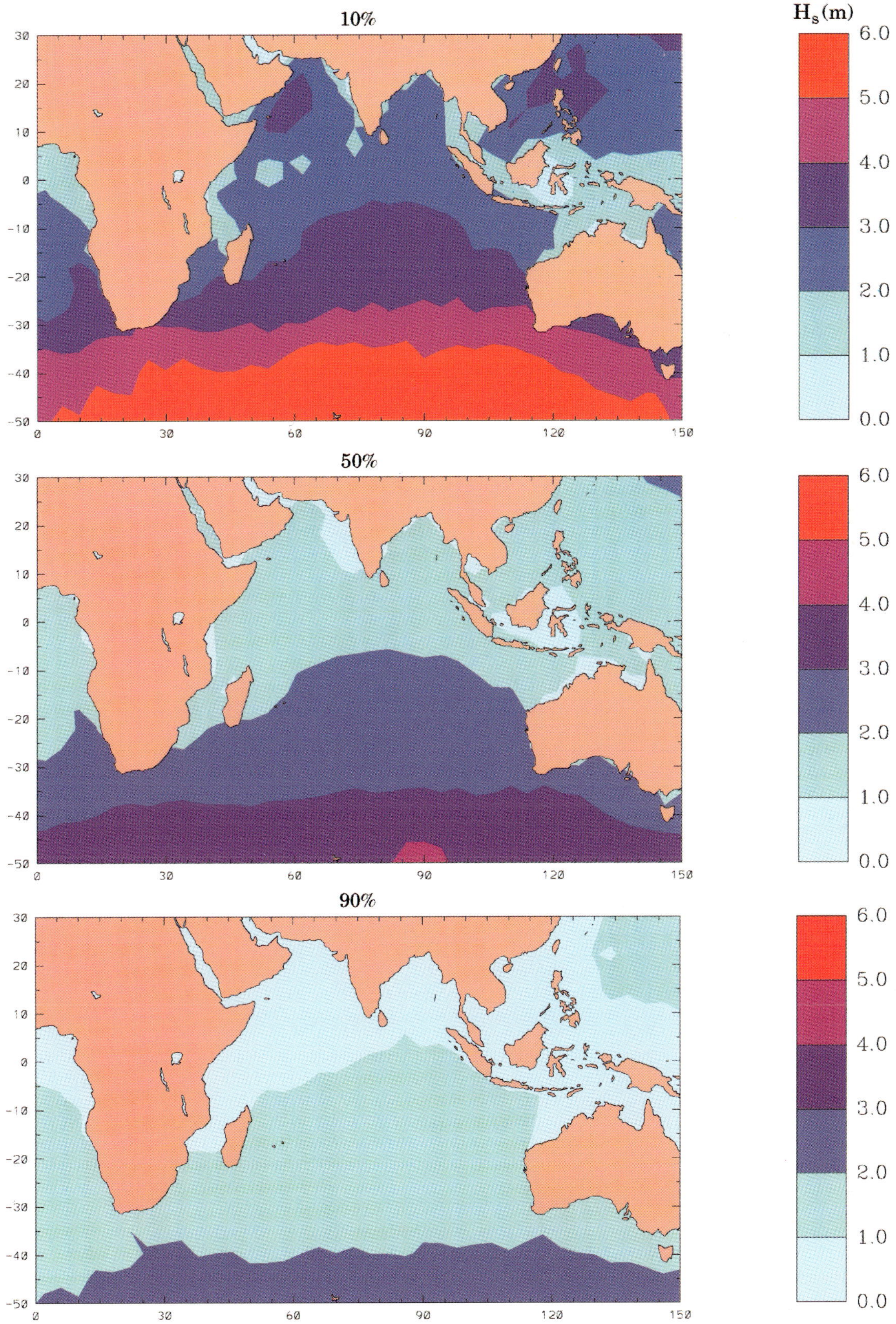

Fig. 7.8(b) Values for the Indian Ocean of the Significant Wave Height which will be exceeded 10% (top), 50% (middle) and 90% (bottom) of the time.

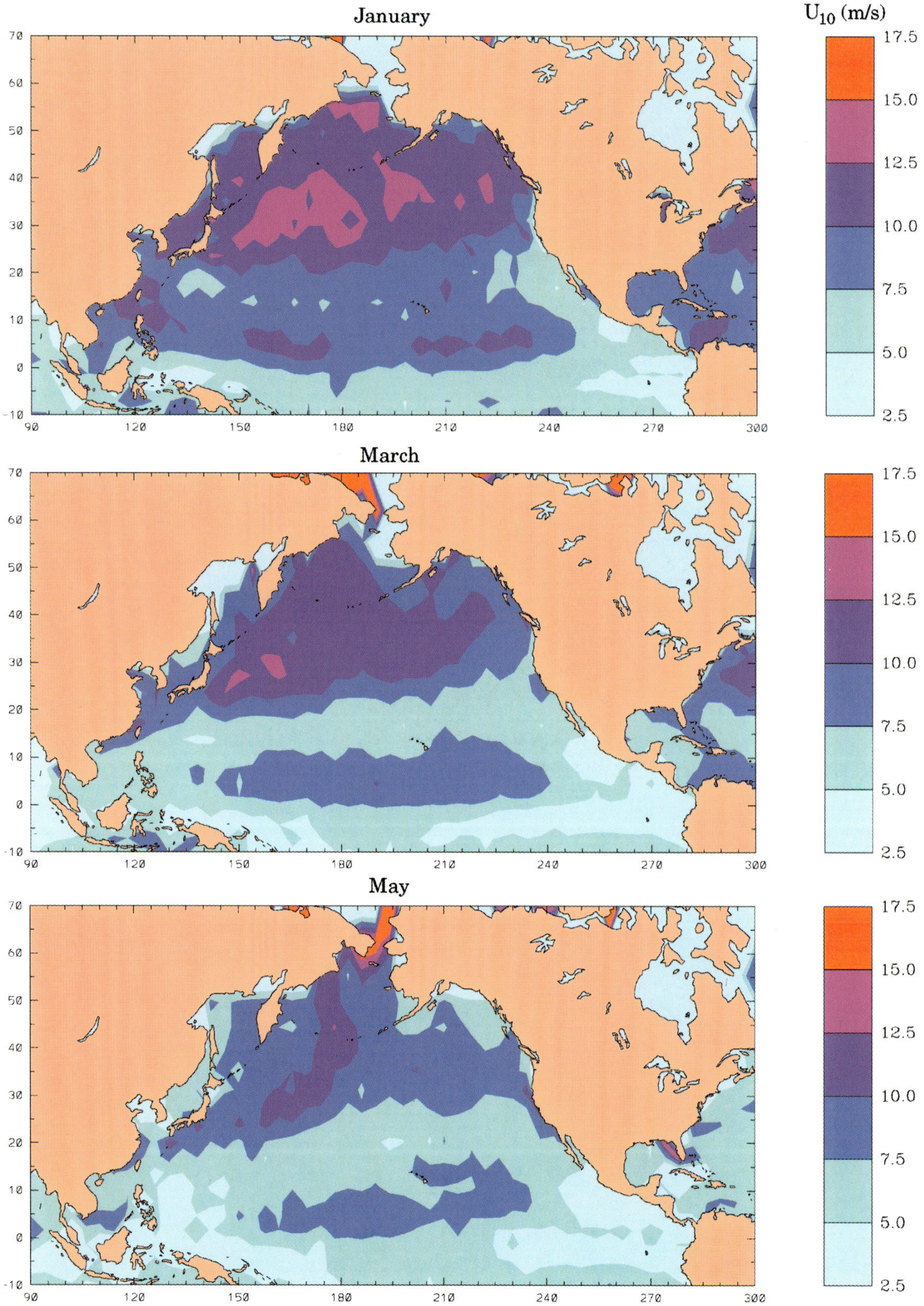

Fig. 7.9(a) Values of Mean Monthly Wind Speed for the North Pacific Ocean for January (top), March (middle) and May (bottom).

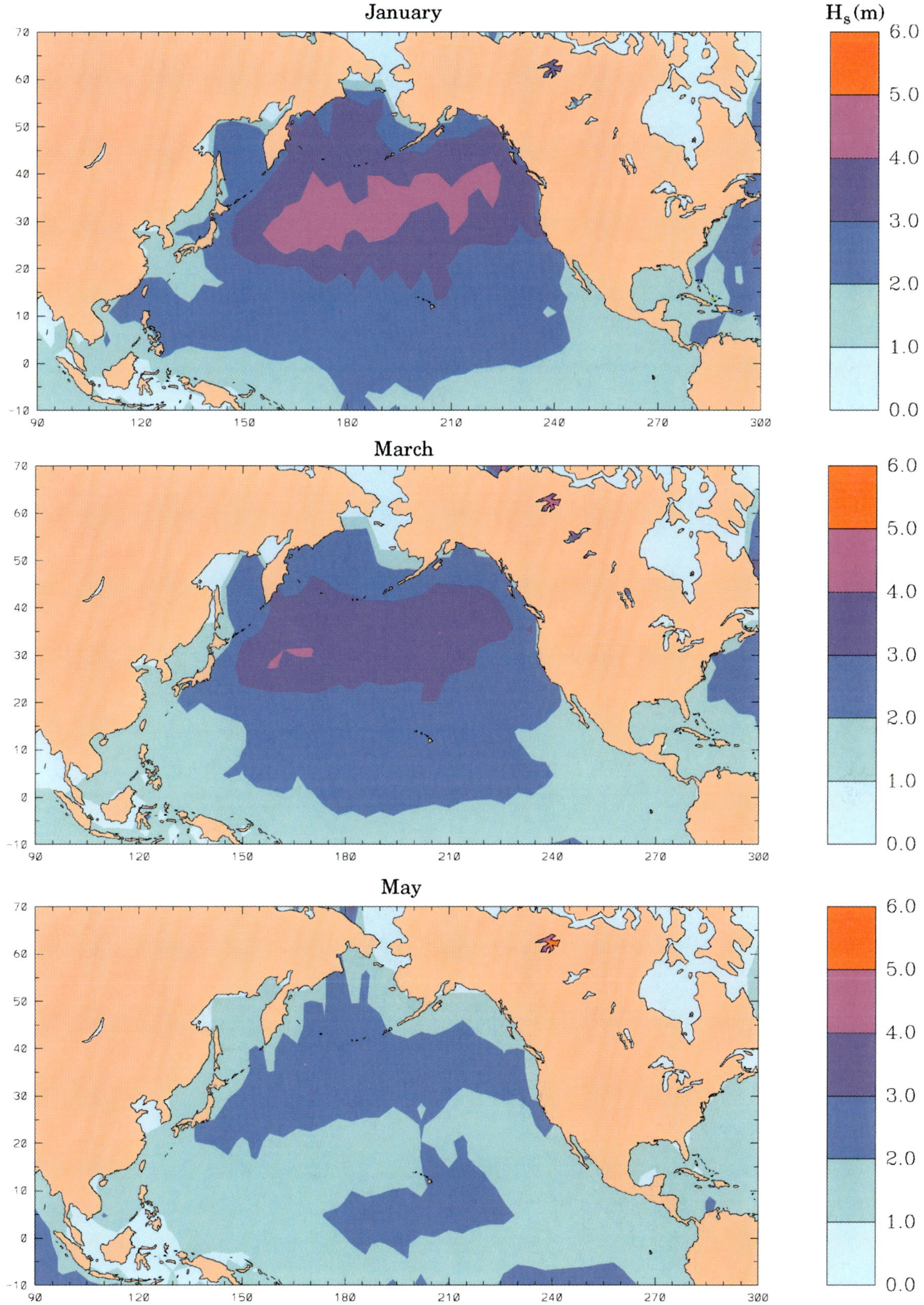

Fig. 7.9(b) Values of Mean Monthly Significant Wave Height for the North Pacific Ocean for January (top), March (middle) and May (bottom).

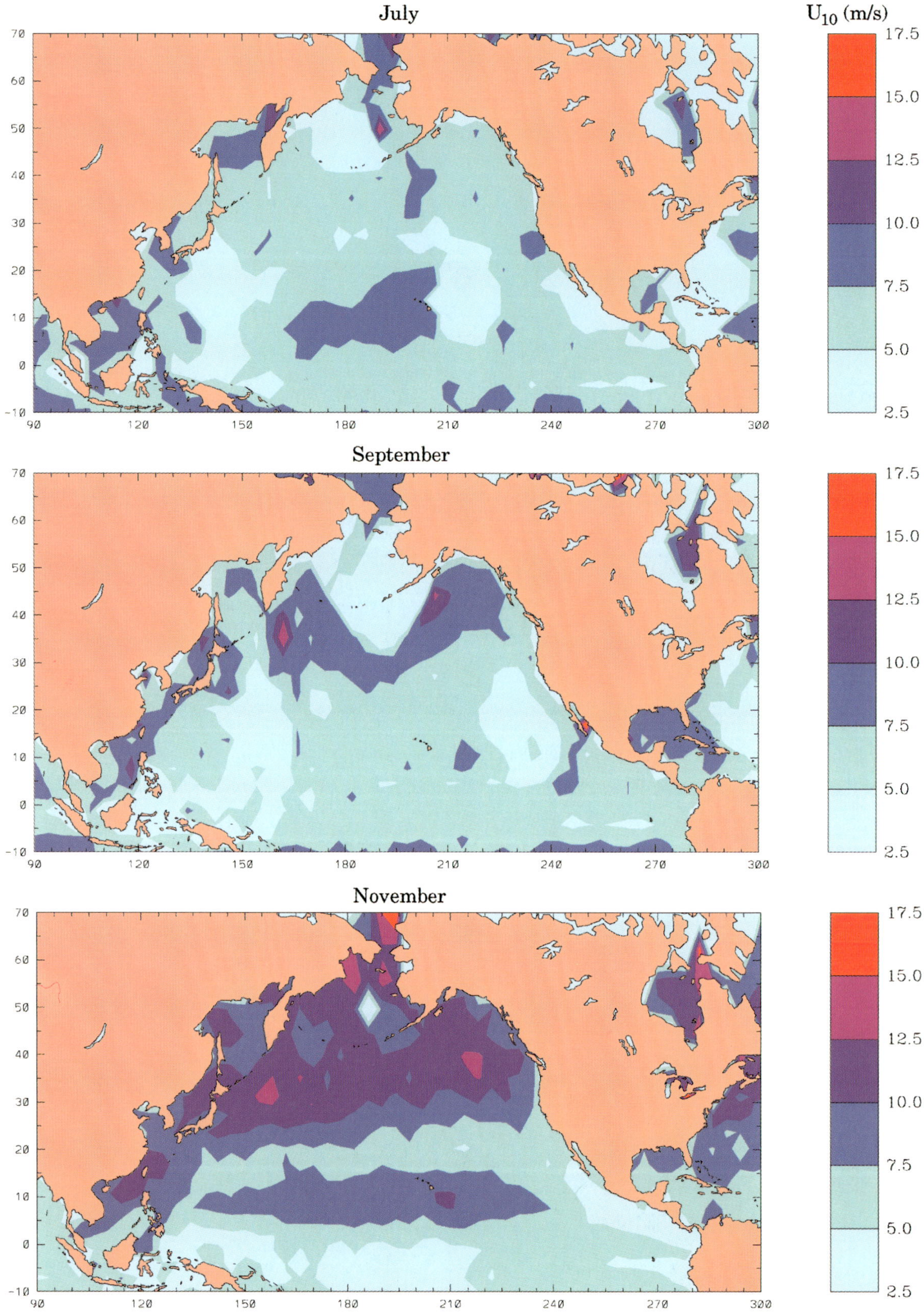

Fig. 7.9(c) Values of Mean Monthly Wind Speed for the North Pacific Ocean for July (top), September (middle) and November (bottom).

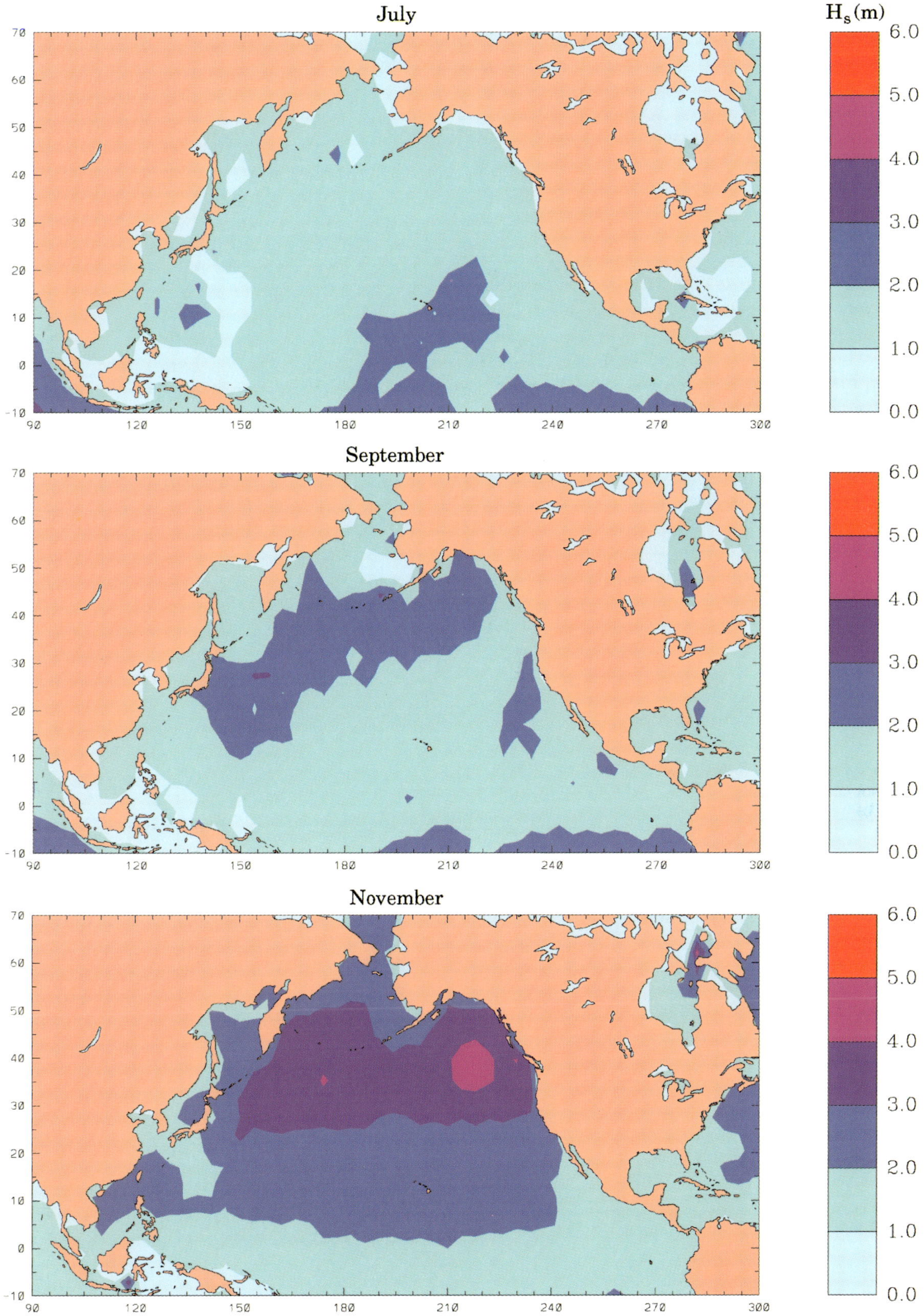

Fig. 7.9(d) Values of Mean Monthly Significant Wave Height for the North Pacific Ocean for July (top), September (middle) and November (bottom).

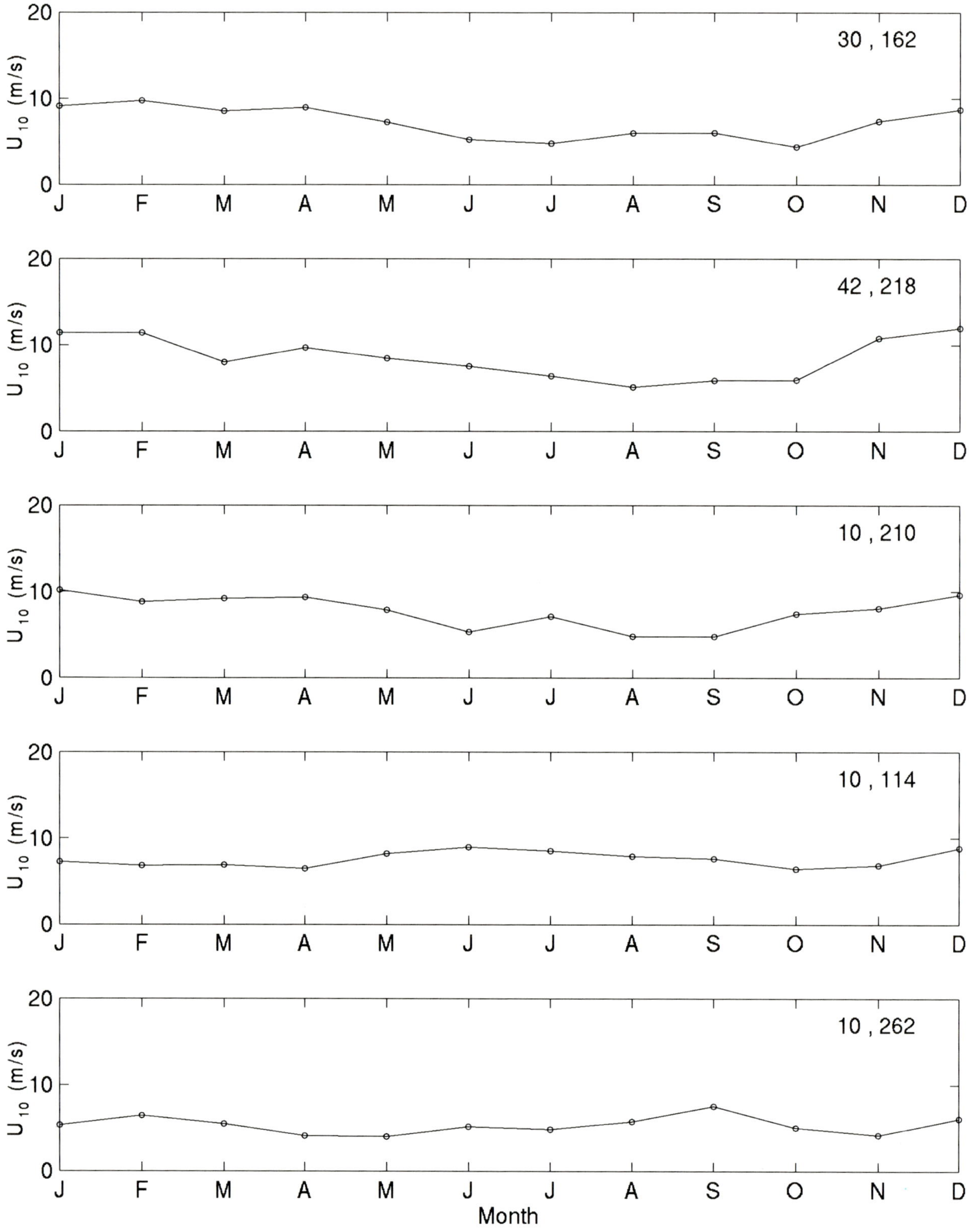

Fig. 7.10(a) Seasonal variability of wind speed at selected sites in the North Pacific Ocean: (i) at the western edge of the westerly winds (30°, 162°), (ii) at the eastern edge of the westerlies (42°, 218°), (iii) in the trade winds of the central Pacific (10°, 210°), (iv) in the summer monsoon of the South China Sea (10°, 114°) and (v) near Central America (10°, 262°).

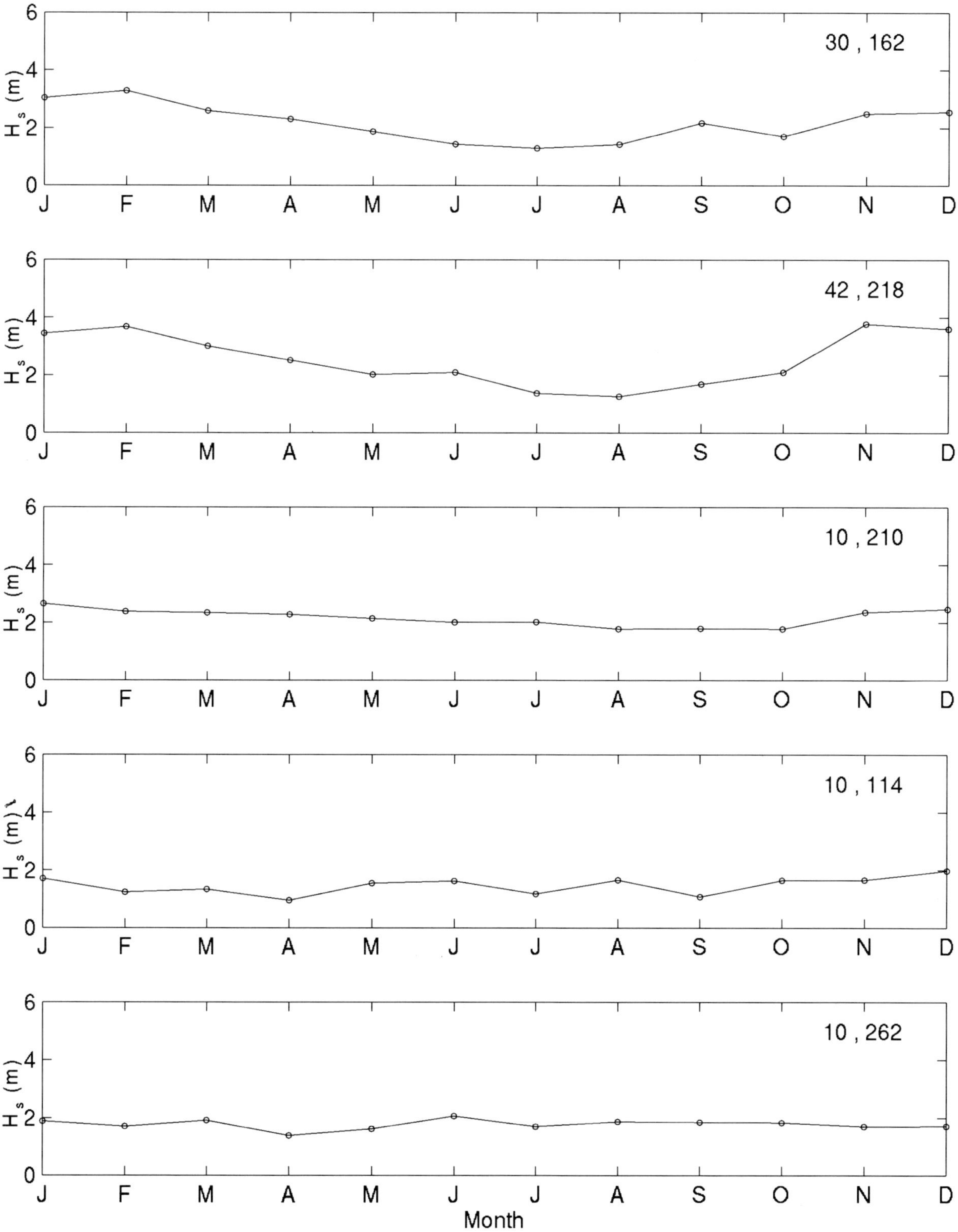

Fig. 7.10(b) Seasonal variability of significant wave height at selected sites in the North Pacific Ocean: (i) at the western edge of the westerly winds (30°, 162°), (ii) at the eastern edge of the westerlies (42°, 218°), (iii) in the trade winds of the central Pacific (10°, 210°), (iv) in the summer monsoon of the South China Sea (10°, 114°) and (v) near Central America (10°, 262°).

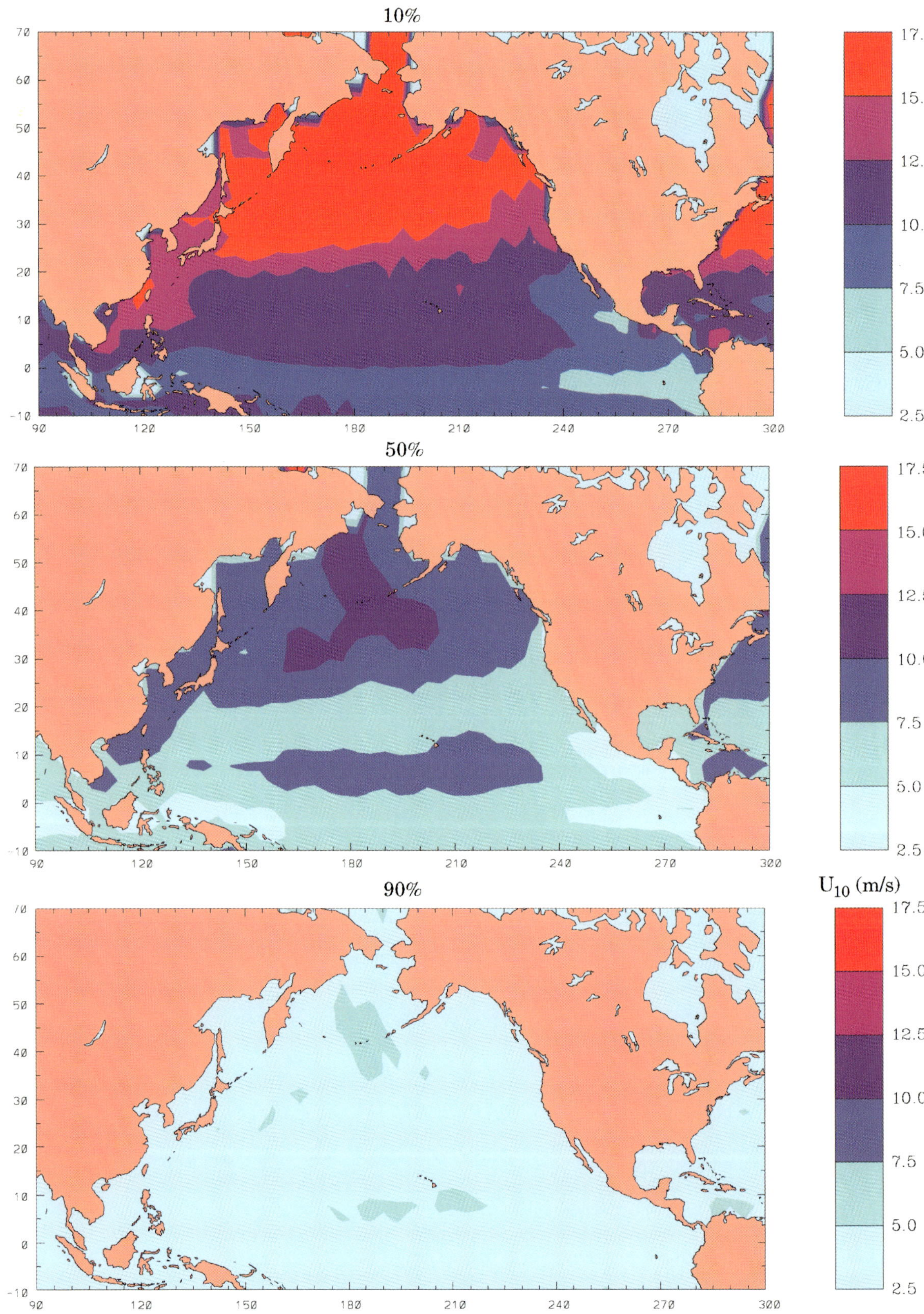

Fig. 7.11(a) Values for the North Pacific Ocean of the Wind Speed which will be exceeded 10% (top), 50% (middle) and 90% (bottom) of the time.

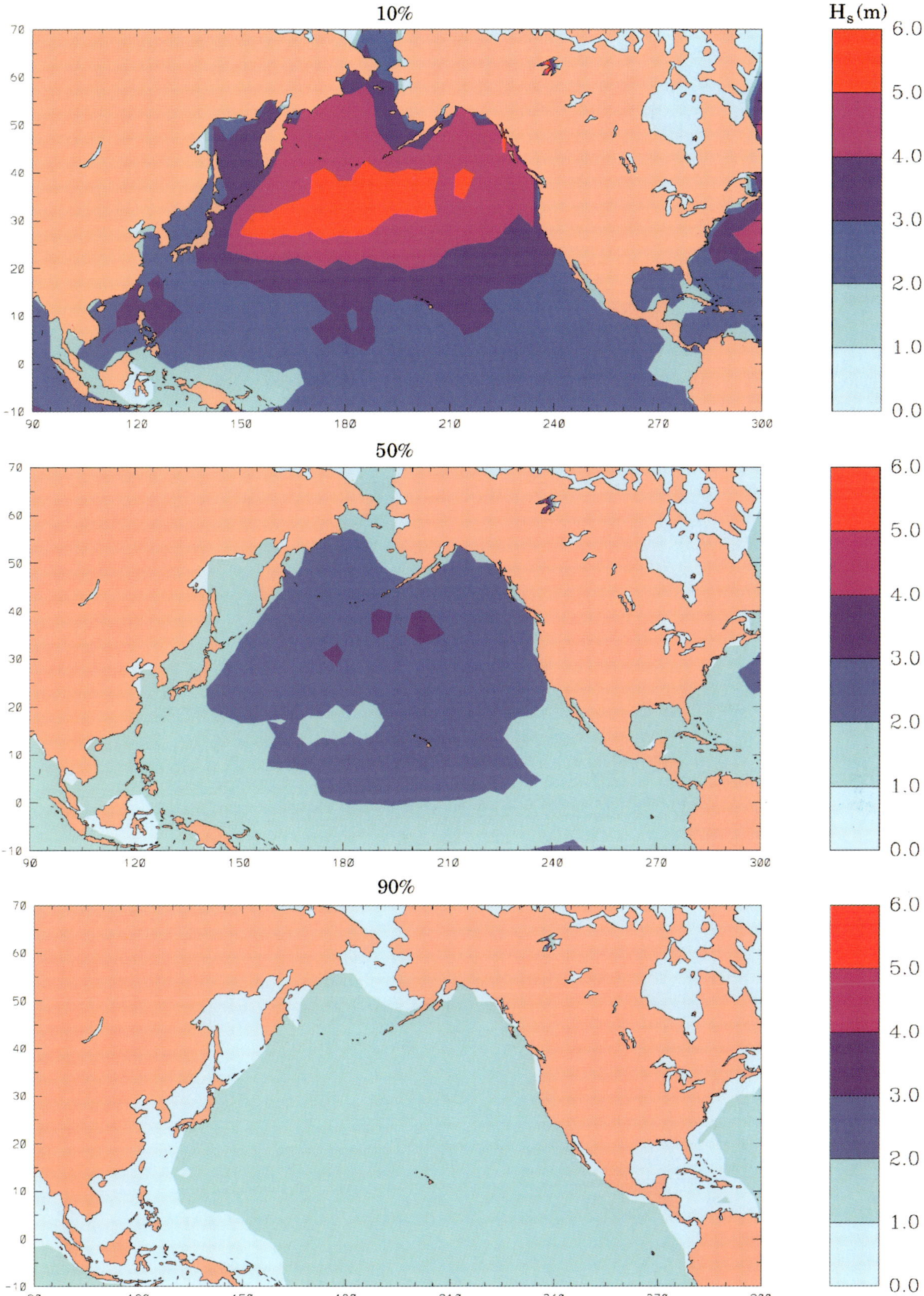

Fig. 7.11(b) Values for the North Pacific Ocean of the Significant Wave Height which will be exceeded 10% (top), 50% (middle) and 90% (bottom) of the time.

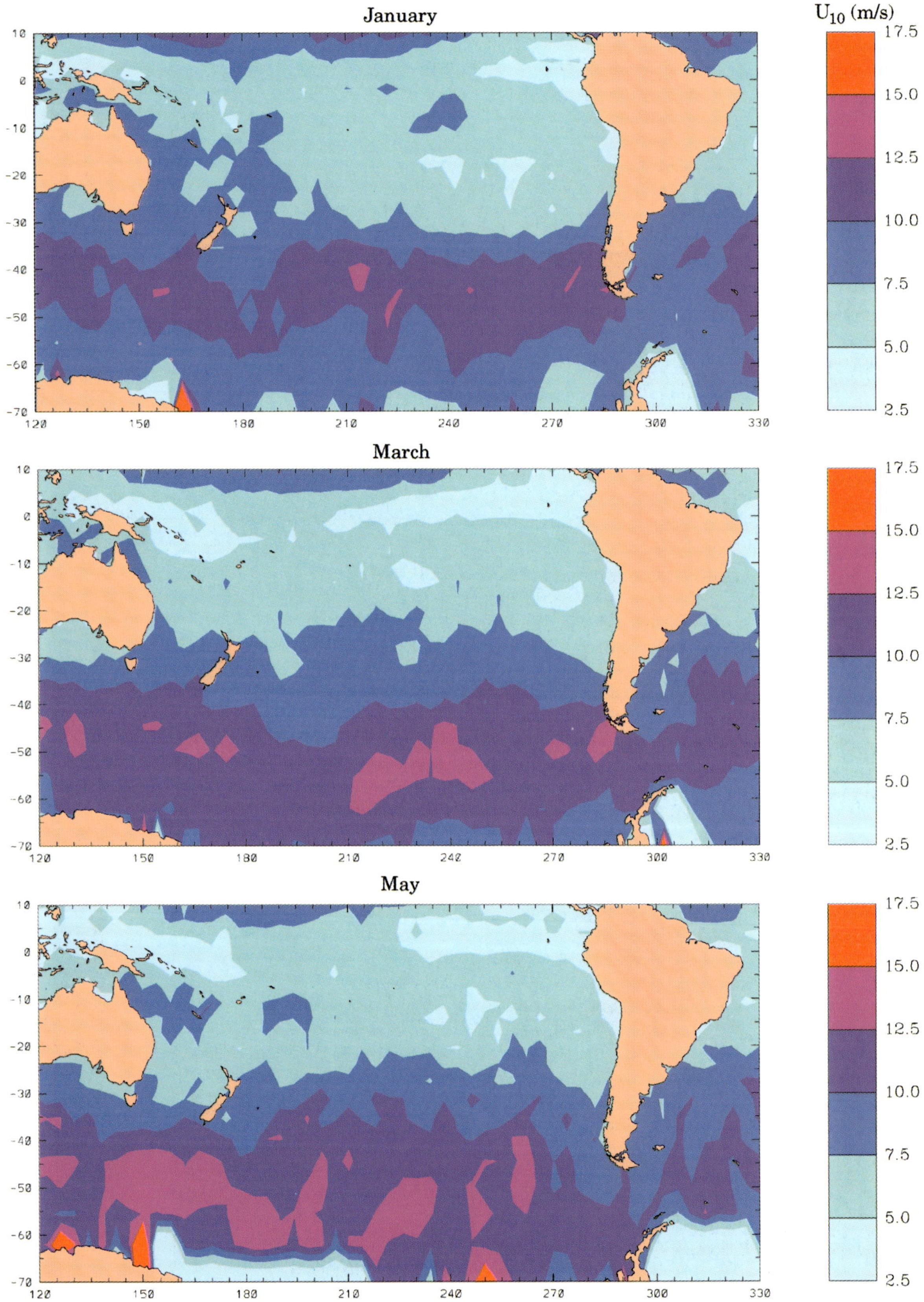

Fig. 7.12(a) Values of Mean Monthly Wind Speed for the South Pacific Ocean for January (top), March (middle) and May (bottom).

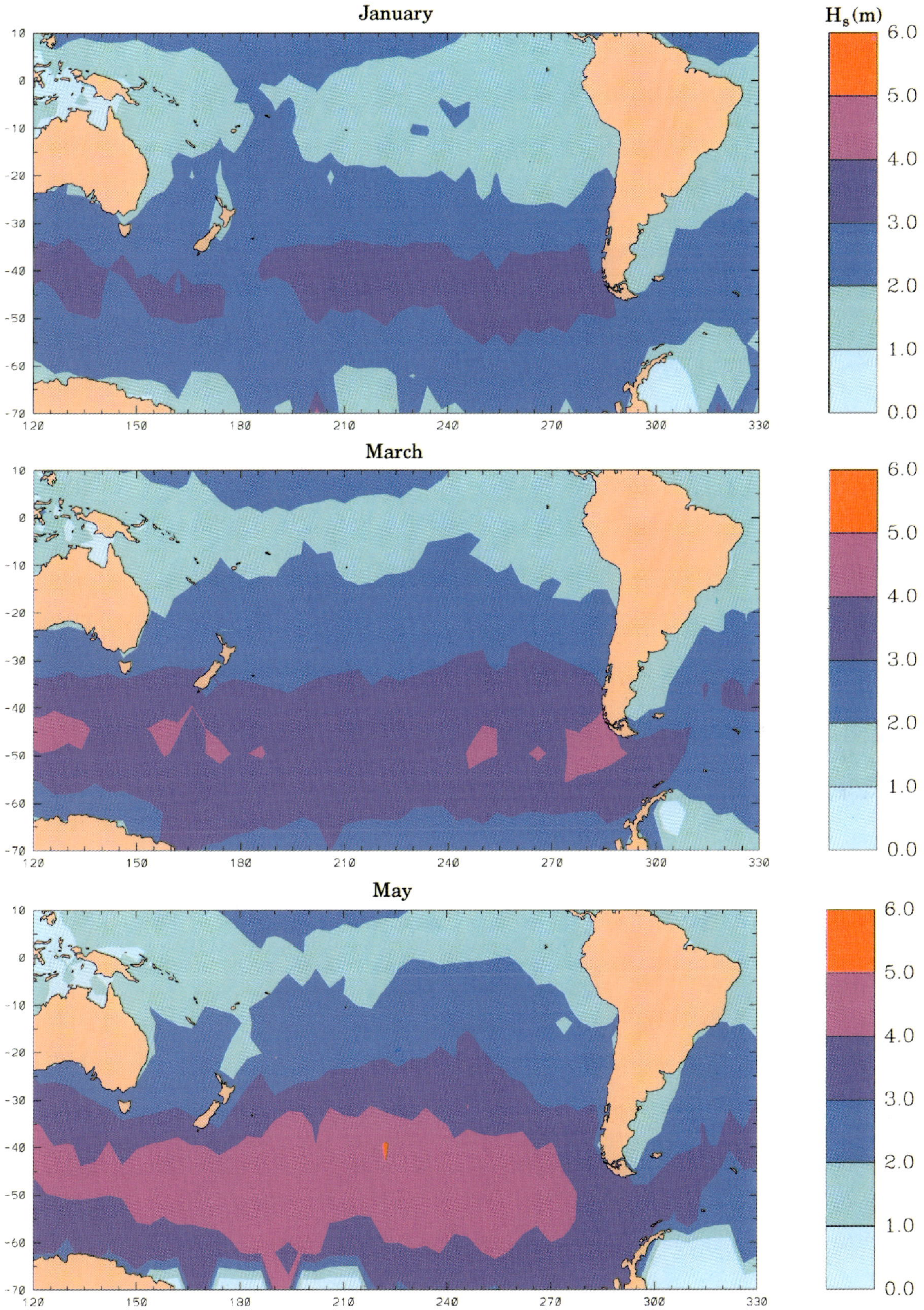

Fig. 7.12(b) Values of Mean Monthly Significant Wave Height for the South Pacific Ocean for January (top), March (middle) and May (bottom).

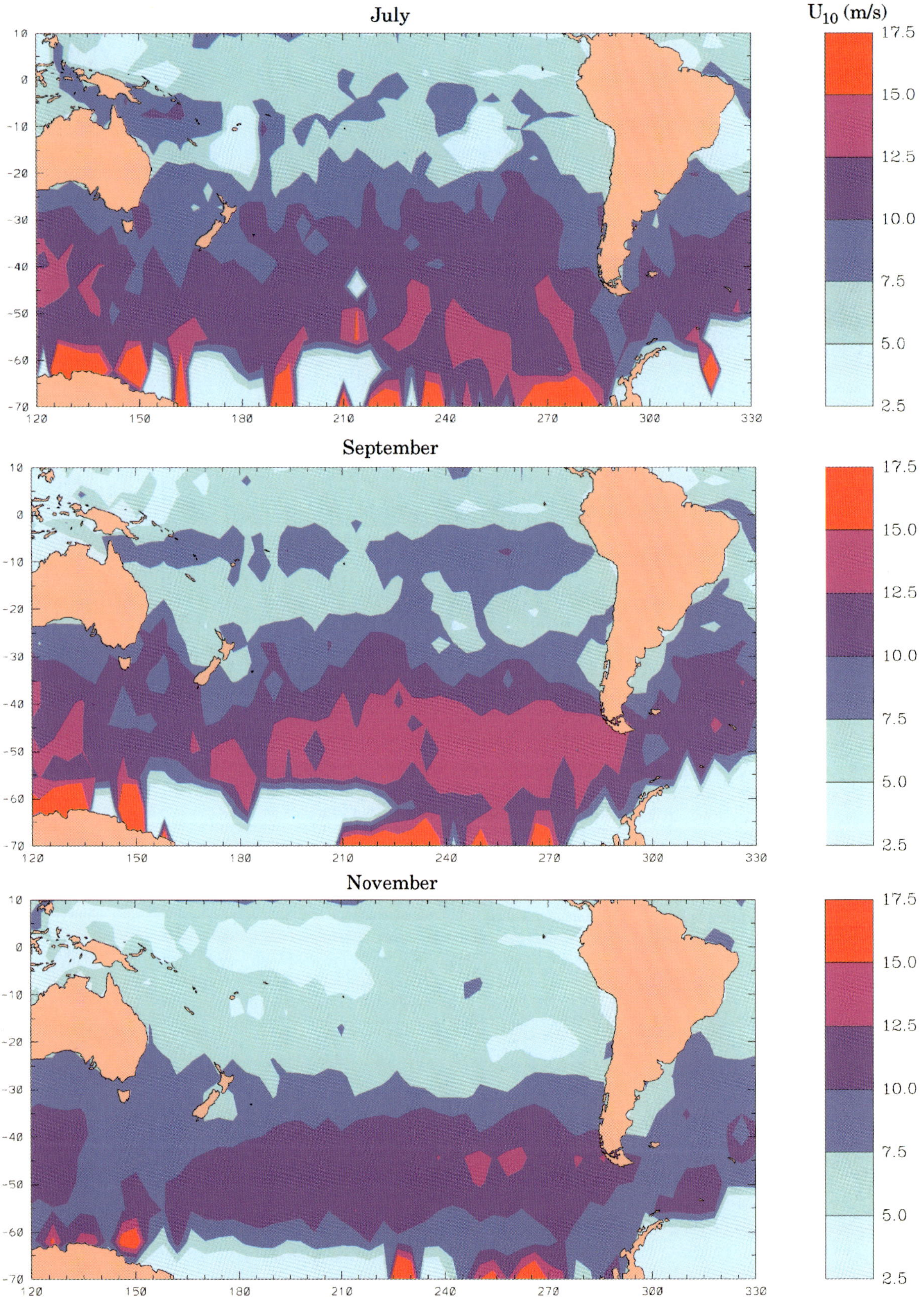

Fig. 7.12(c) Values of Mean Monthly Wind Speed for the South Pacific Ocean for July (top), September (middle) and November (bottom).

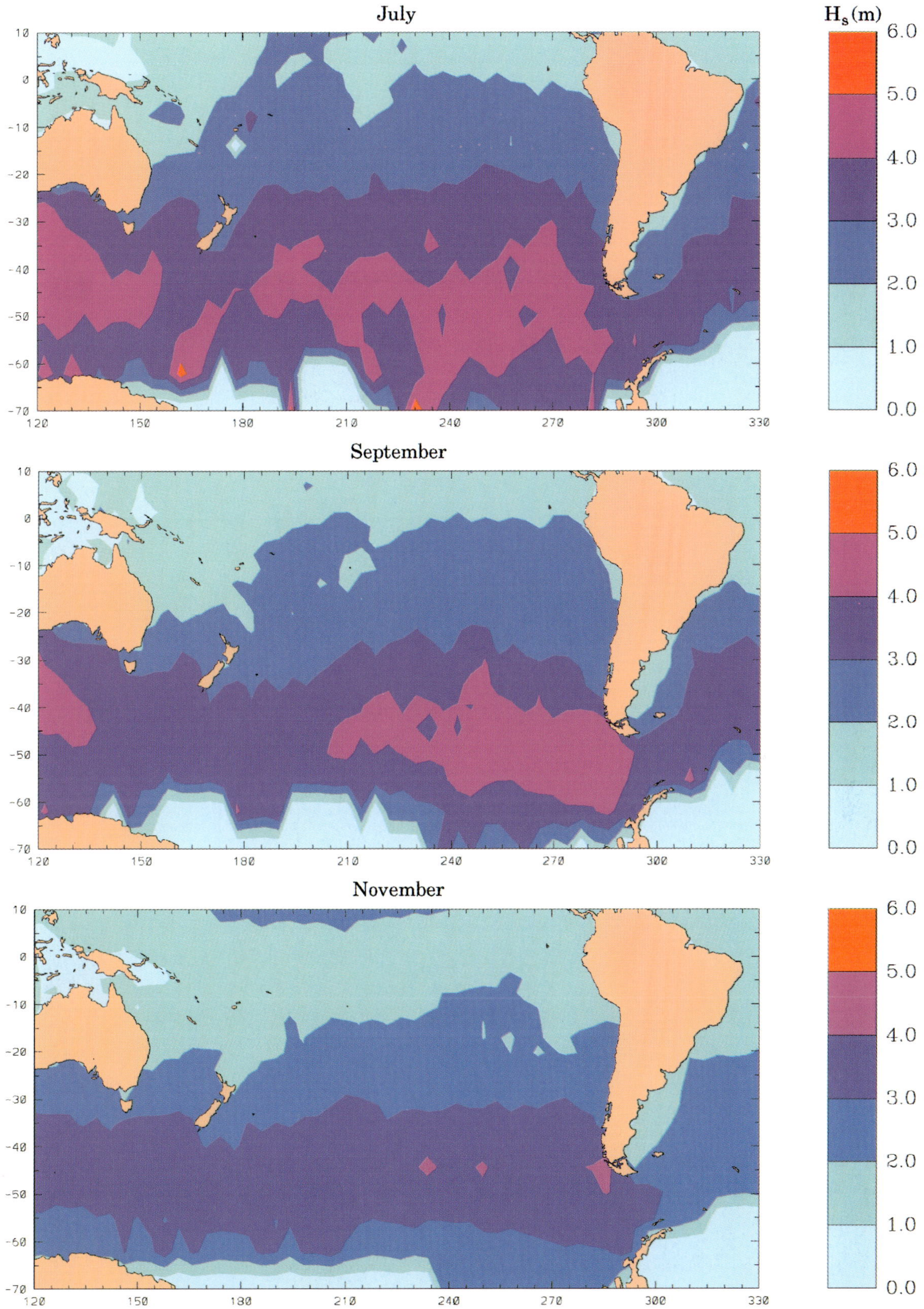

Fig. 7.12(d) Values of Mean Monthly Significant Wave Height for the South Pacific Ocean for July (top), September (middle) and November (bottom).

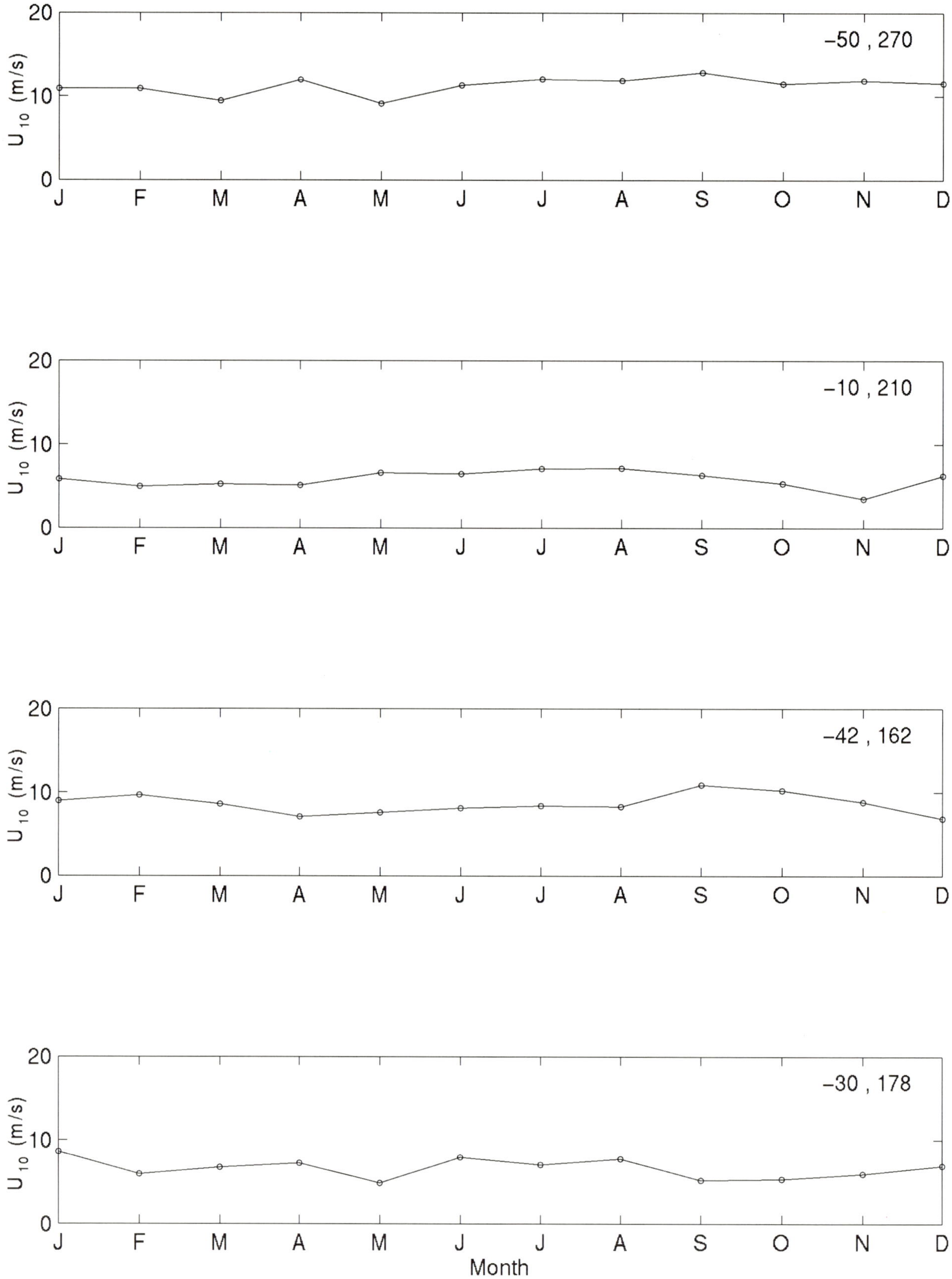

Fig. 7.13(a) Seasonal variability of wind speed at selected sites in the South Pacific Ocean: (i) in the belt of high latitude westerlies (−50°, 270°), (ii) in the extensive trade wind flow (−10°, 210°), (iii) in the Tasman Sea (−42°, 162°) and (iv) north-west of New Zealand (−30°, 178°).

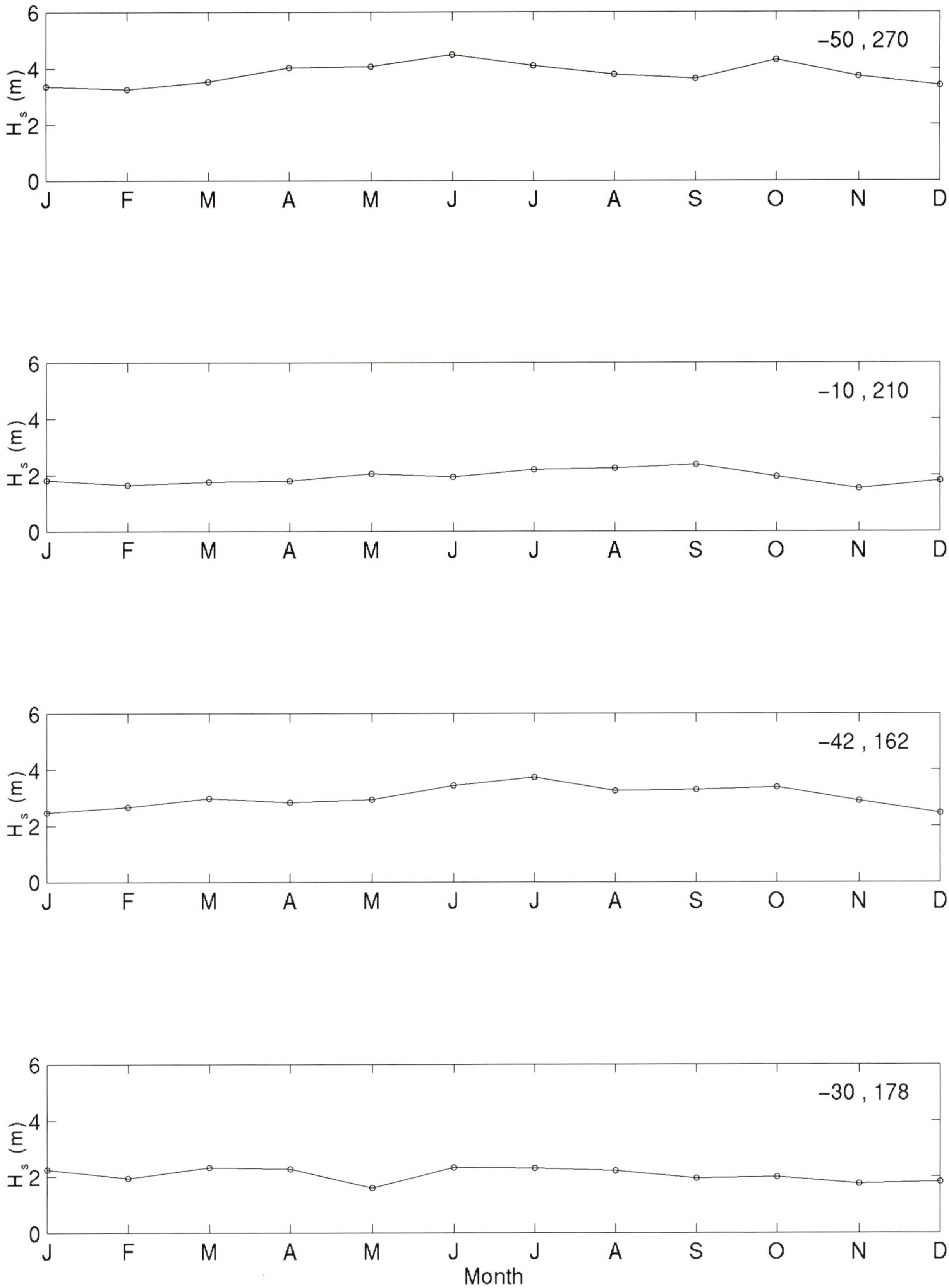

Fig. 7.13(b) Seasonal variability of significant wave height at selected sites in the South Pacific Ocean: (i) in the belt of high latitude westerlies ($-50^\circ, 270^\circ$), (ii) in the extensive trade wind flow ($-10^\circ, 210^\circ$), (iii) in the Tasman Sea ($-42^\circ, 162^\circ$) and (iv) north-west of New Zealand ($-30^\circ, 178^\circ$).

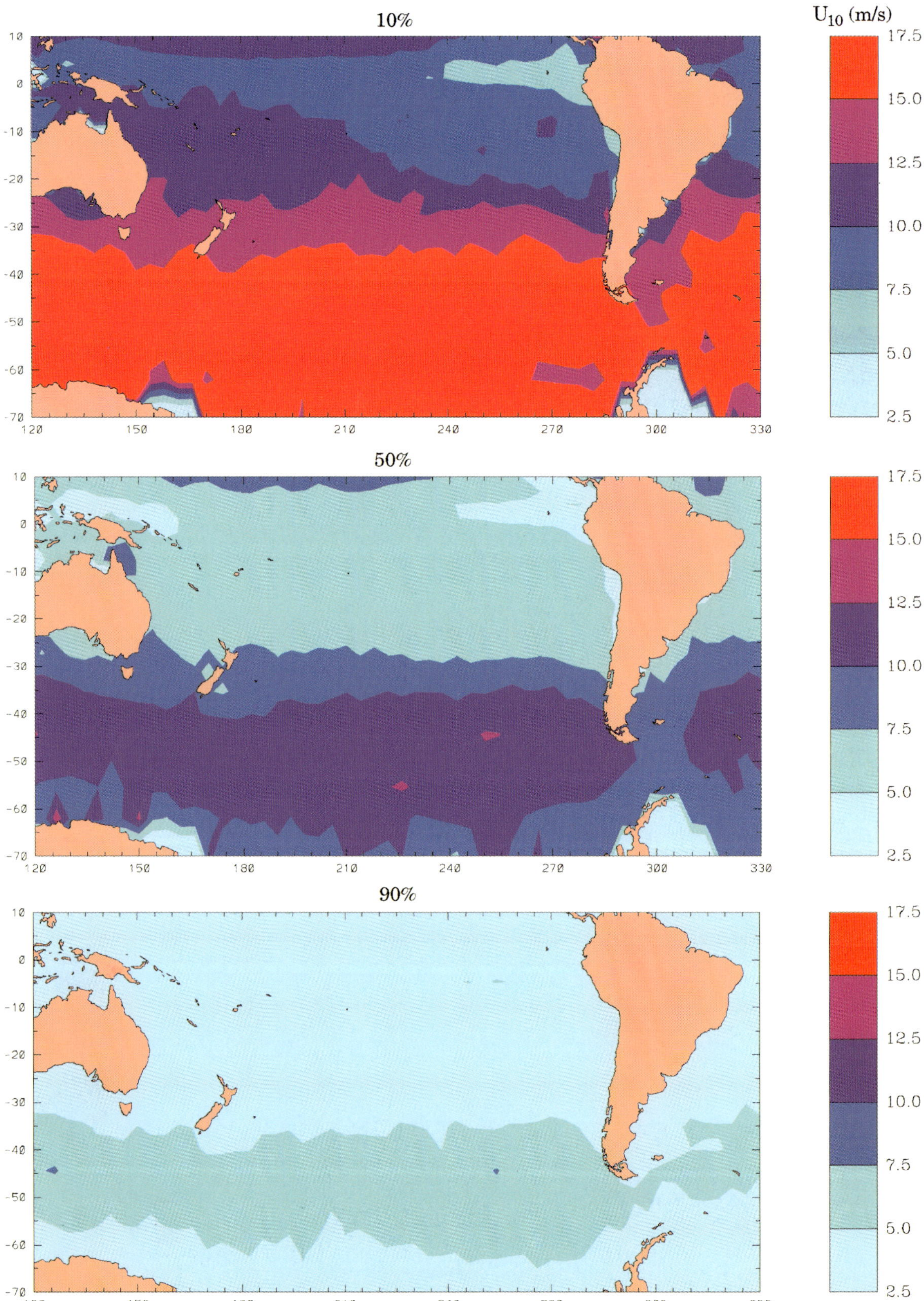

Fig. 7.14(a) Values for the South Pacific Ocean of the Wind Speed which will be exceeded 10% (top), 50% (middle) and 90% (bottom) of the time.

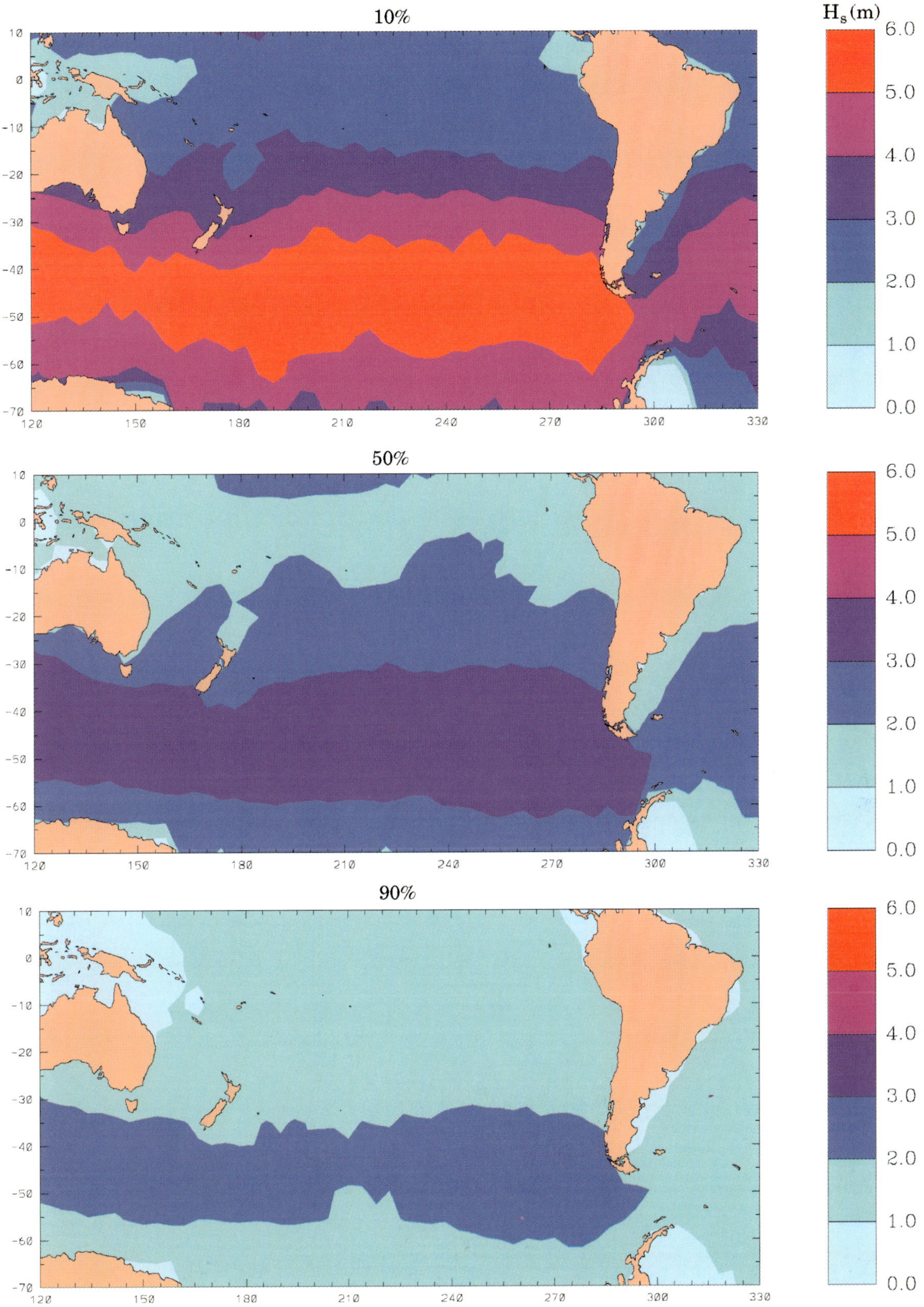

Fig. 7.14(b) Values for the South Pacific Ocean of the Significant Wave Height which will be exceeded 10% (top), 50% (middle) and 90% (bottom) of the time.

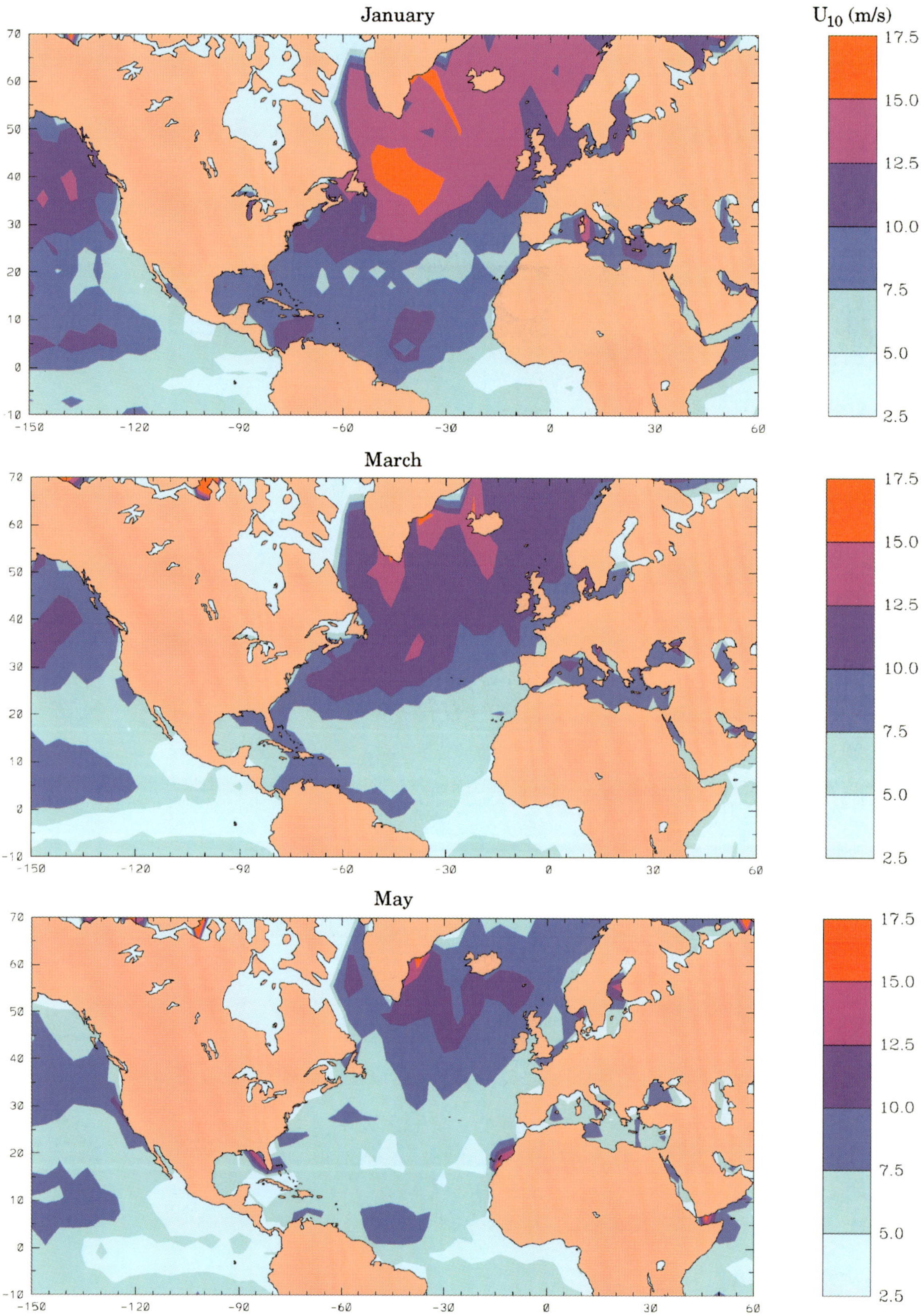

Fig. 7.15(a) Values of Mean Monthly Wind Speed for the North Atlantic Ocean for January (top), March (middle) and May (bottom).

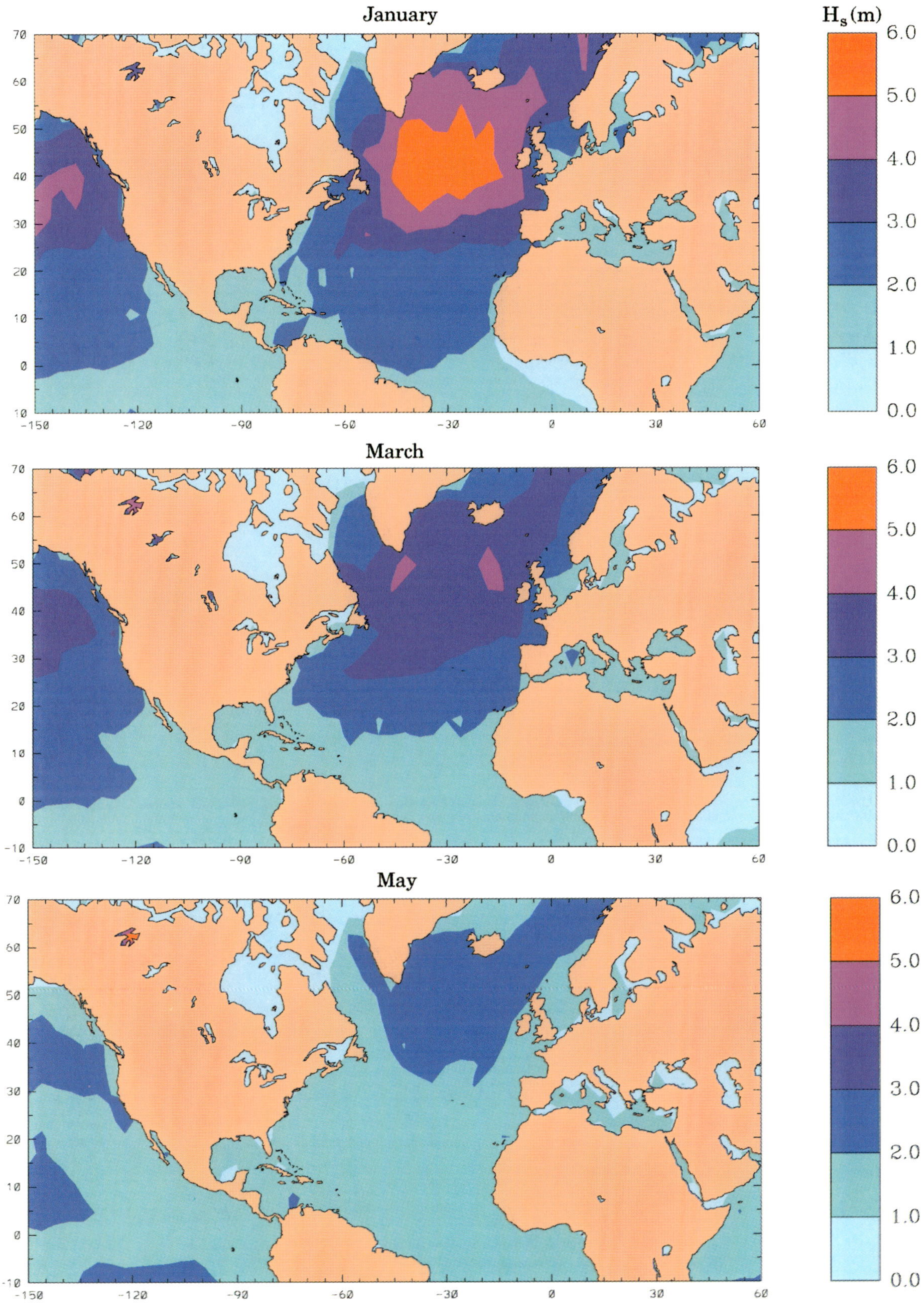

Fig. 7.15(b) Values of Mean Monthly Significant Wave Height for the North Atlantic Ocean for January (top), March (middle) and May (bottom).

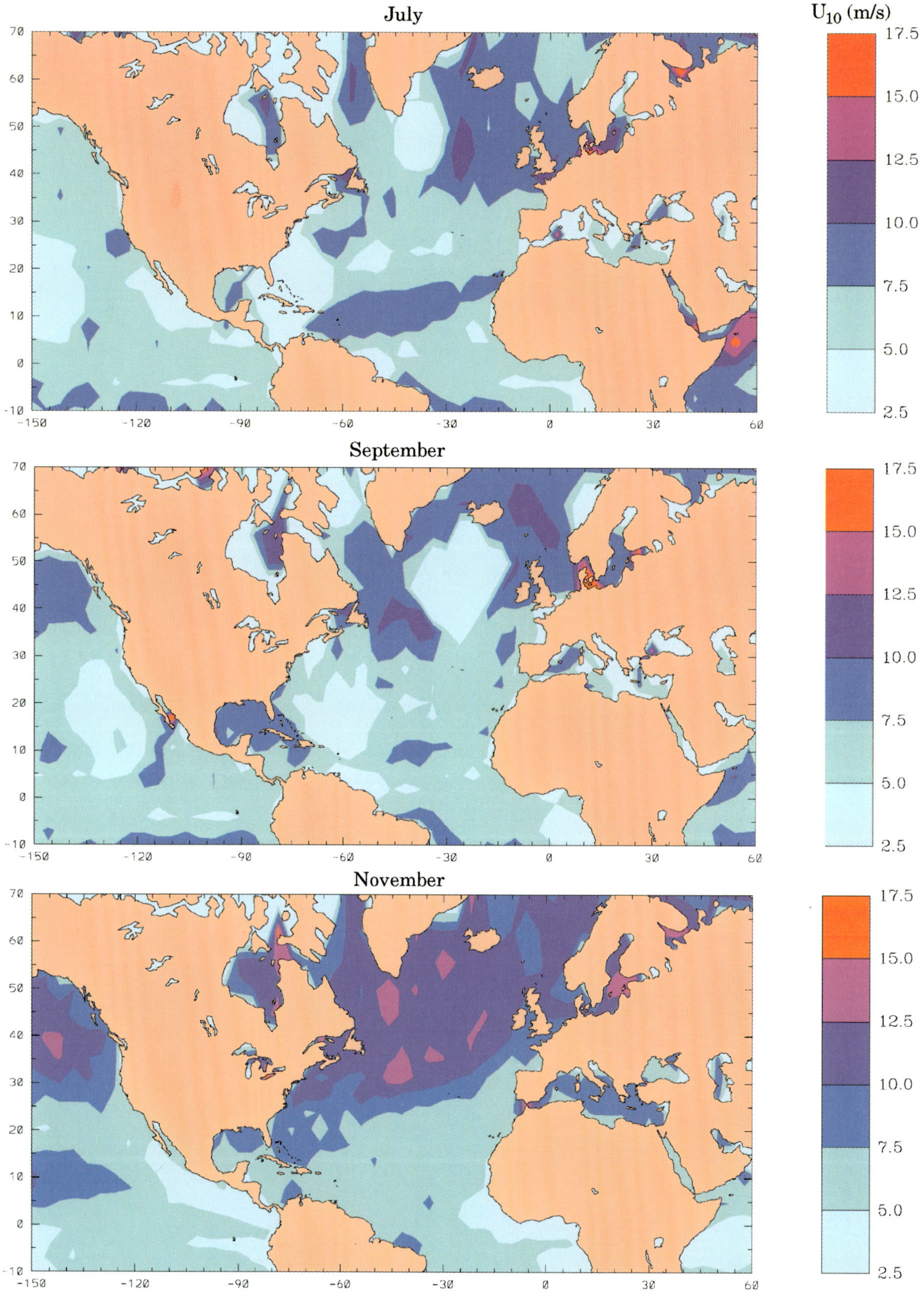

Fig. 7.15(c) Values of Mean Monthly Wind Speed for the North Atlantic Ocean for July (top), September (middle) and November (bottom).

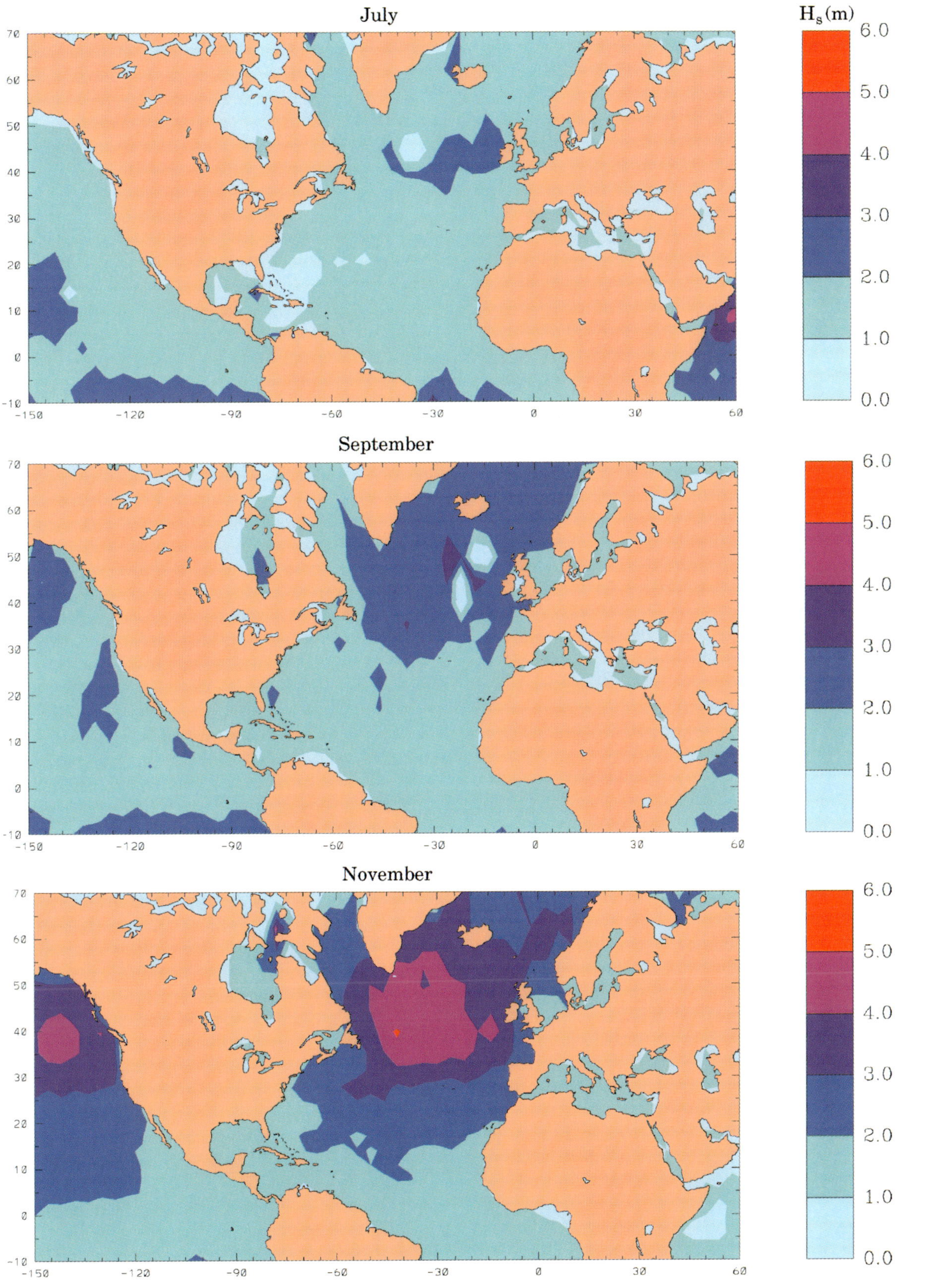

Fig. 7.15(d) Values of Mean Monthly Significant Wave Height for the North Atlantic Ocean for July (top), September (middle) and November (bottom).

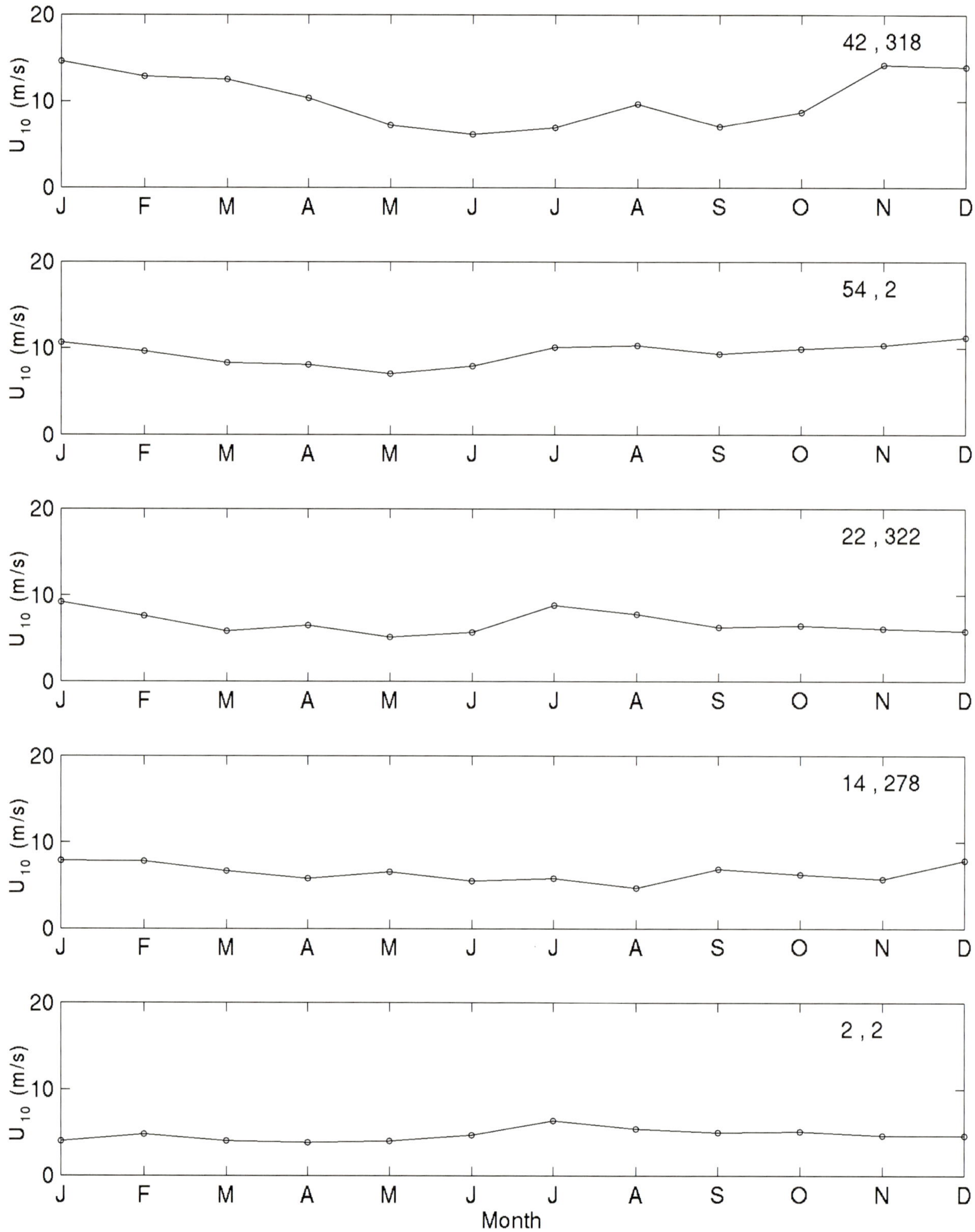

Fig. 7.16(a) Seasonal Variability of Wind Speed at selected sites in the North Atlantic Ocean: (i) in the strong westerly flow (42°, 318°), (ii) in the North Sea (54°, 2°), (iii) in the trade wind regime (22°, 322°), (iv) in the Caribbean Sea (14°, 278°) and (v) in the monsoonal Gulf of Guinea (2°, 2°).

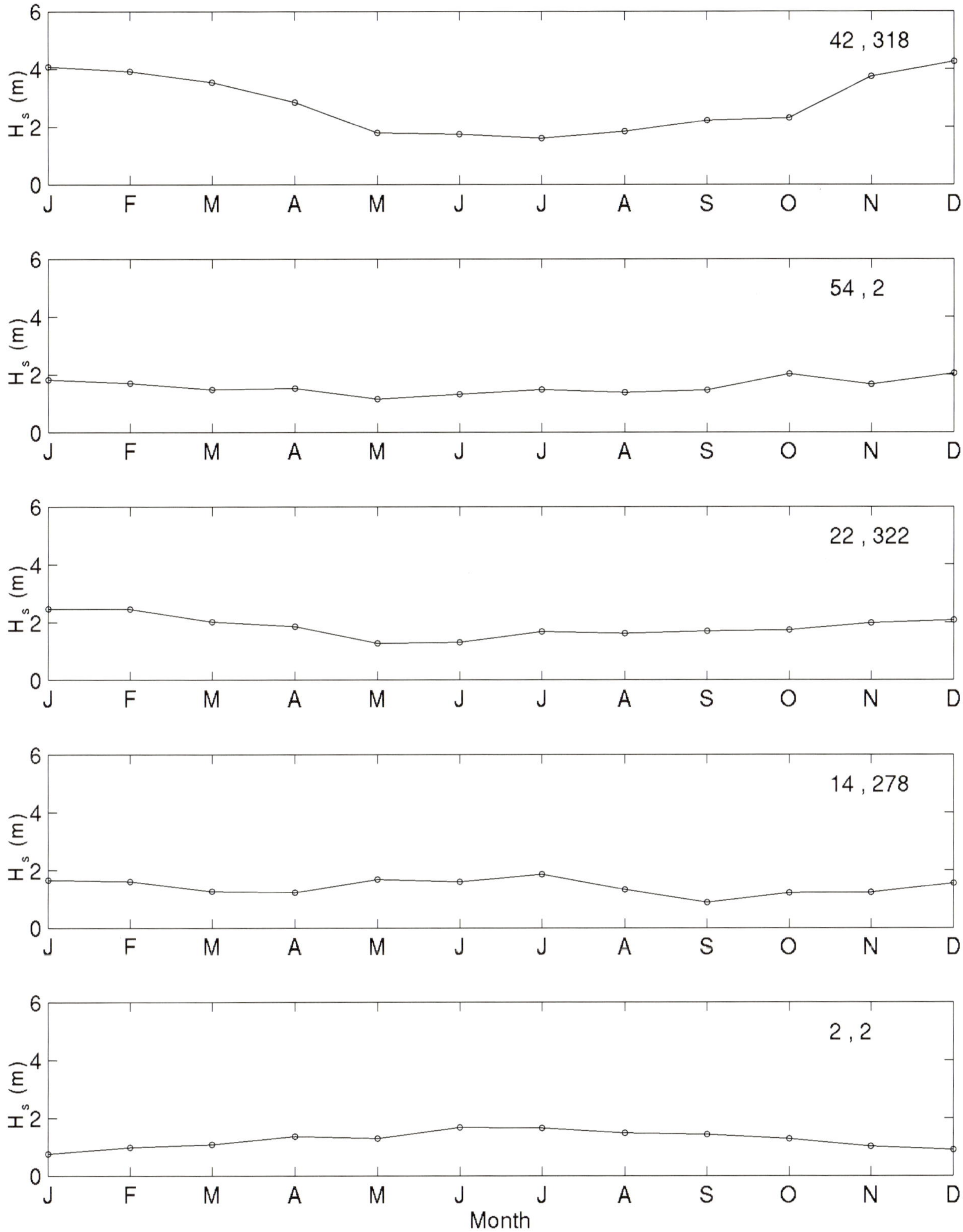

Fig. 7.16(b) Seasonal variability of Significant Wave Height at selected sites in the North Atlantic Ocean: (i) in the strong westerly flow (42°, 318°), (ii) in the North Sea (54°, 2°), (iii) in the trade wind regime (22°, 322°), (iv) in the Caribbean Sea (14°, 278°) and (v) in the monsoonal Gulf of Guinea (2°, 2°).

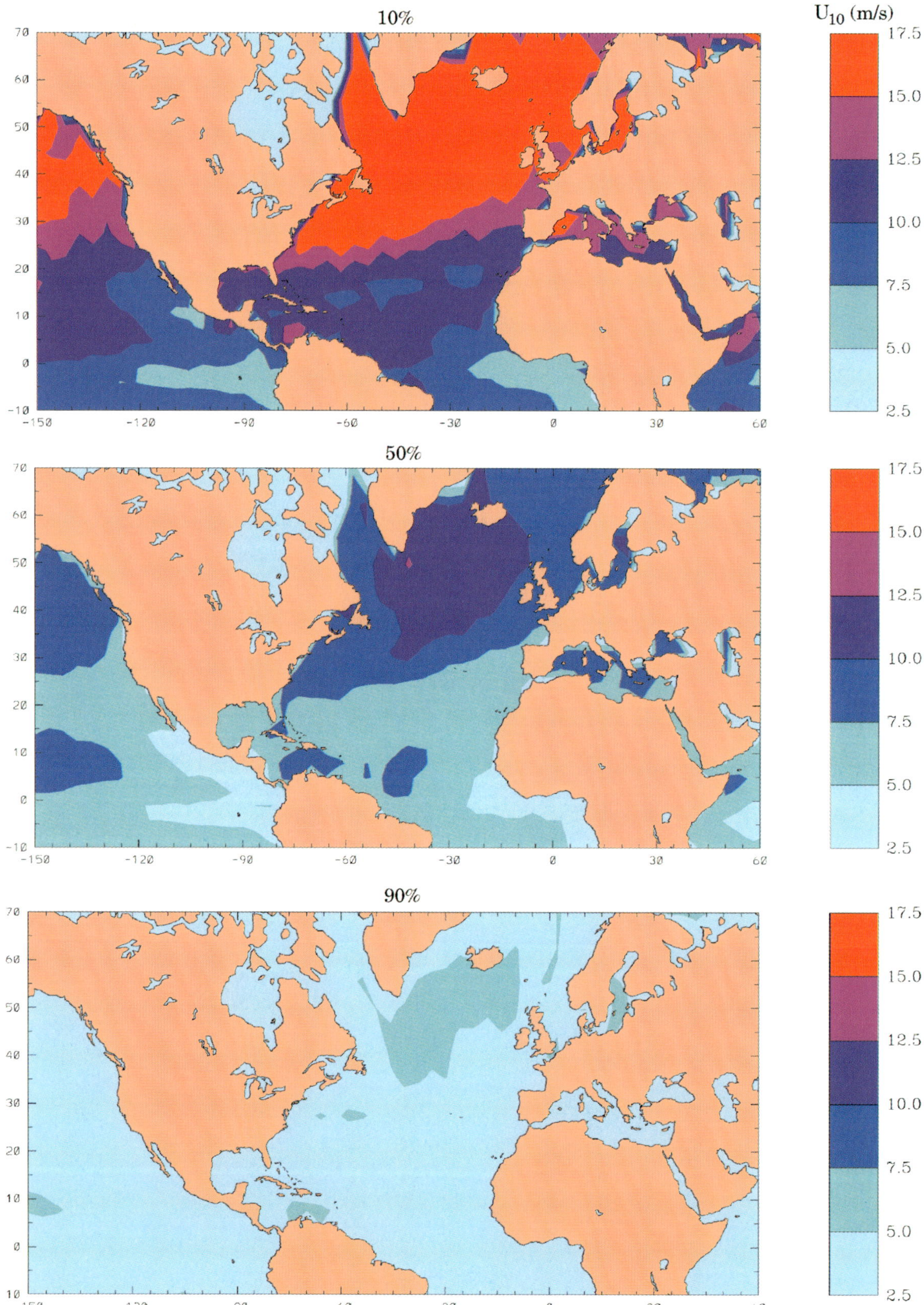

Fig. 7.17(a) Values for the North Atlantic Ocean of the Wind Speed which will be exceeded 10% (top), 50% (middle) and 90% (bottom) of the time.

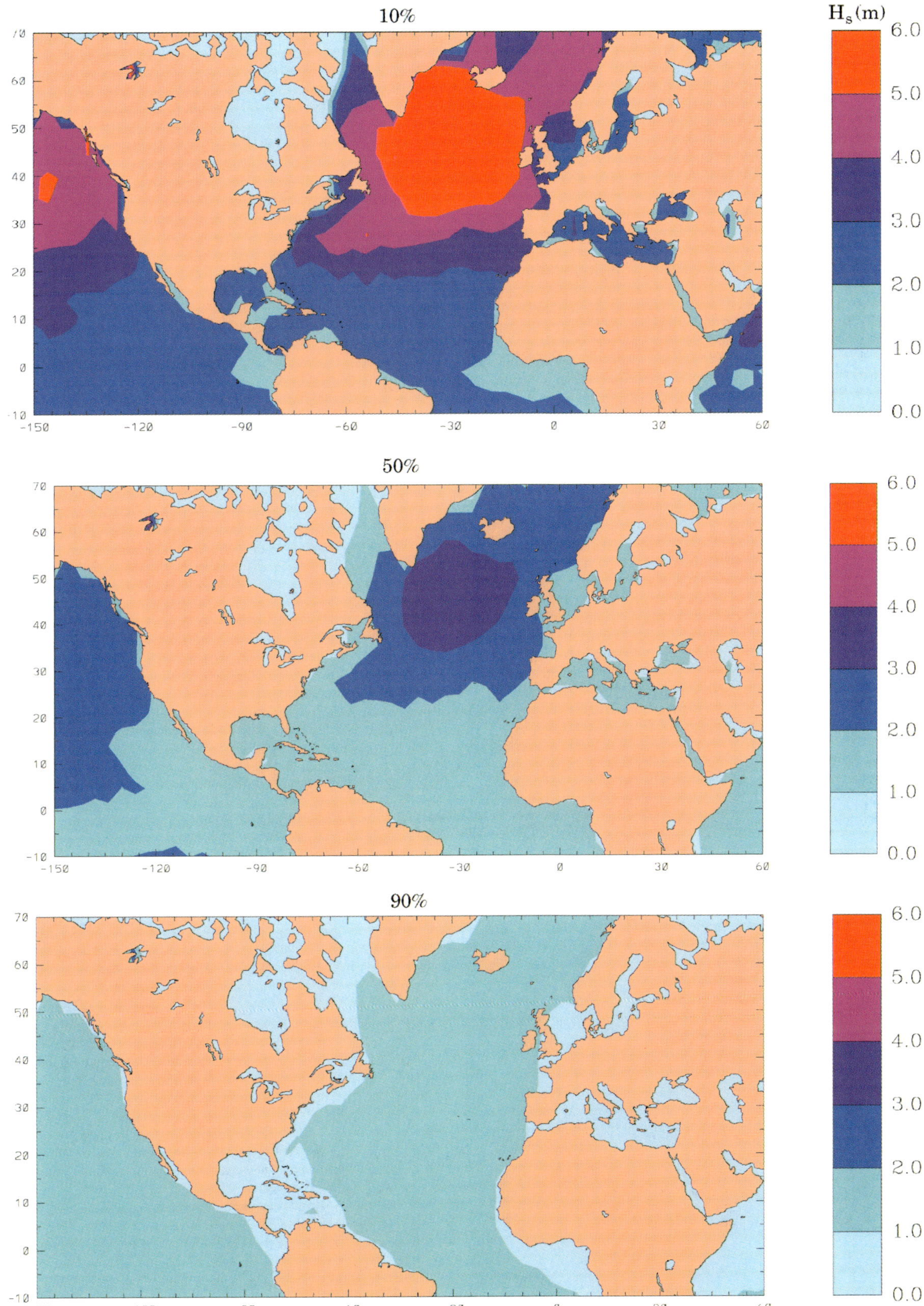

Fig. 7.17(b) Values for the North Atlantic Ocean of the Significant Wave Height which will be exceeded 10% (top), 50% (middle) and 90% (bottom) of the time.

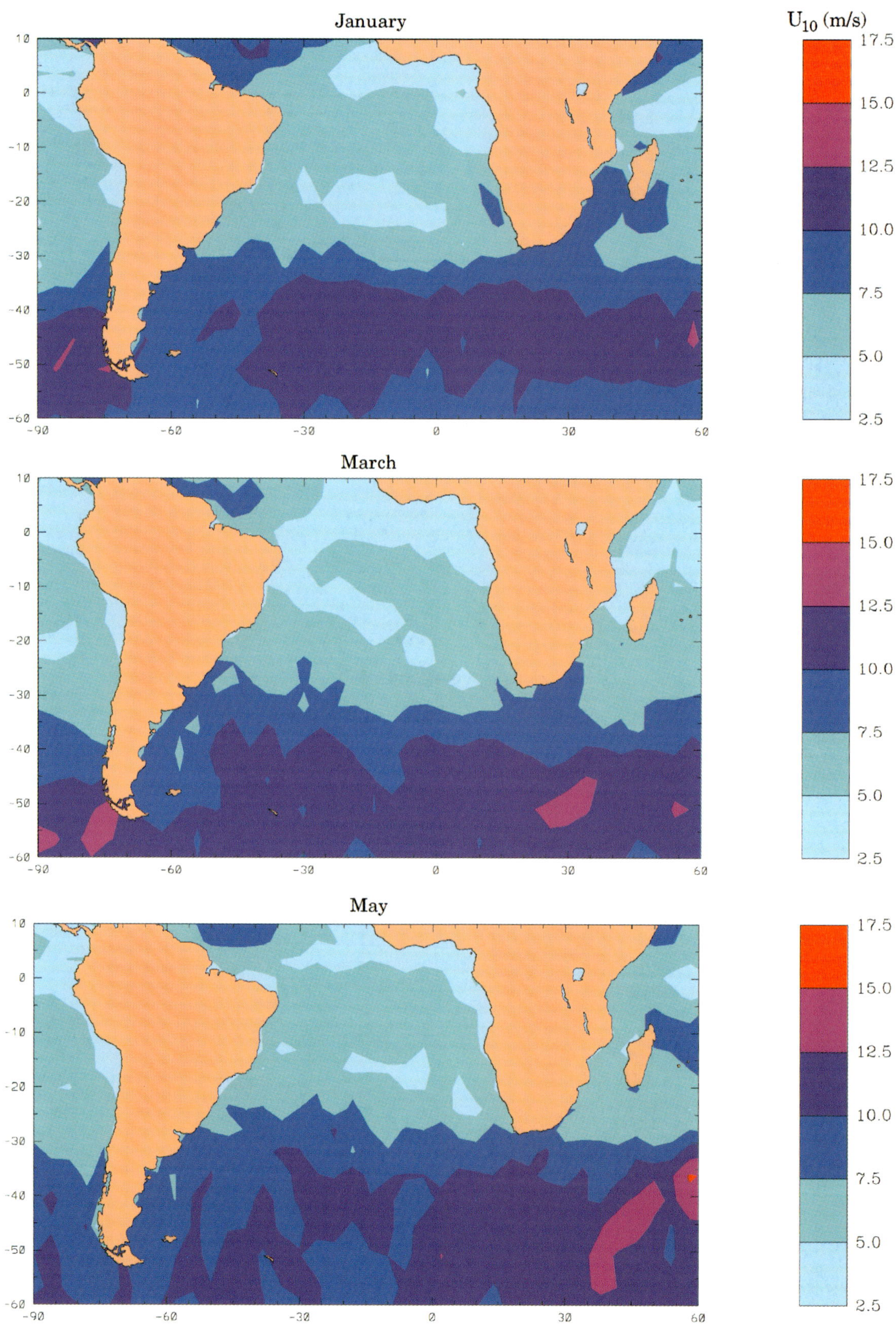

Fig. 7.18(a) Values of Mean Monthly Wind Speed for the South Atlantic Ocean for January (top), March (middle) and May (bottom).

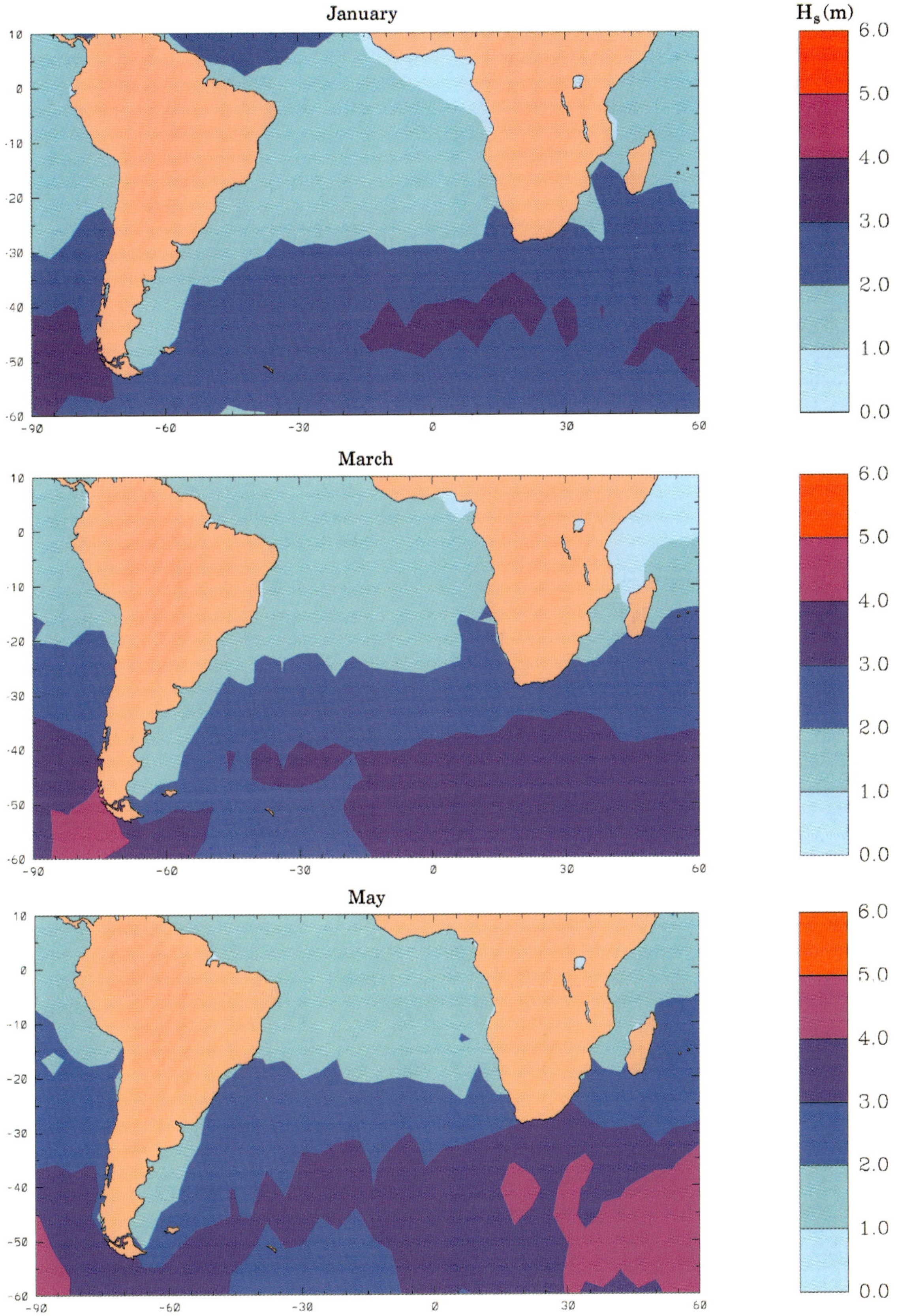

Fig. 7.18(b) Values of Mean Monthly Significant Wave Height for the South Atlantic Ocean for January (top), March (middle) and May (bottom).

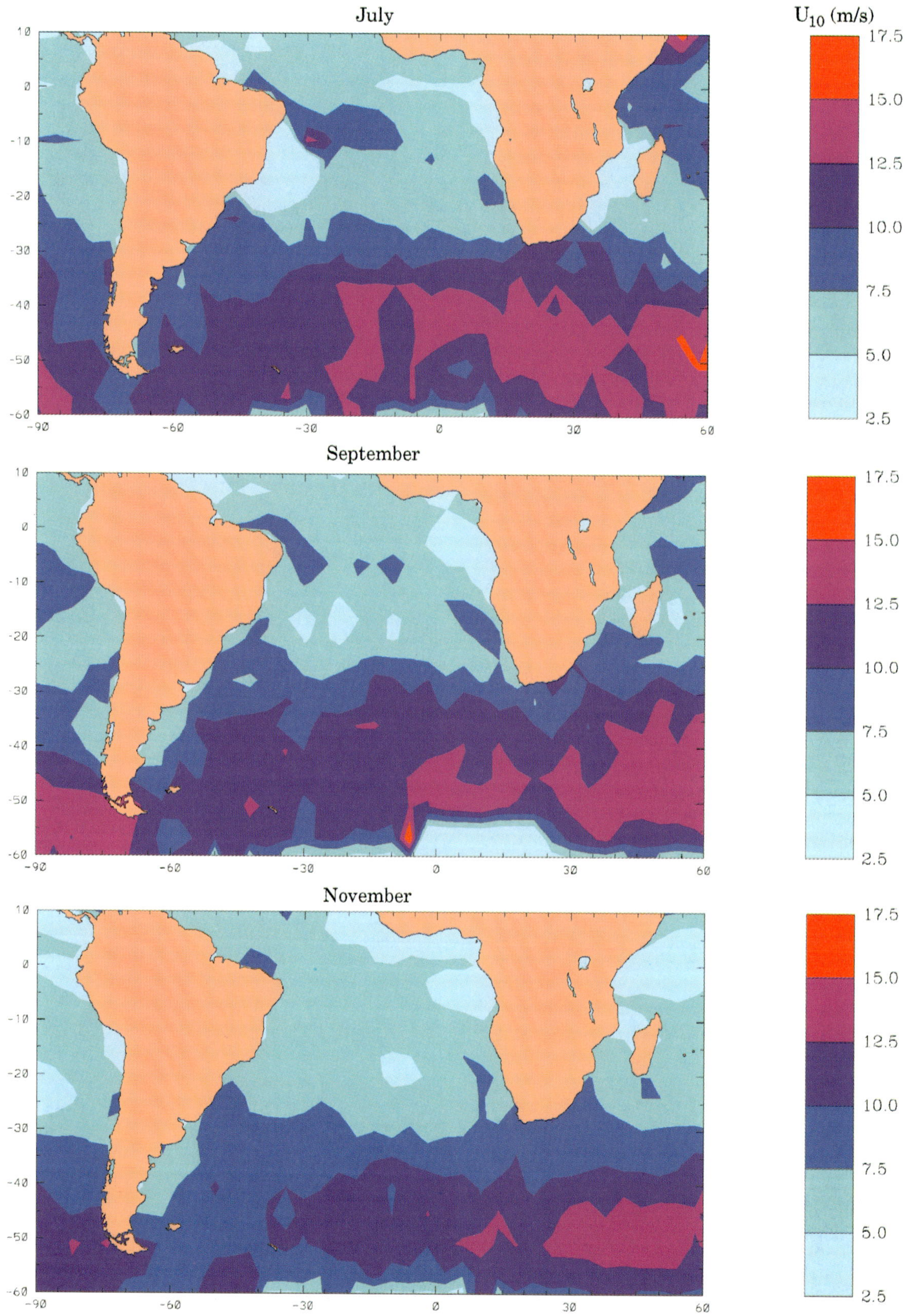

Fig. 7.18(c) Values of Mean Monthly Wind Speed for the South Atlantic Ocean for July (top), September (middle) and November (bottom).

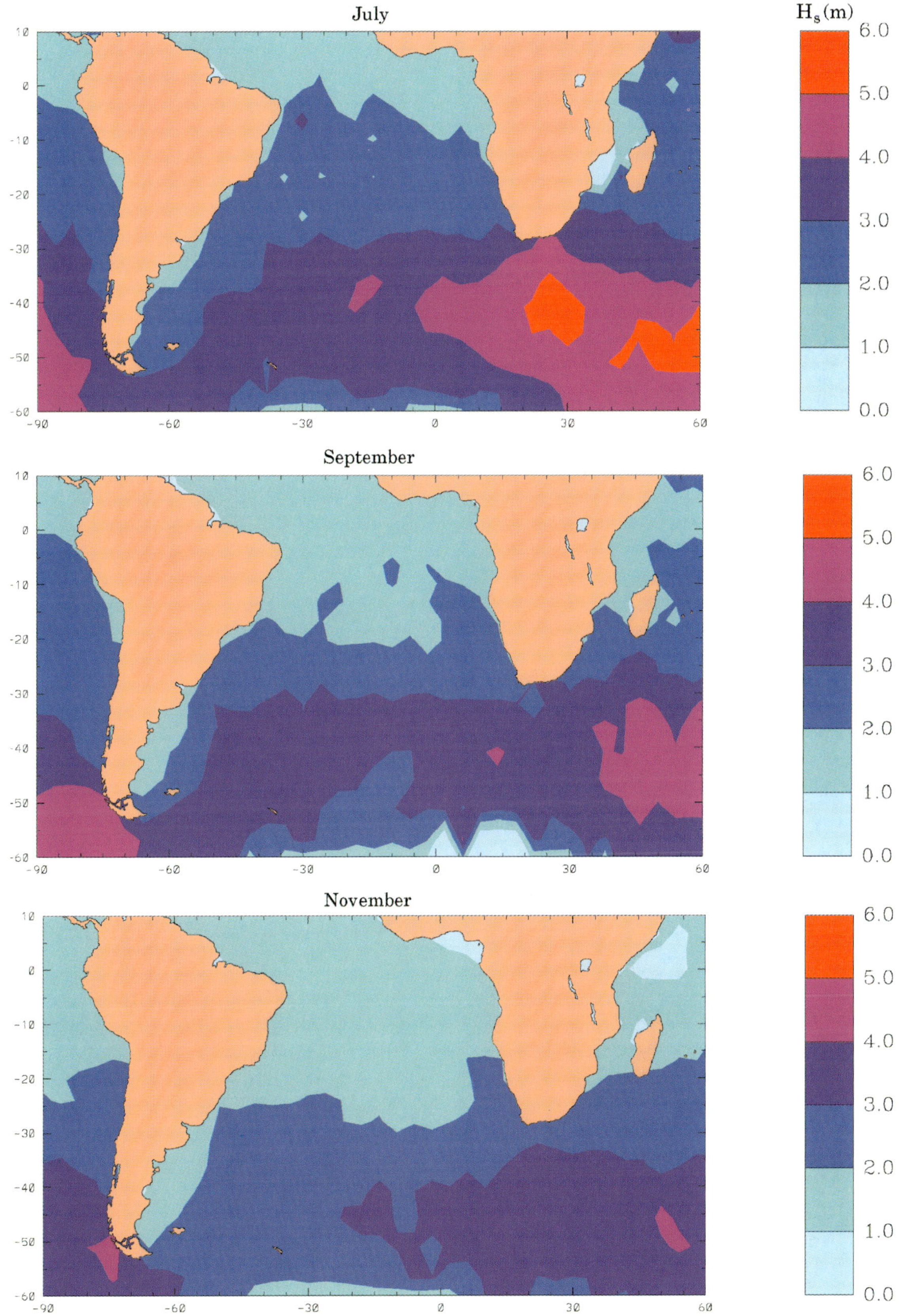

Fig. 7.18(d) Values of Mean Monthly Significant Wave Height for the South Atlantic Ocean for July (top), September (middle) and November (bottom).

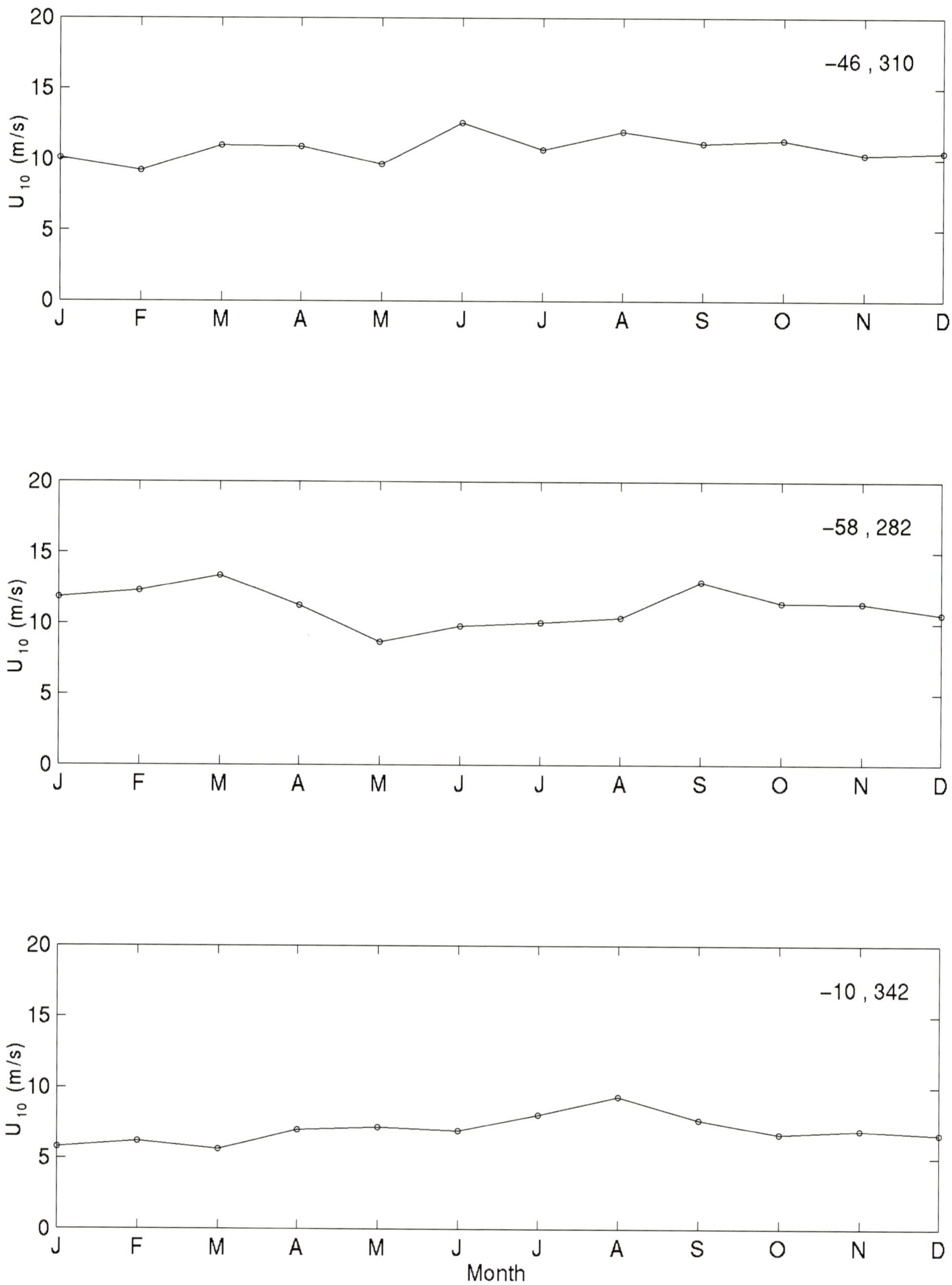

Fig. 7.19(a) Seasonal Variability of Wind Speed for selected sites in the South Atlantic Ocean: (i) in the westerlies in the lee of South America ($-46°, 310°$) with (ii) the entrance to Drake Passage ($-58°, 282°$) for comparison and (iii) in the trade flow ($-10°, 342°$).

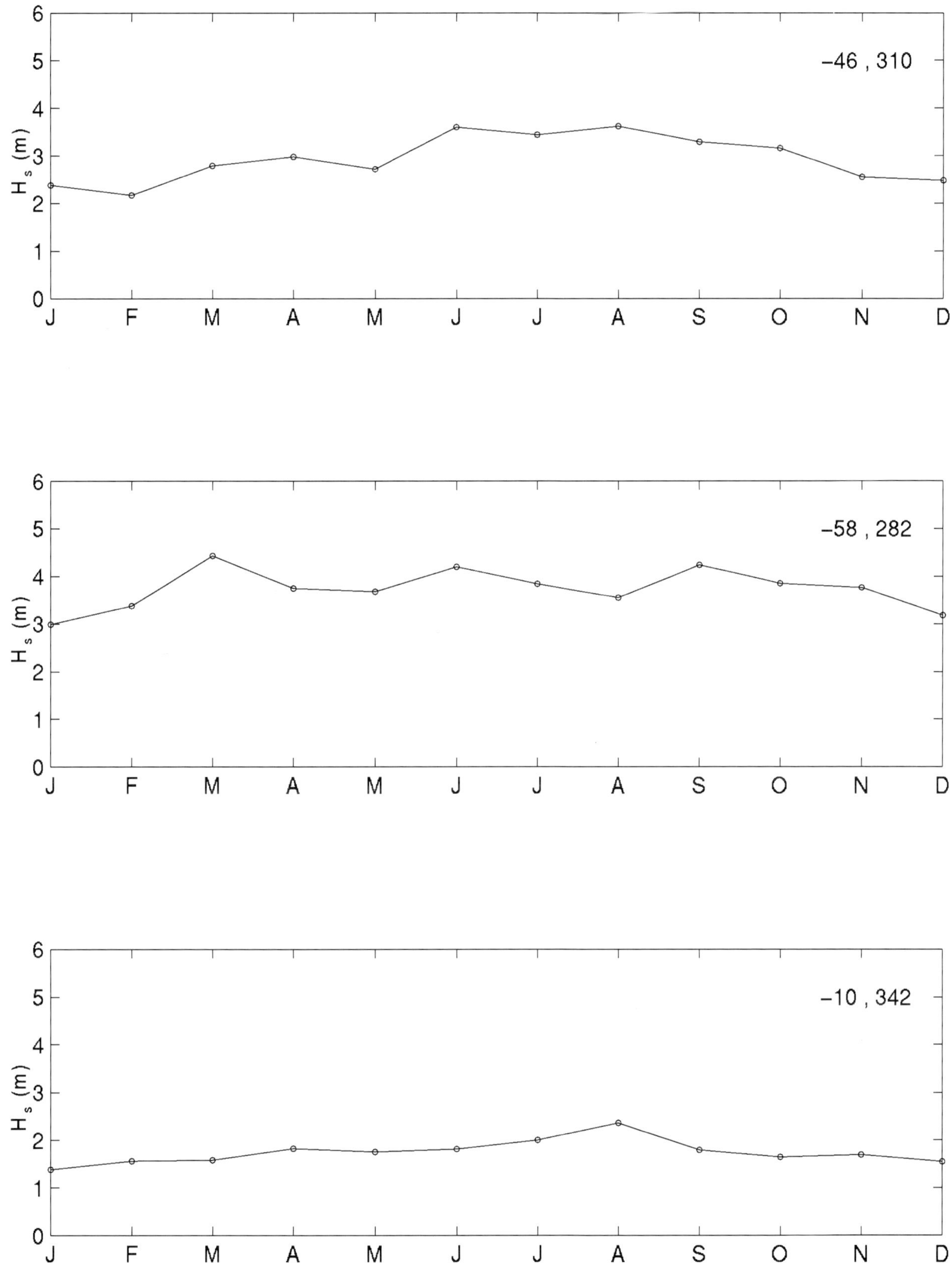

Fig. 7.19(b) Seasonal Variability of Significant Wave Height for selected sites in the South Atlantic Ocean: (i) in the westerlies in the lee of South America ($-46°, 310°$) with (ii) the entrance to Drake Passage ($-58°, 282°$) for comparison and (iii) in the trade flow ($-10°, 342°$).

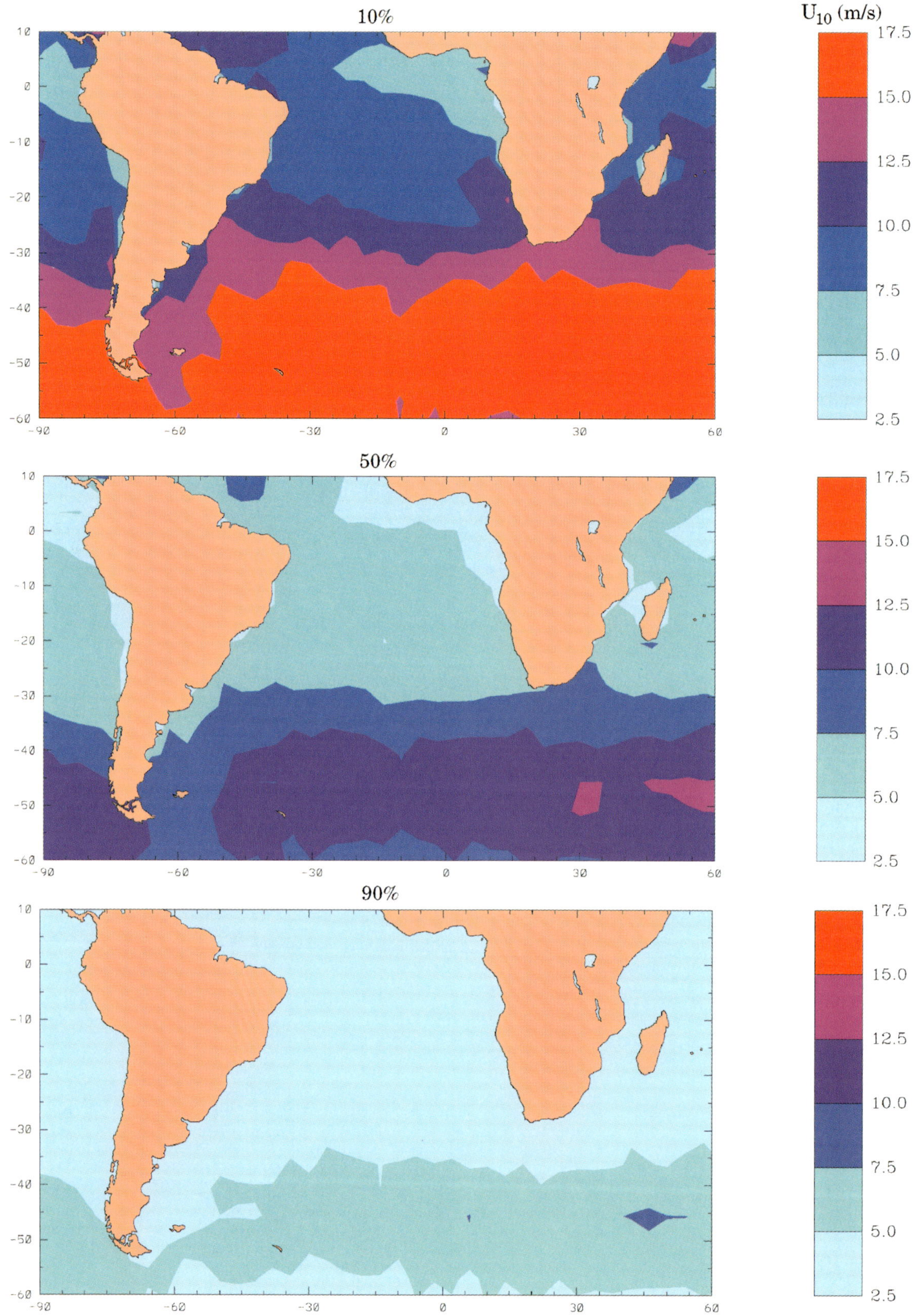

Fig. 7.20(a) Values for the South Atlantic Ocean of the Wind Speed which will be exceeded 10% (top), 50% (middle) and 90% (bottom) of the time.

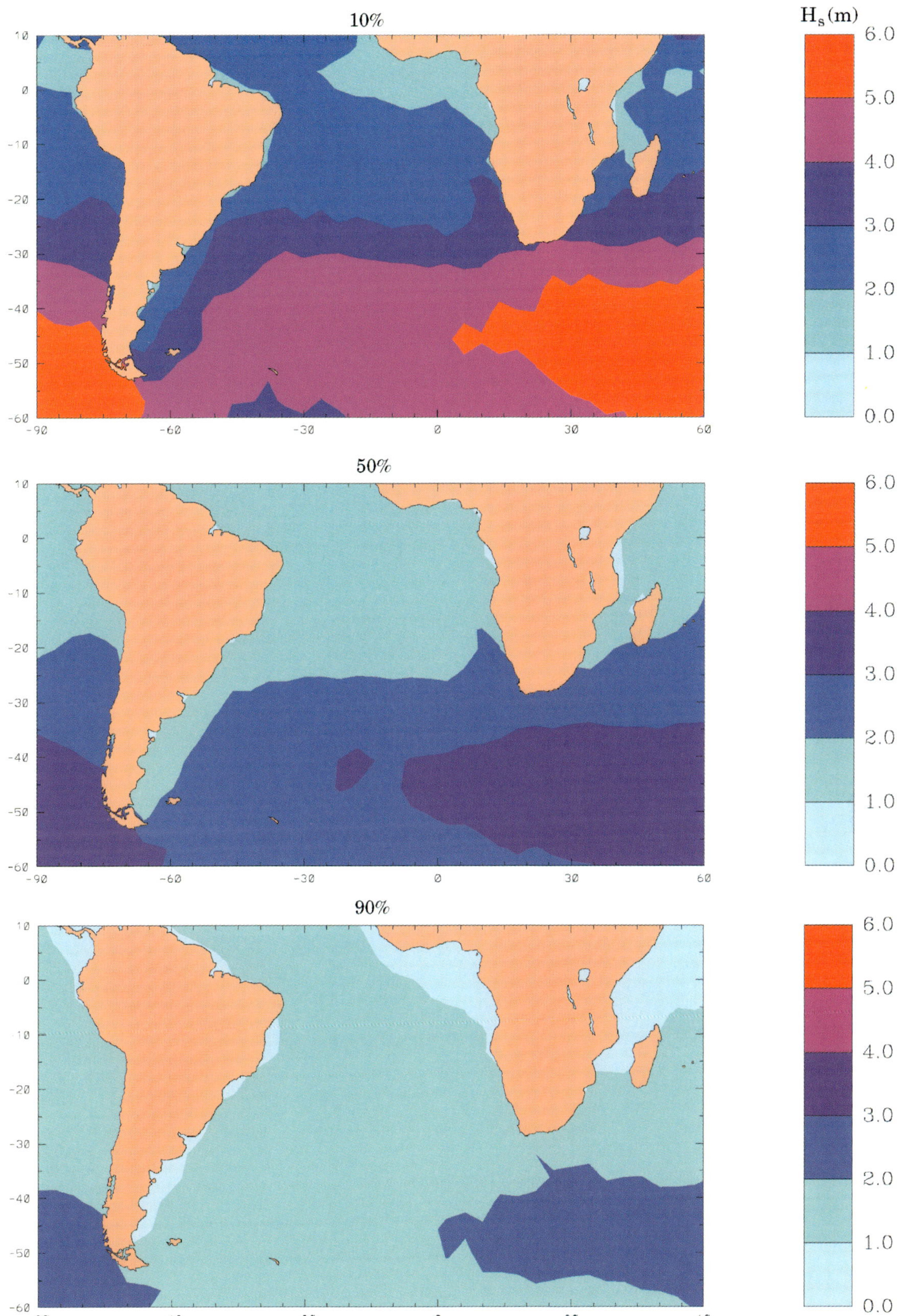

Fig. 7.20(b) Values for the South Atlantic Ocean of the Significant Wave Height which will be exceeded 10% (top), 50% (middle) and 90% (bottom) of the time.

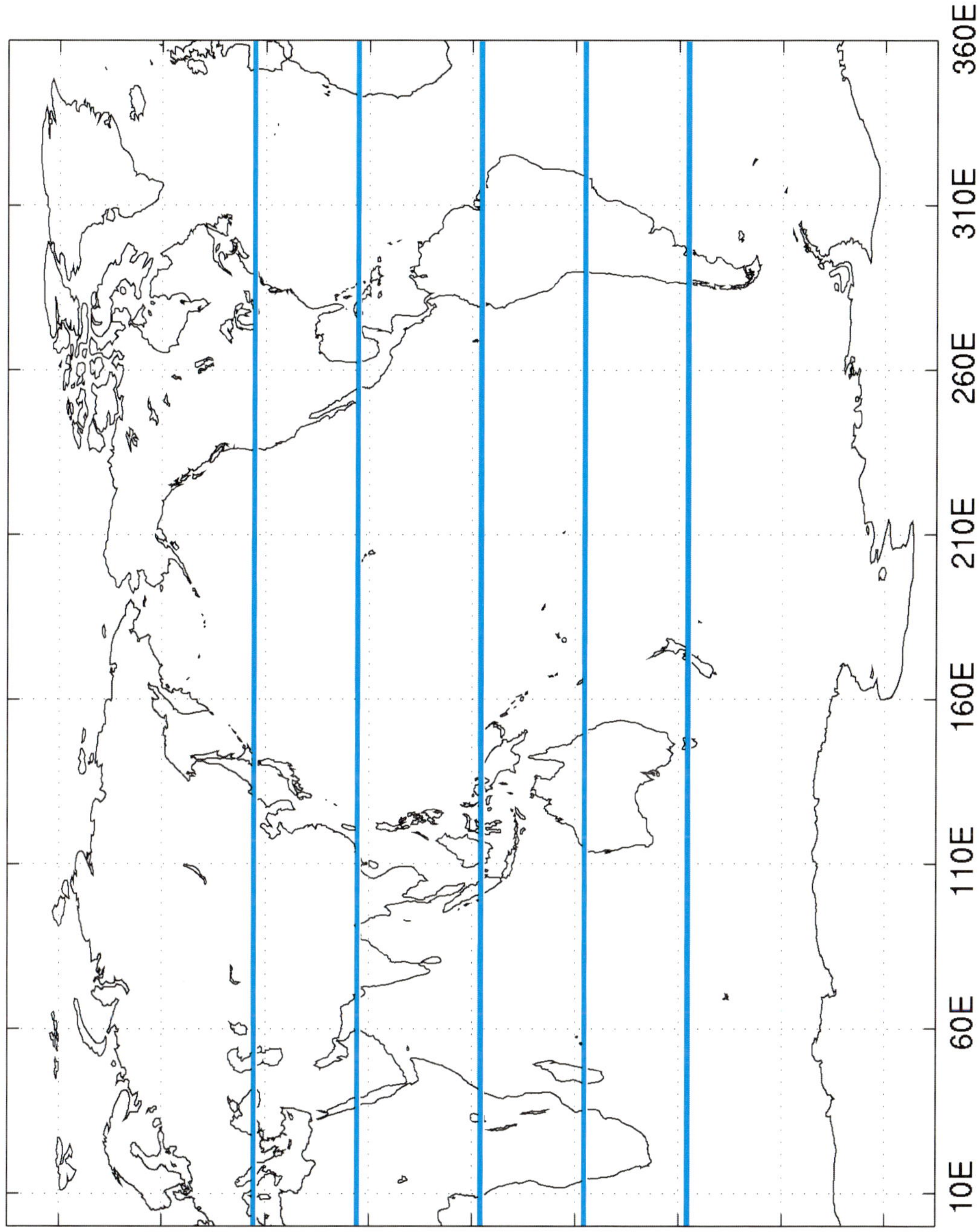

Fig. 7.21 Lines of constant latitude along which results are presented in Figs 7.22 and 7.23.

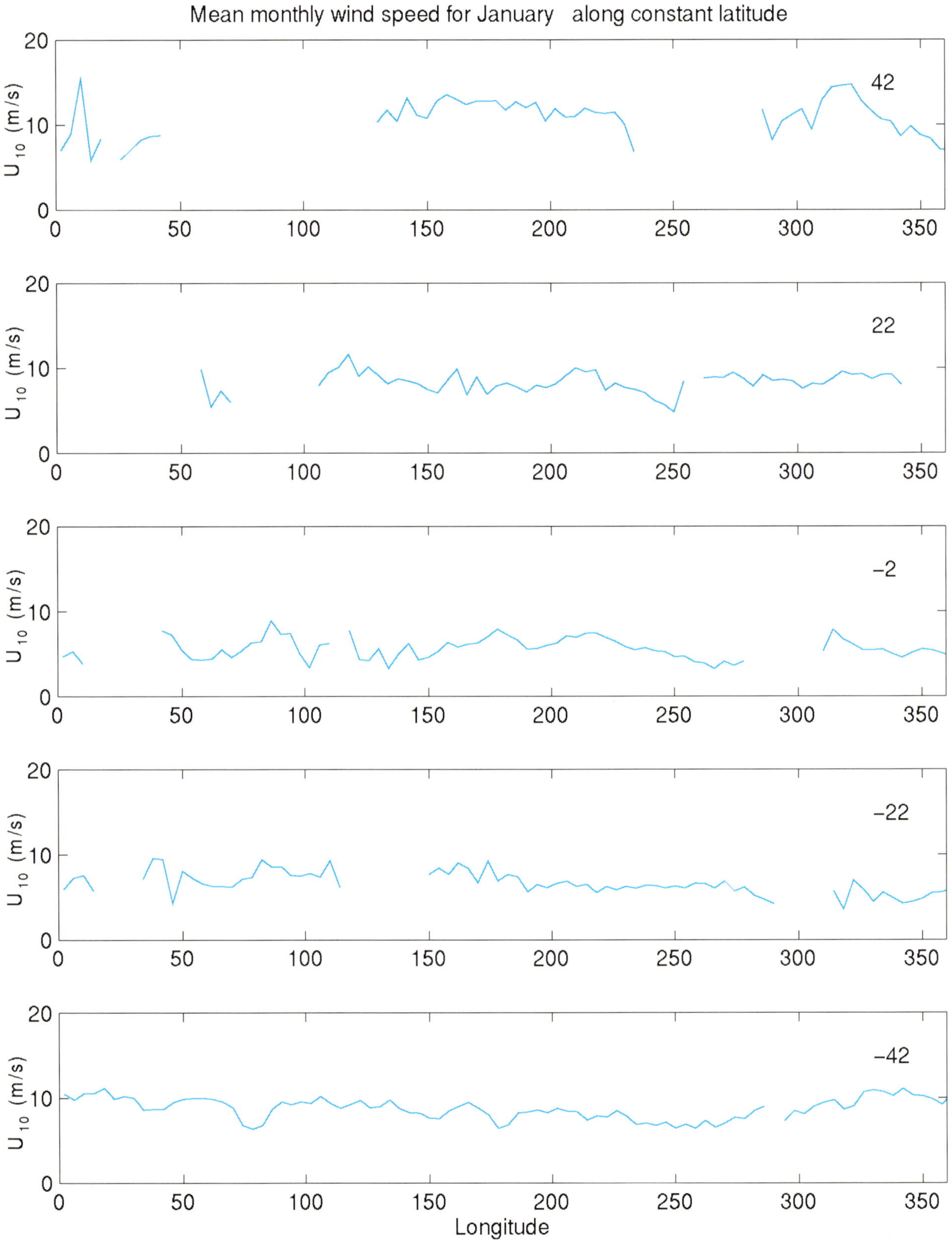

Fig. 7.22(a) Values of Mean Monthly Wind Speed for the month of January along the lines of constant latitude shown in Fig. 7.21. The values of latitude are shown in the top right corner of each panel.

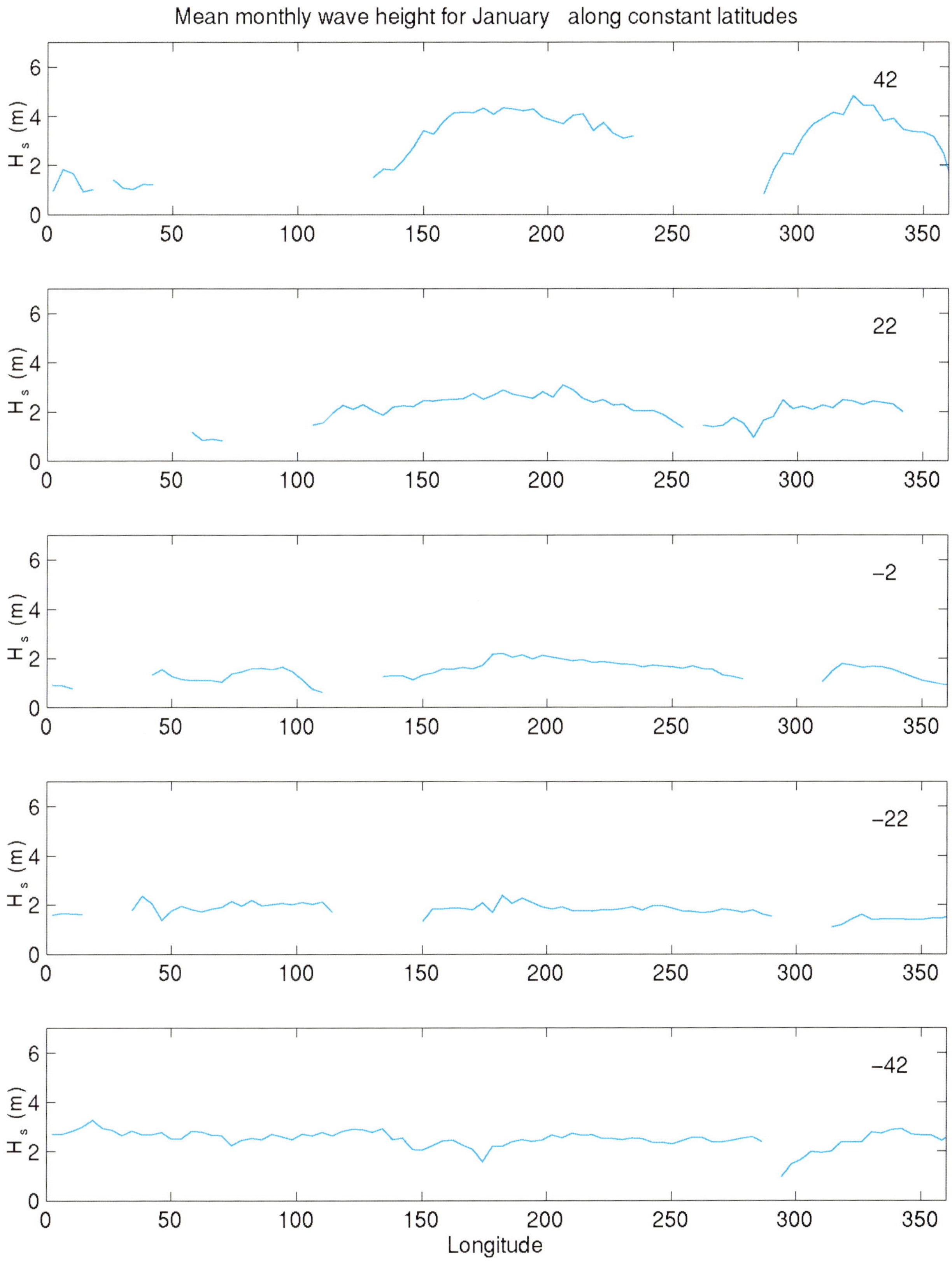

Fig. 7.22(b) Values of Mean Monthly Significant Wave Height for the month of January along the lines of constant latitude shown in Fig. 7.21. The values of latitude are shown in the top right corner of each panel.

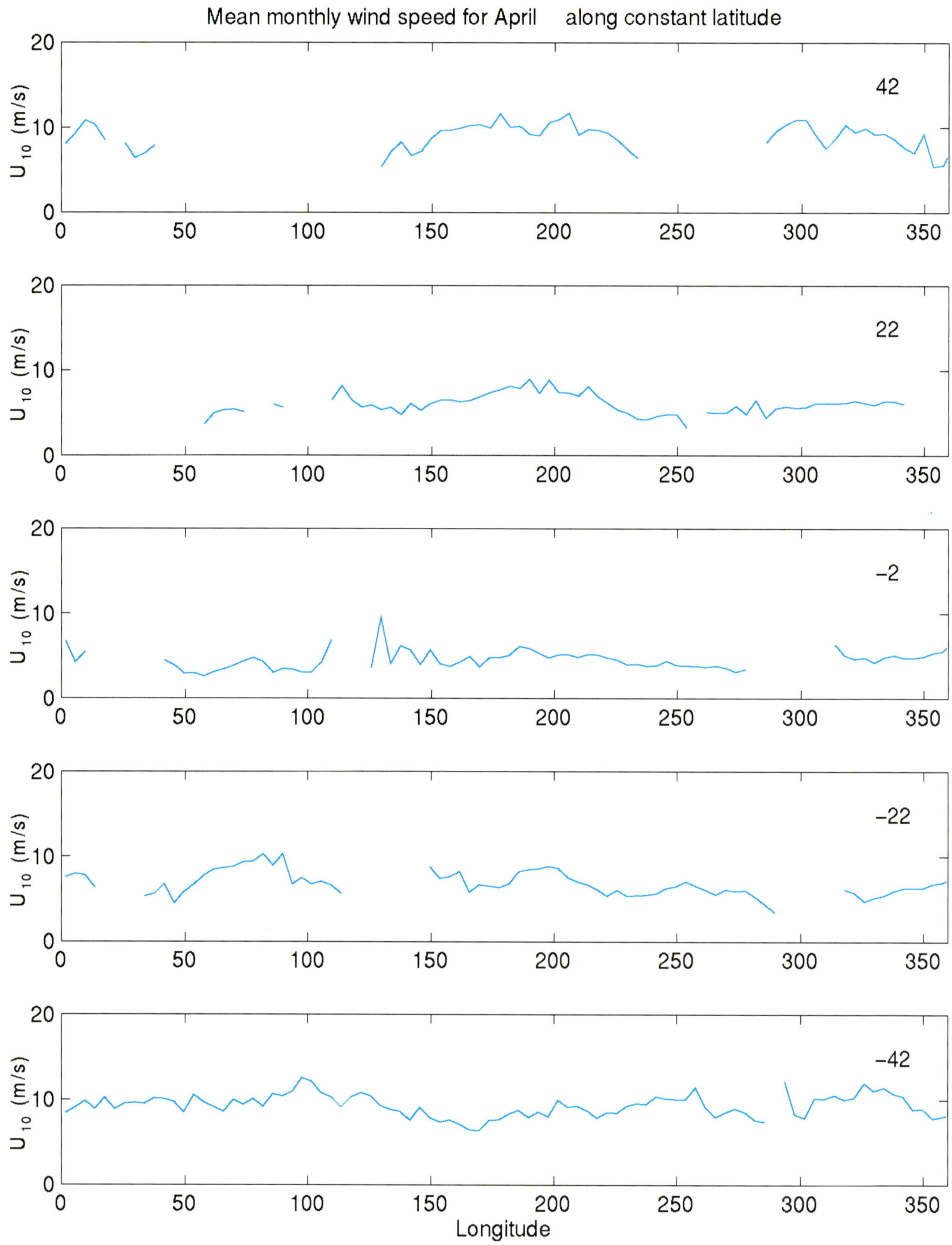

Fig. 7.22(c) Values of Mean Monthly Wind Speed for the month of April along the lines of constant latitude shown in Fig. 7.21. The values of latitude are shown in the top right corner of each panel.

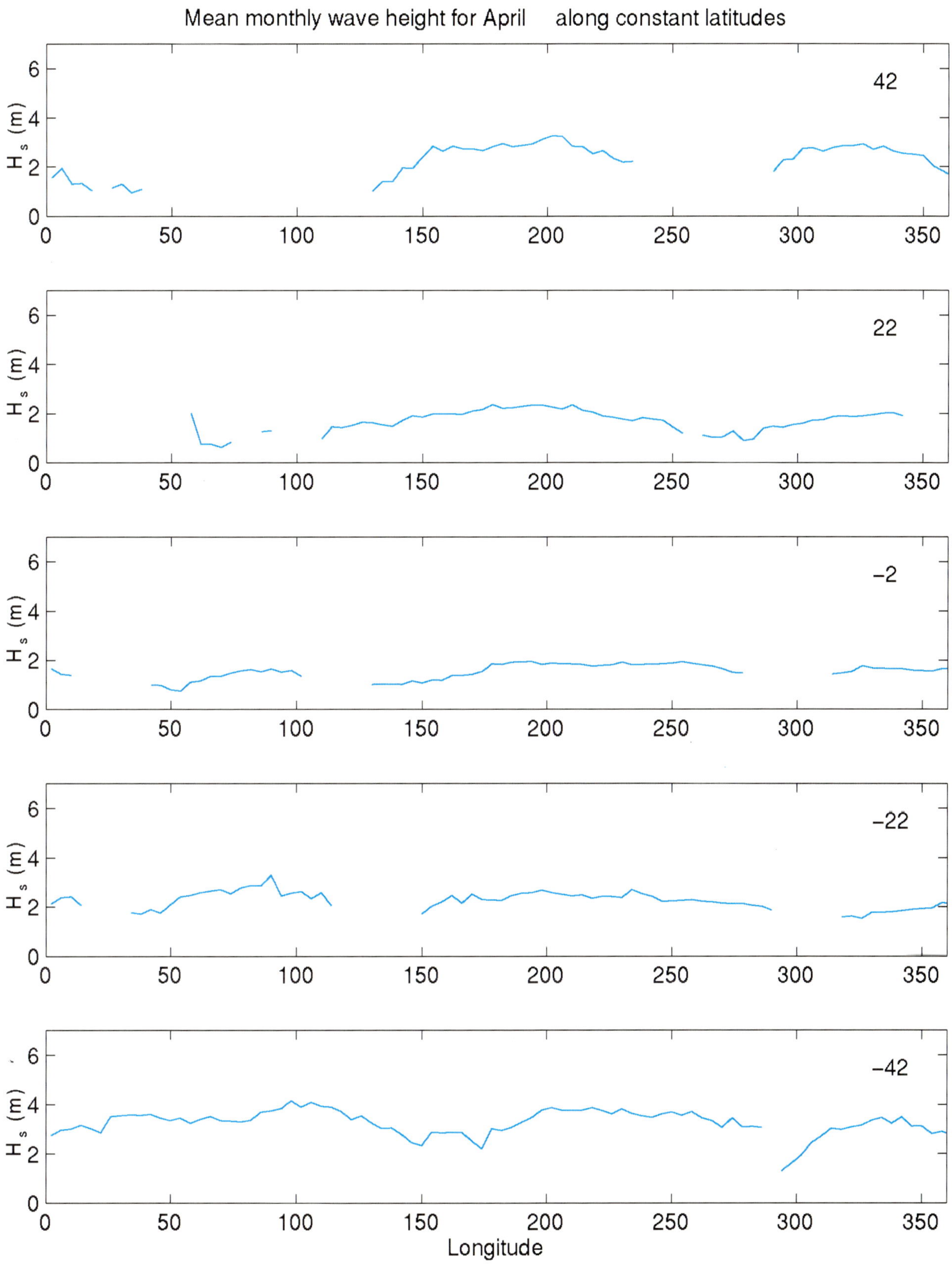

Fig. 7.22(d) Values of Mean Monthly Significant Wave Height for the month of April along the lines of constant latitude shown in Fig. 7.21. The values of latitude are shown in the top right corner of each panel.

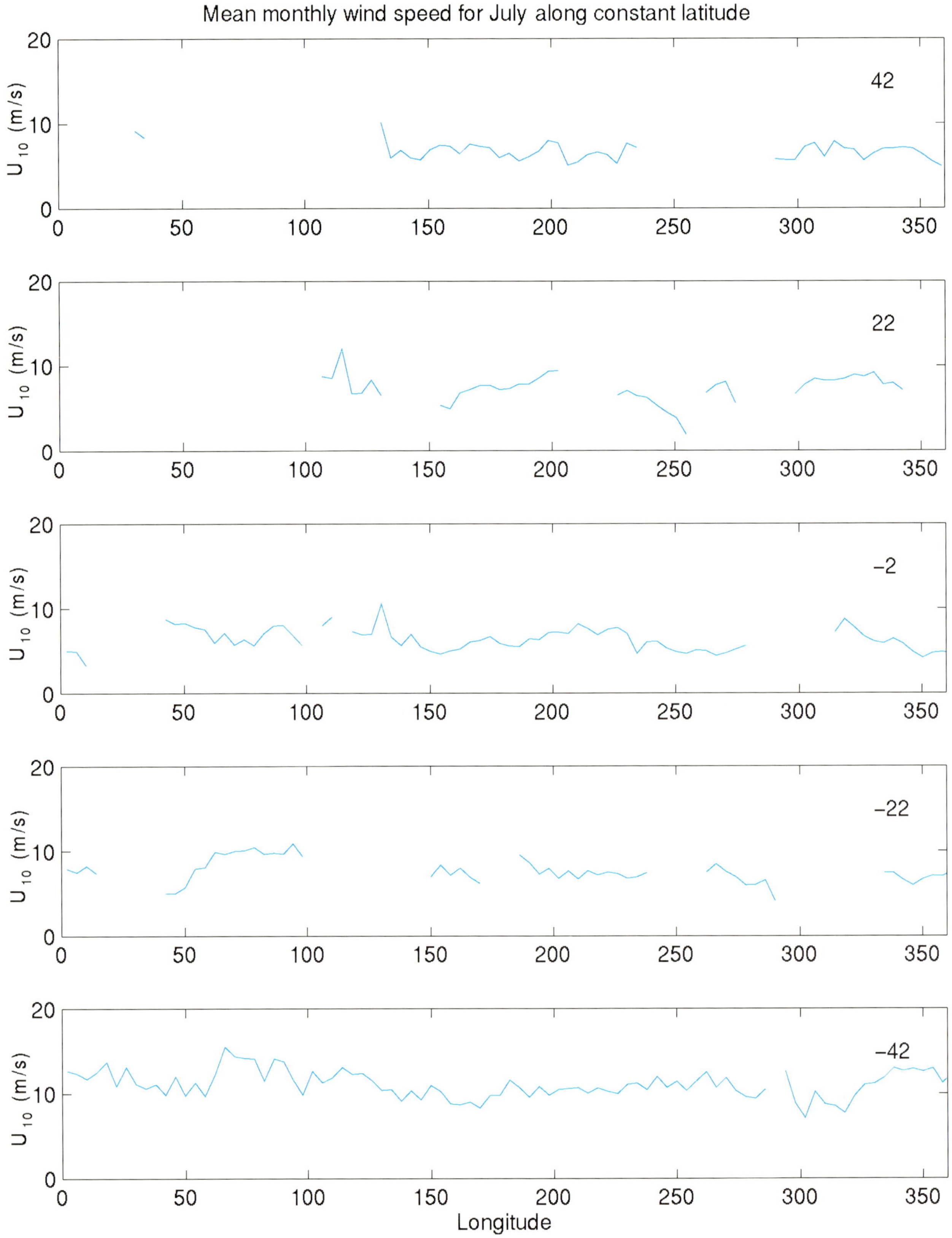

Fig. 7.22(e) Values of Mean Monthly Wind Speed for the month of July along the lines of constant latitude shown in Fig. 7.21. The values of latitude are shown in the top right corner of each panel.

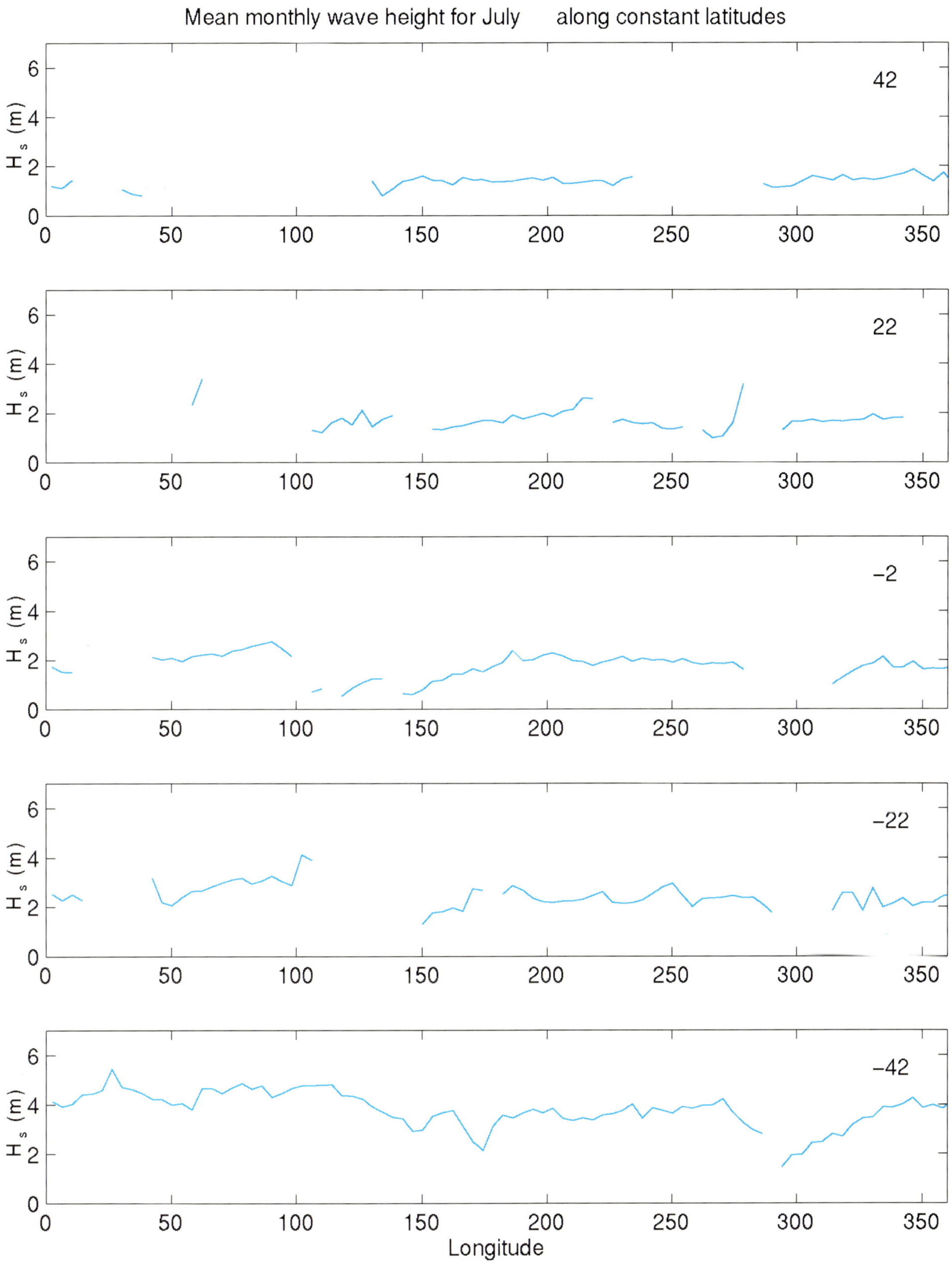

Fig. 7.22(f) Values of Mean Monthly Significant Wave Height for the month of July along the lines of constant latitude shown in Fig. 7.21. The values of latitude are shown in the top right corner of each panel.

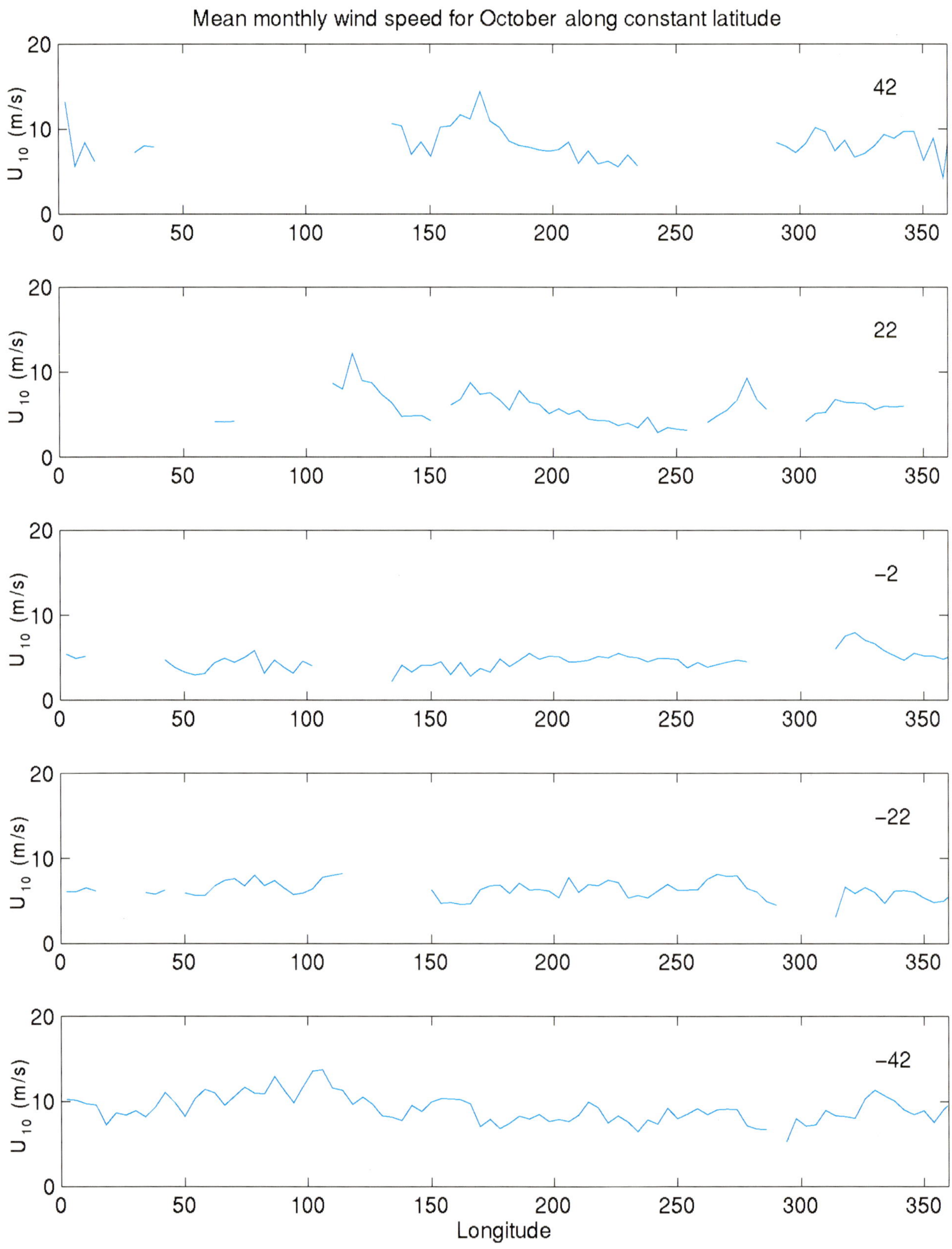

Fig. 7.22(g) Values of Mean Monthly Wind Speed for the month of October along the lines of constant latitude shown in Fig. 7.21. The values of latitude are shown in the top right corner of each panel.

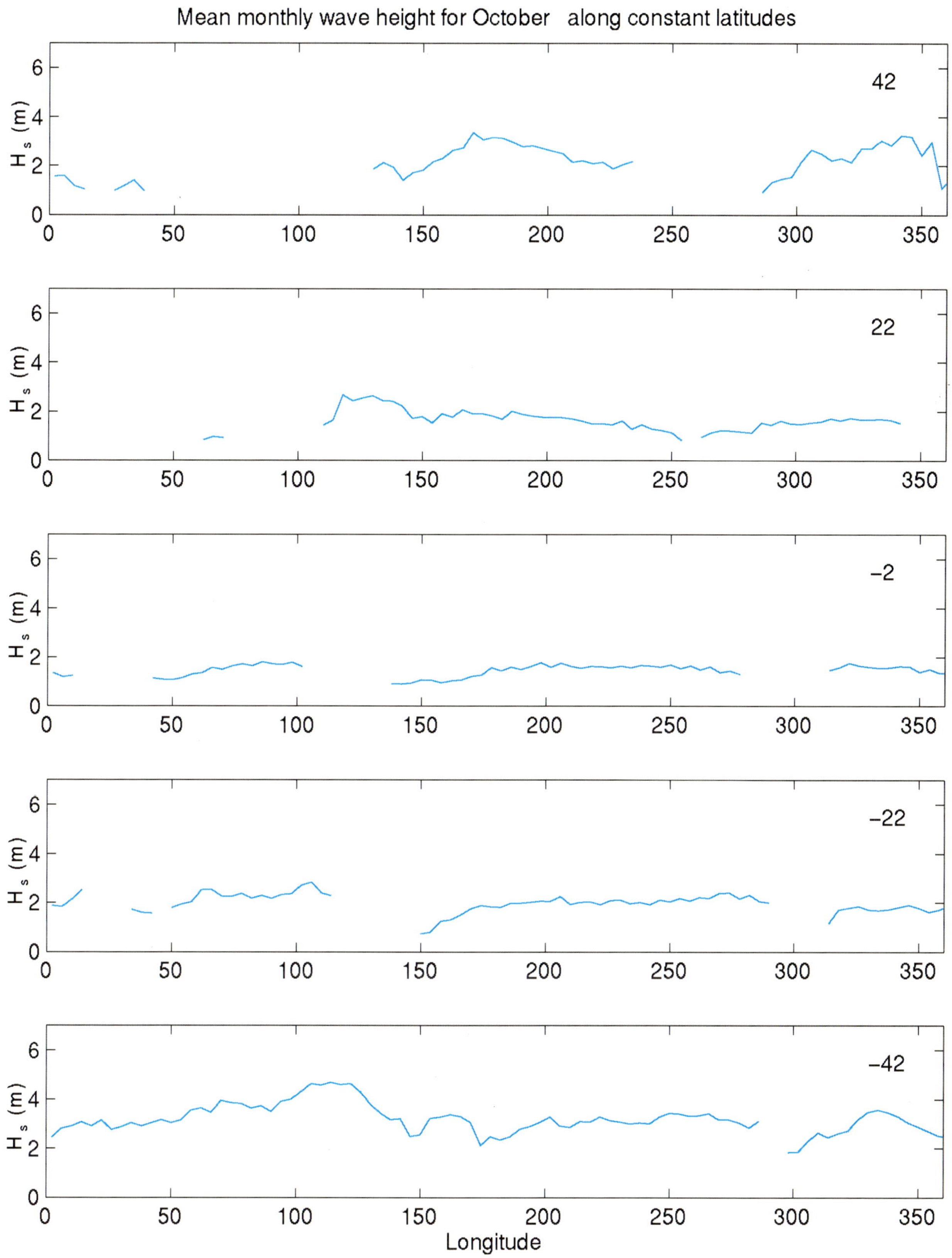

Fig. 7.22(h) Values of Mean Monthly Significant Wave Height for the month of October along the lines of constant latitude shown in Fig. 7.21. The values of latitude are shown in the top right corner of each panel.

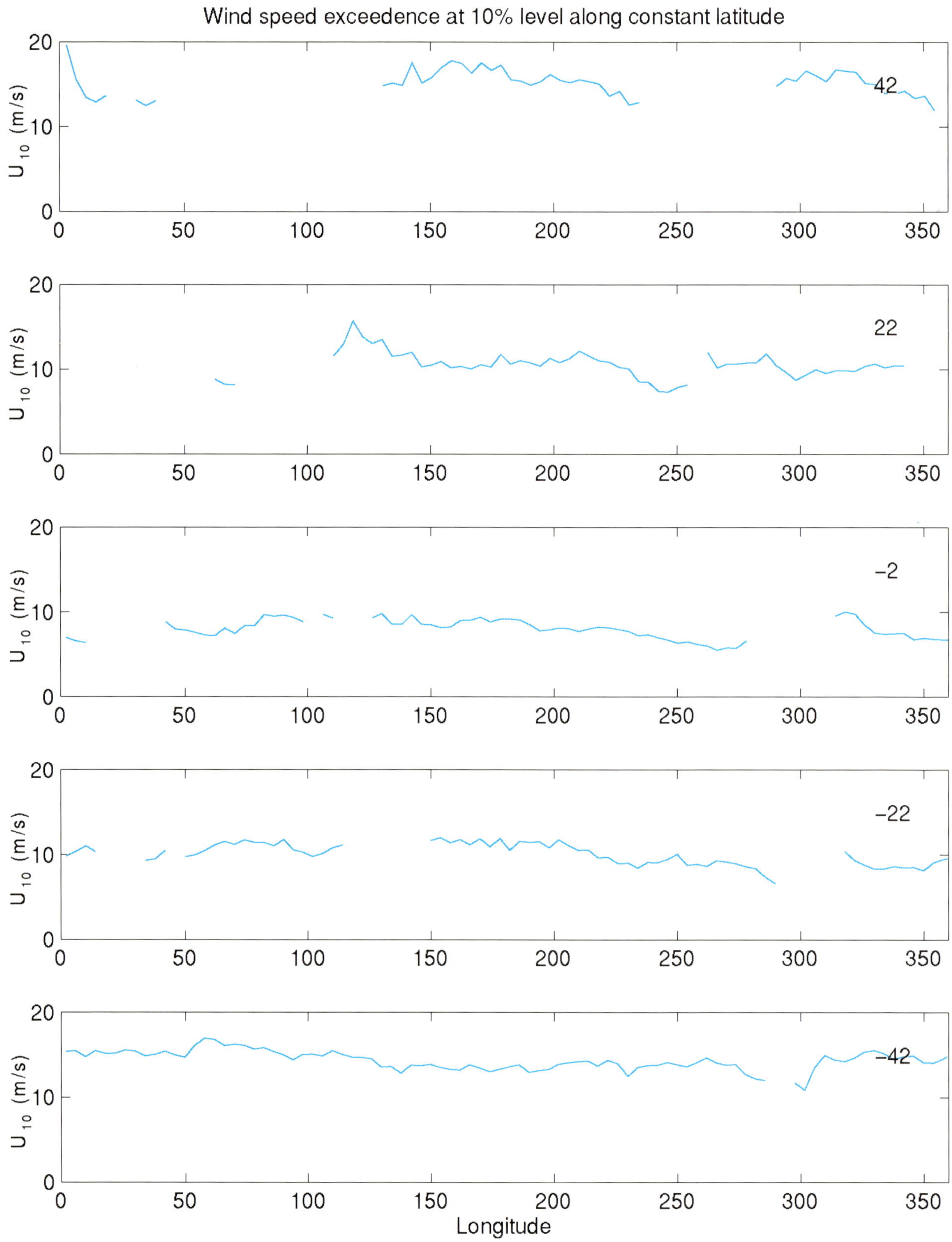

Fig. 7.23(a) Values of Wind Speed which will be exceeded 10% of the time along the lines of constant latitude shown in Fig. 7.21. The values of latitude are shown in the top right corner of each panel.

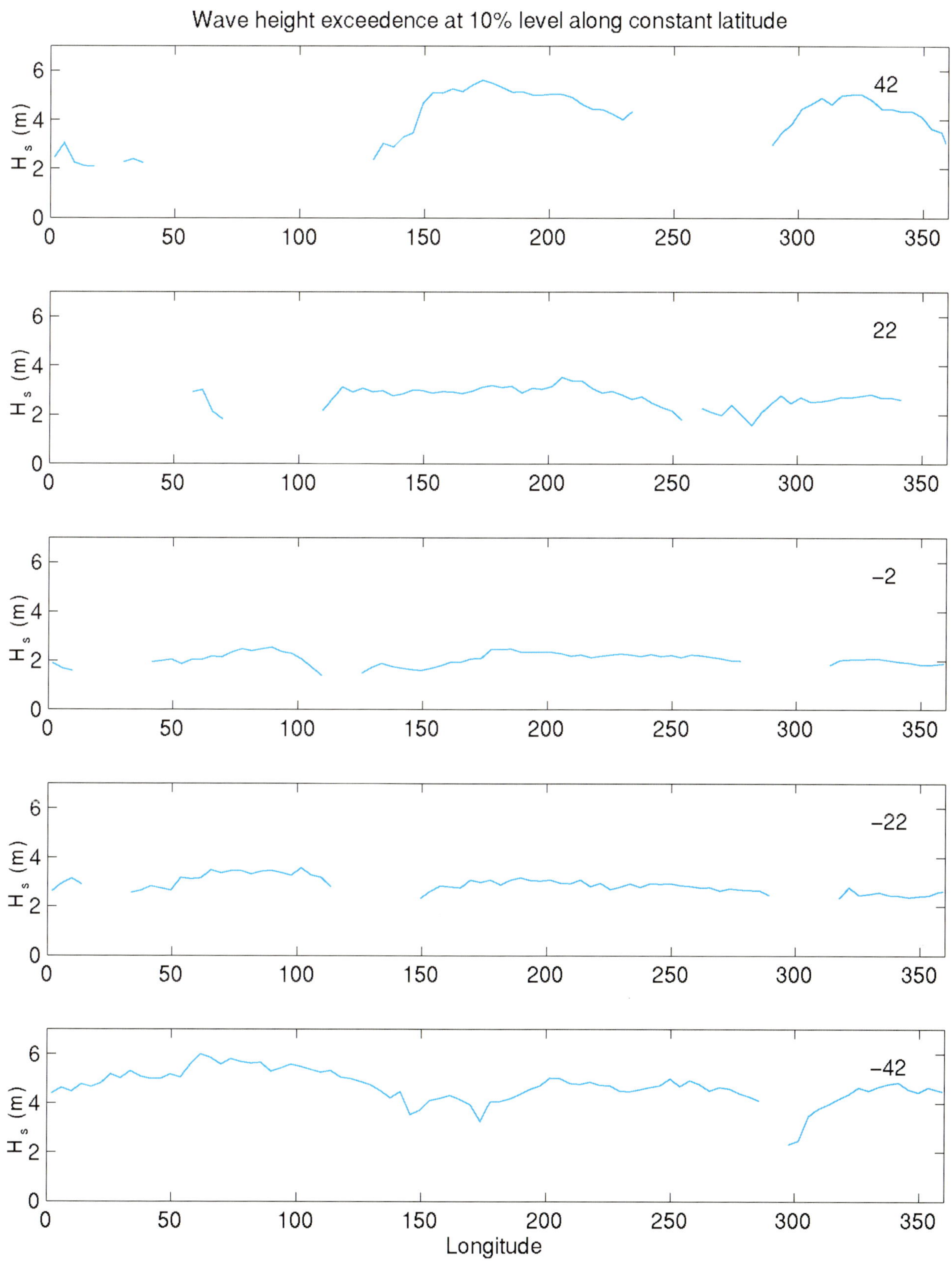

Fig. 7.23(b) Values of Significant Wave Height which will be exceeded 10% of the time along the lines of constant latitude shown in Fig. 7.21. The values of latitude are shown in the top right corner of each panel.

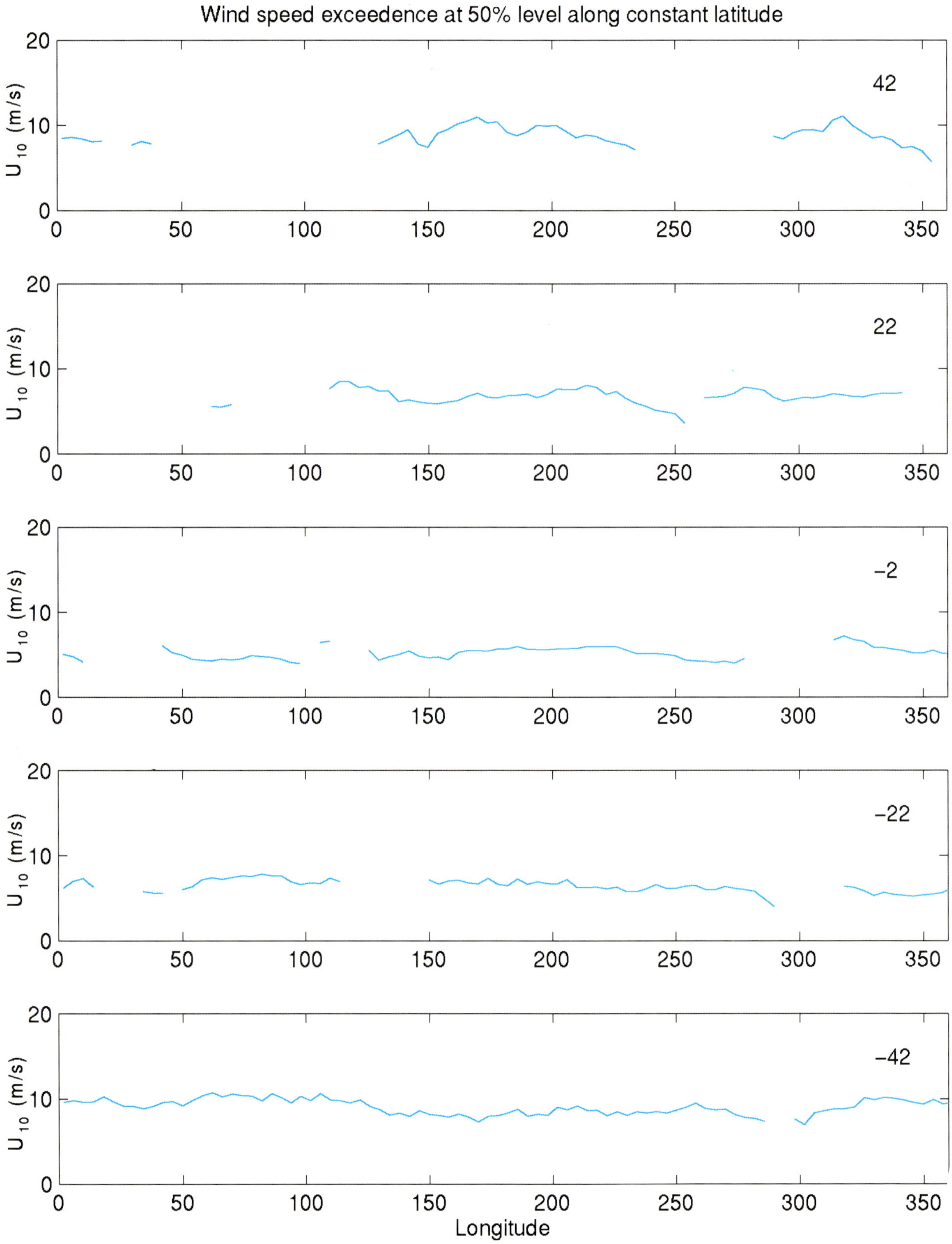

Fig. 7.23(c) Values of Wind Speed which will be exceeded 50% of the time along the lines of constant latitude shown in Fig. 7.21. The values of latitude are shown in the top right corner of each panel.

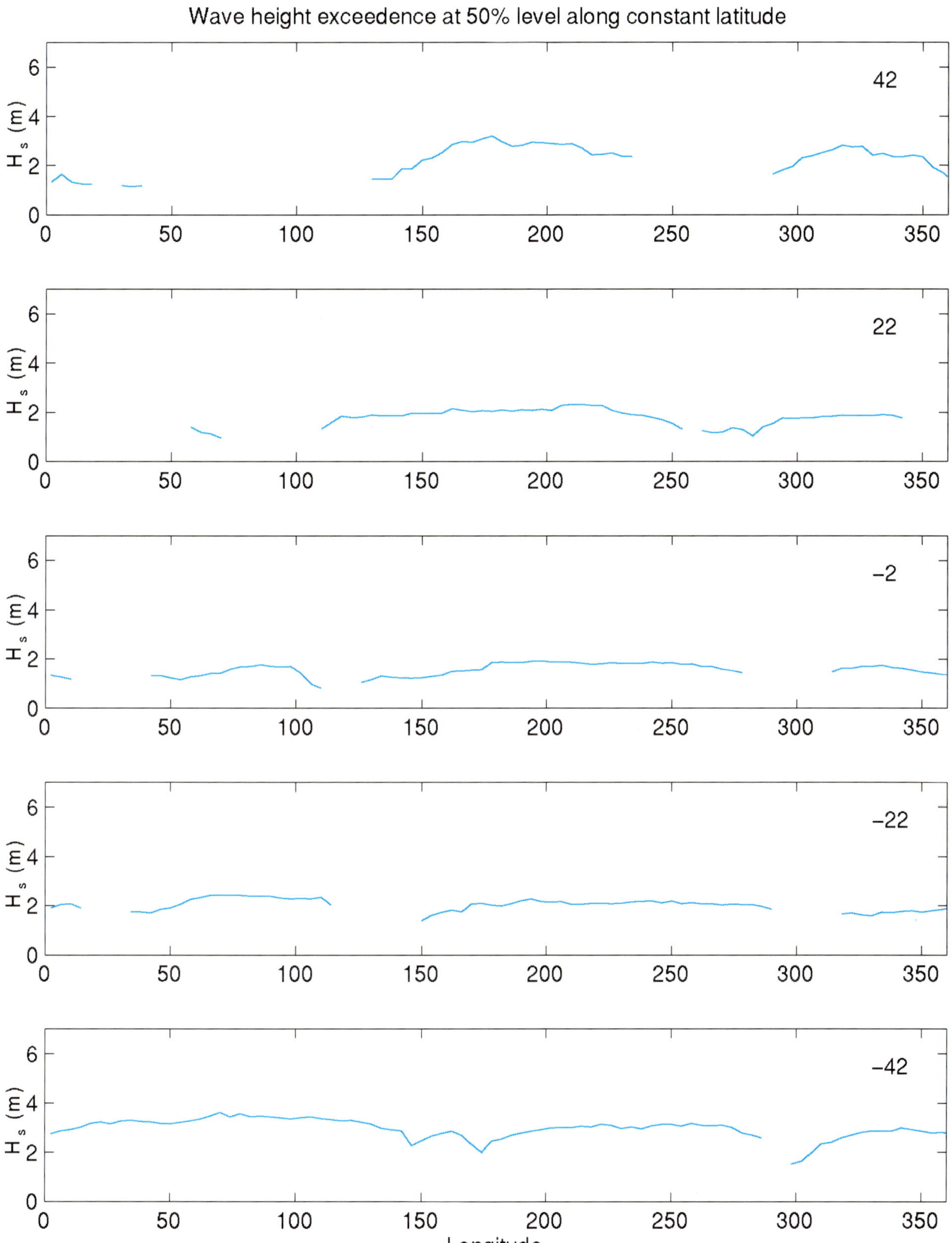

Fig. 7.23(d) Values of Significant Wave Height which will be exceeded 50% of the time along the lines of constant latitude shown in Fig. 7.21. The values of latitude are shown in the top right corner of each panel.

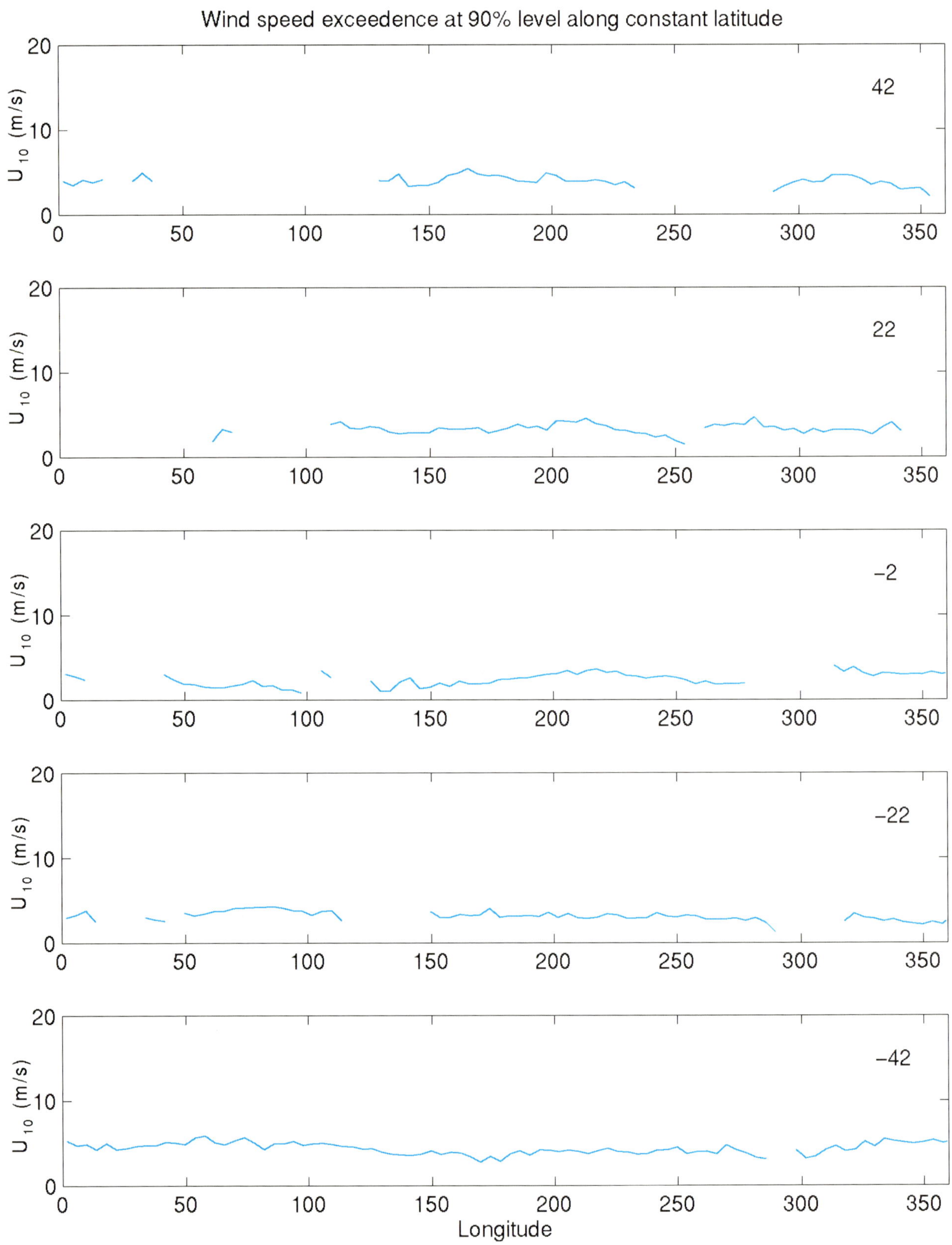

Fig. 7.23(e) Values of Wind Speed which will be exceeded 90% of the time along the lines of constant latitude shown in Fig. 7.21. The values of latitude are shown in the top right corner of each panel.

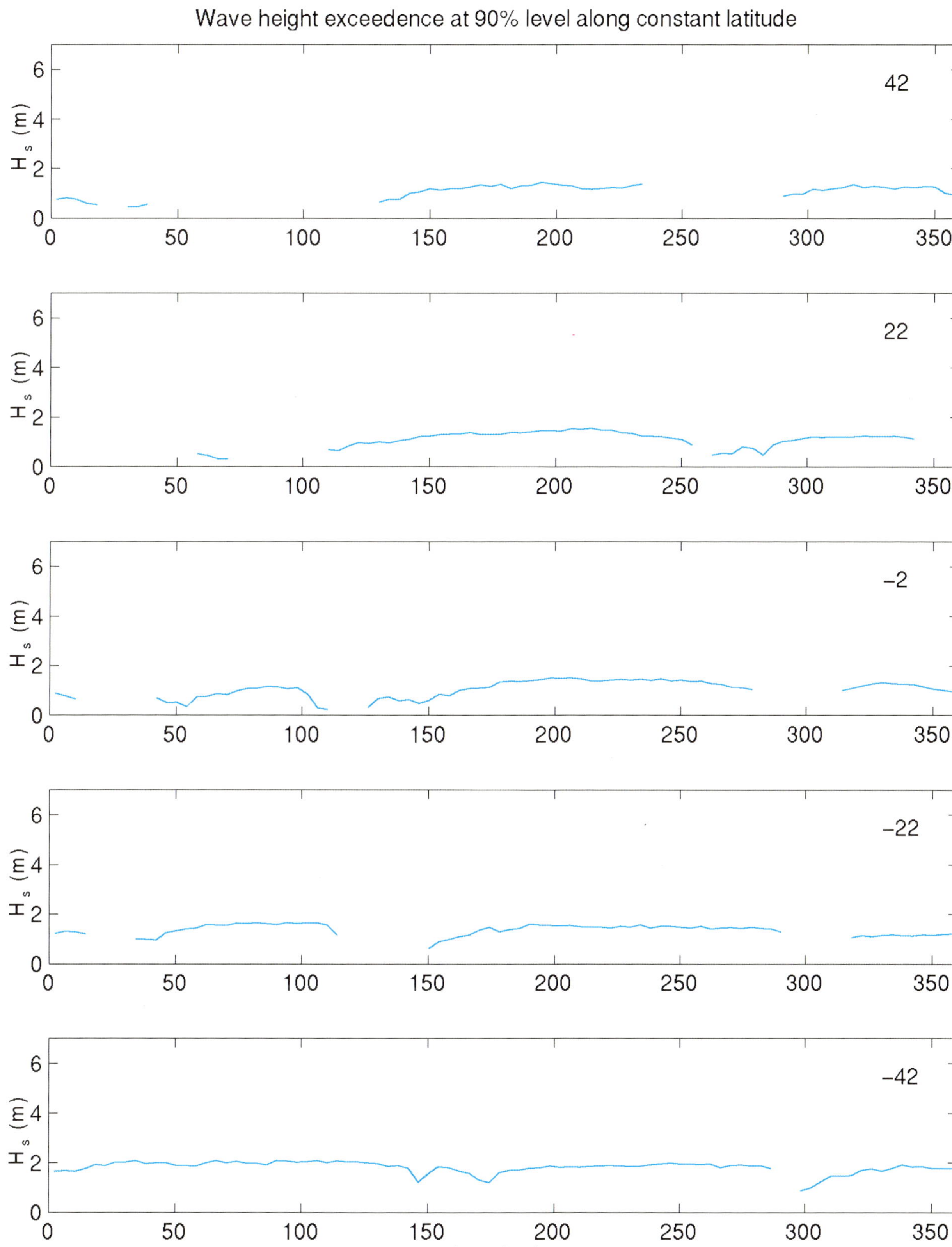

Fig. 7.23(f) Values of Significant Wave Height which will be exceeded 90% of the time along the lines of constant latitude shown in Fig. 7.21. The values of latitude are shown in the top right corner of each panel.

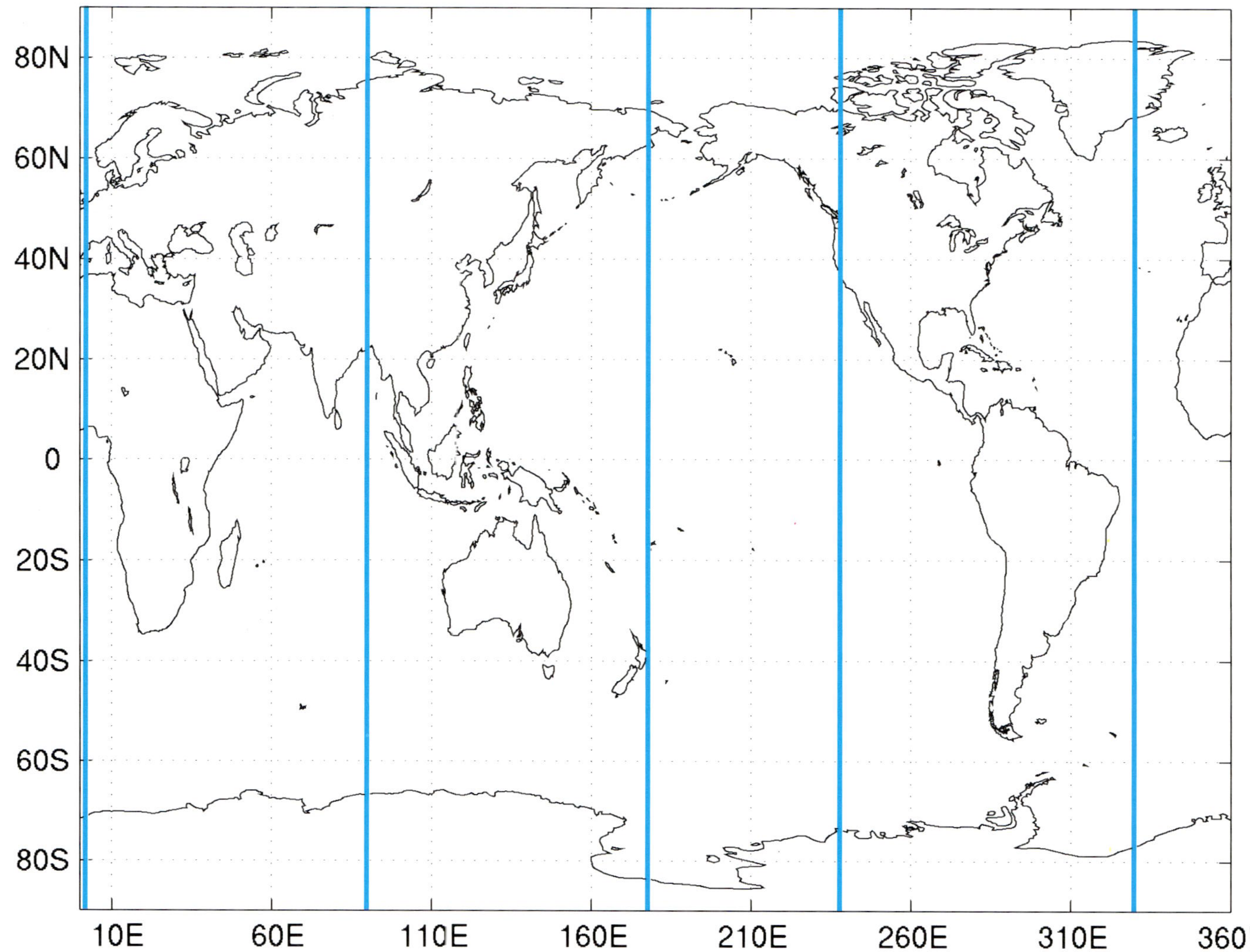

Fig. 7.24 Lines of constant longitude along which results are presented in Figs 7.25 and 7.26.

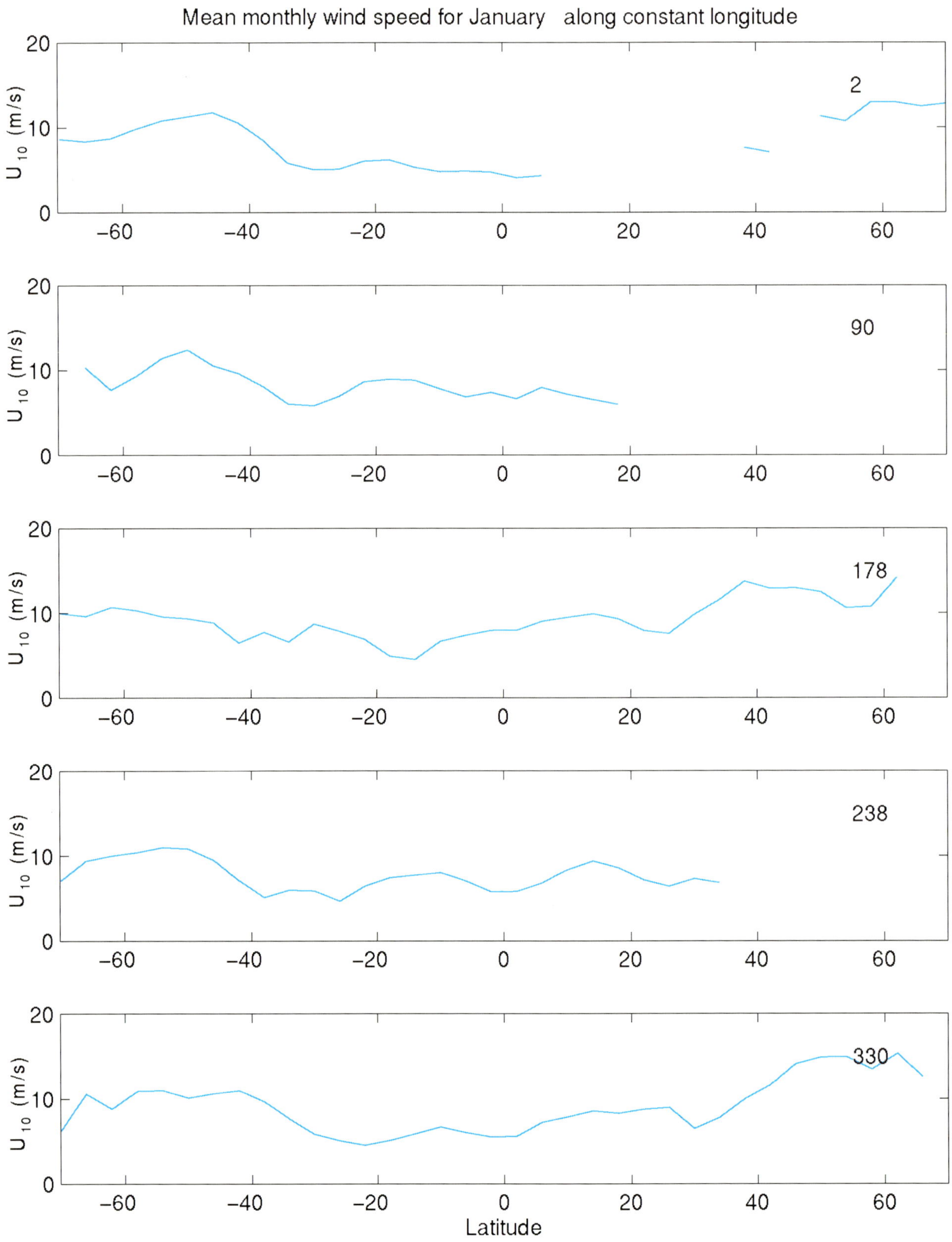

Fig. 7.25(a) Values of Mean Monthly Wind Speed for the month of January along the lines of constant longitude shown in Fig. 7.24. The values of longitude are shown in the top right corner of each panel.

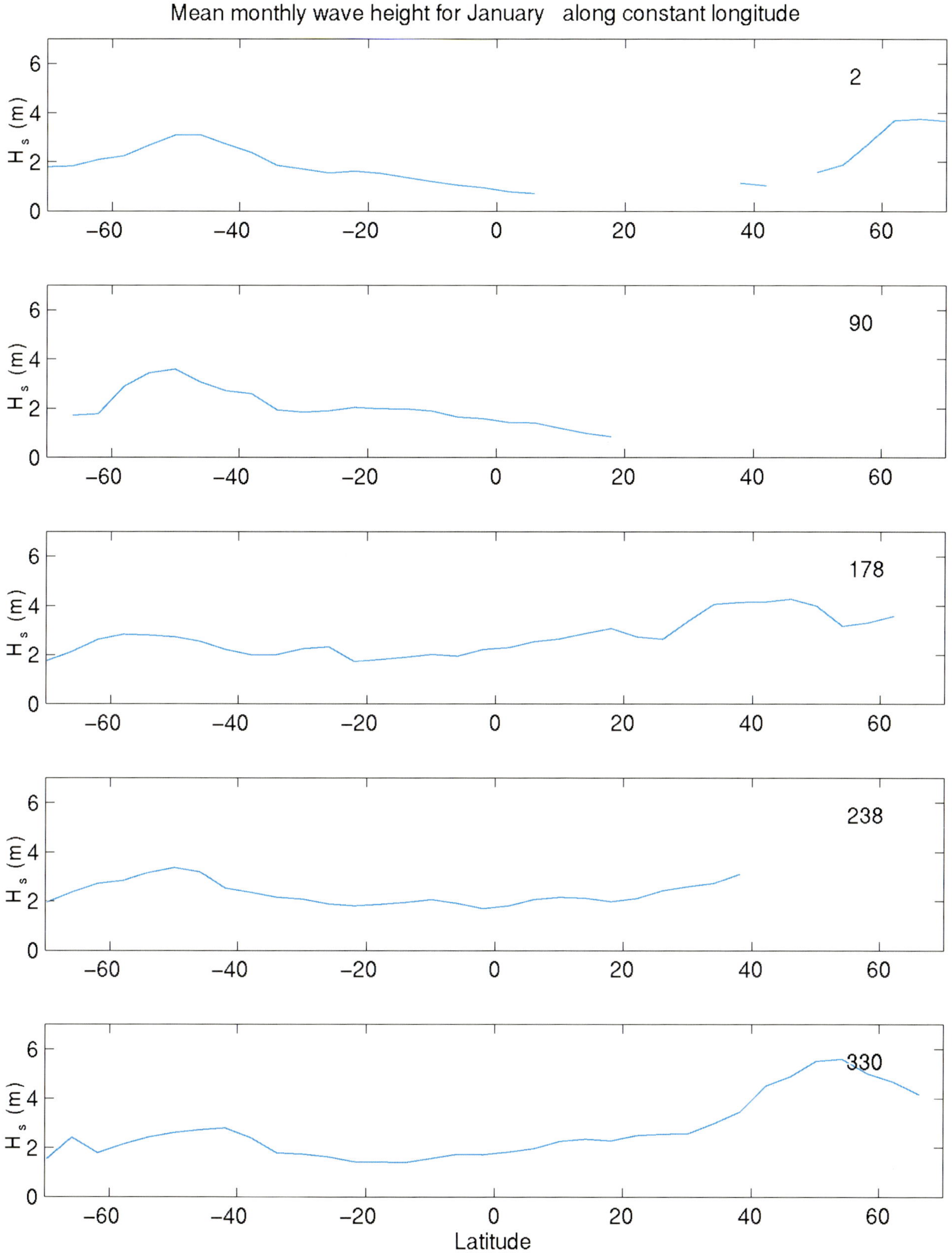

Fig. 7.25(b) Values of Mean Monthly Significant Wave Height for the month of January along the lines of constant longitude shown in Fig. 7.24. The values of longitude are shown in the top right corner of each panel.

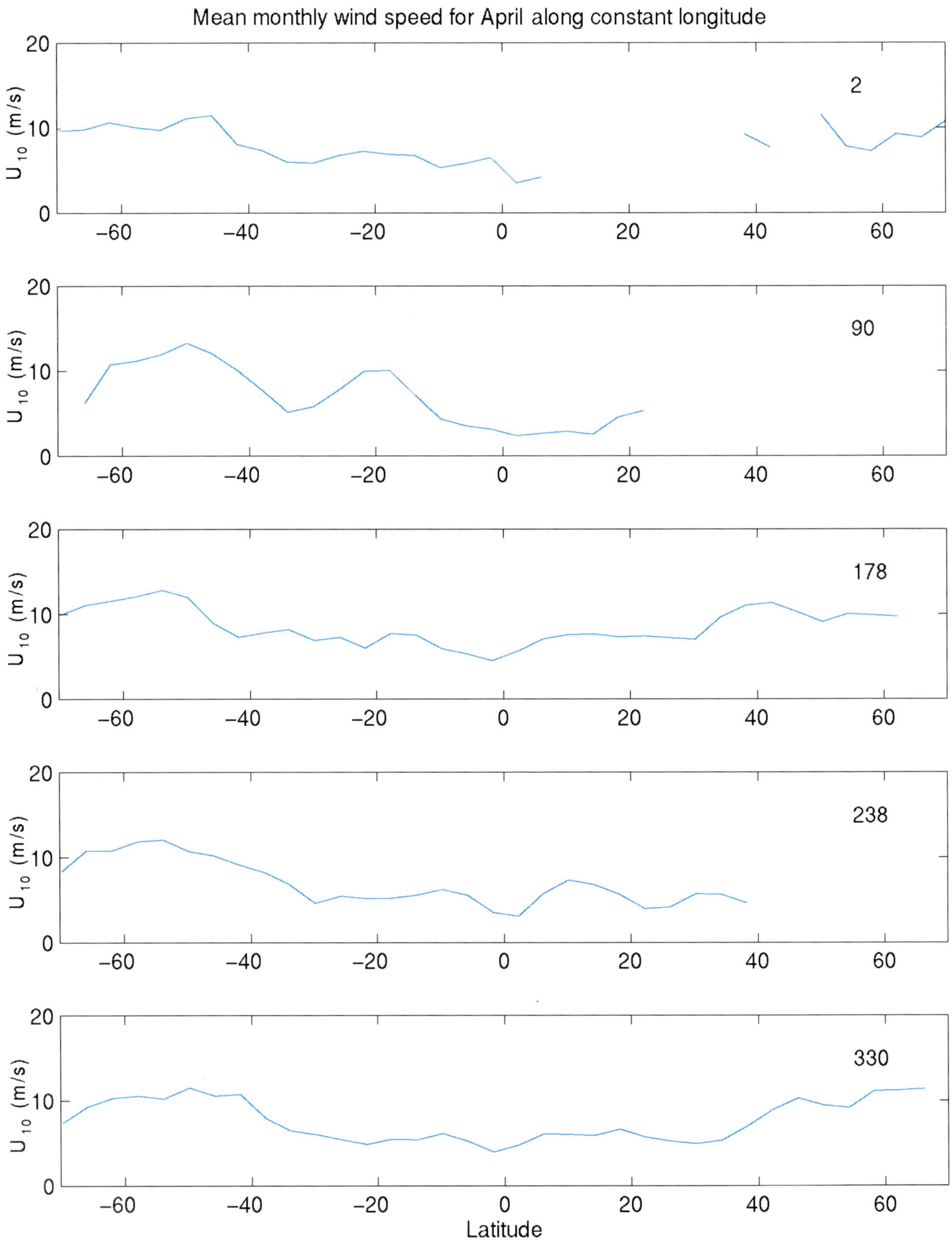

Fig. 7.25(c) Values of Mean Monthly Wind Speed for the month of April along the lines of constant longitude shown in Fig. 7.24. The values of longitude are shown in the top right corner of each panel.

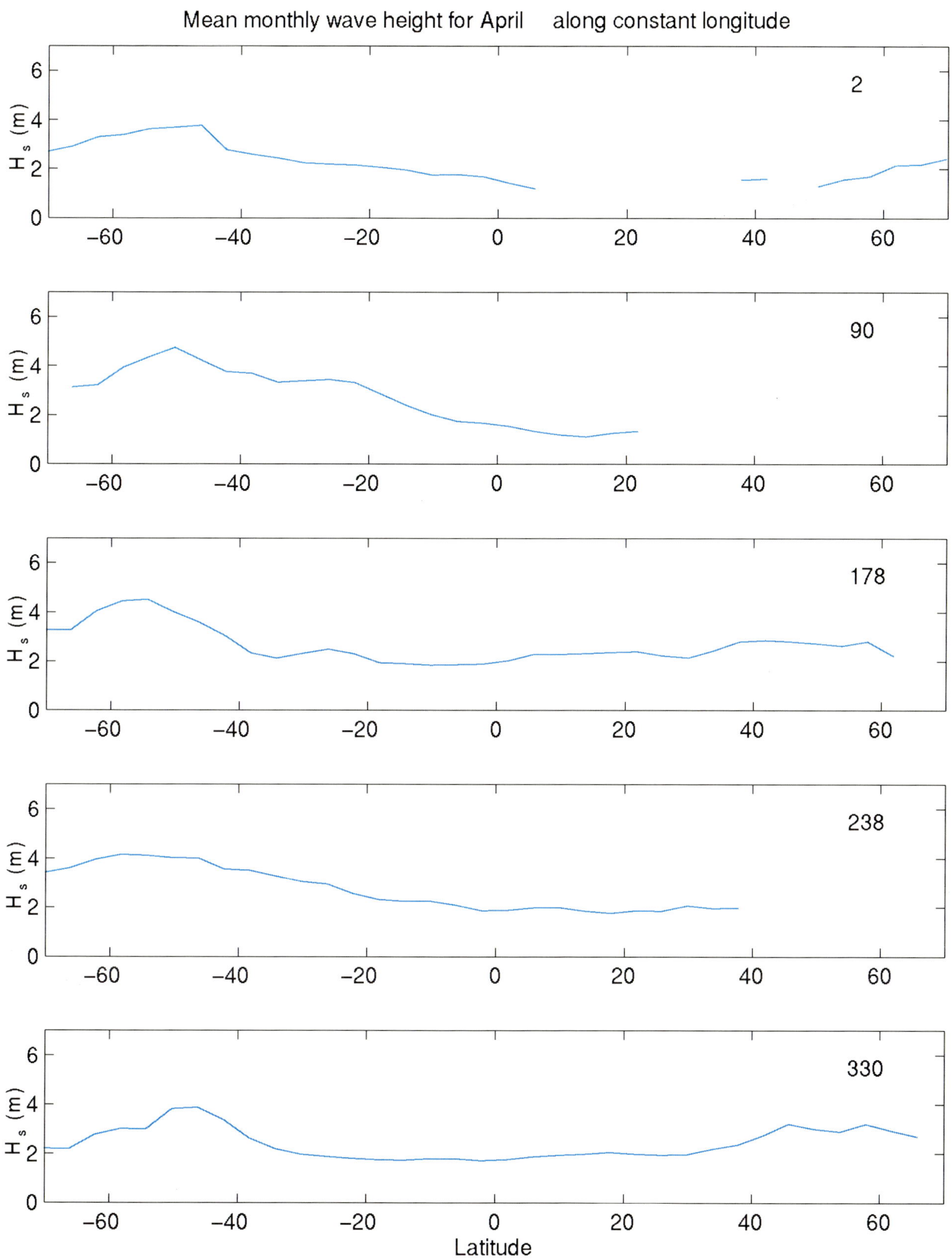

Fig. 7.25(d) Values of Mean Monthly Significant Wave Height for the month of April along the lines of constant longitude shown in Fig. 7.24. The values of longitude are shown in the top right corner of each panel.

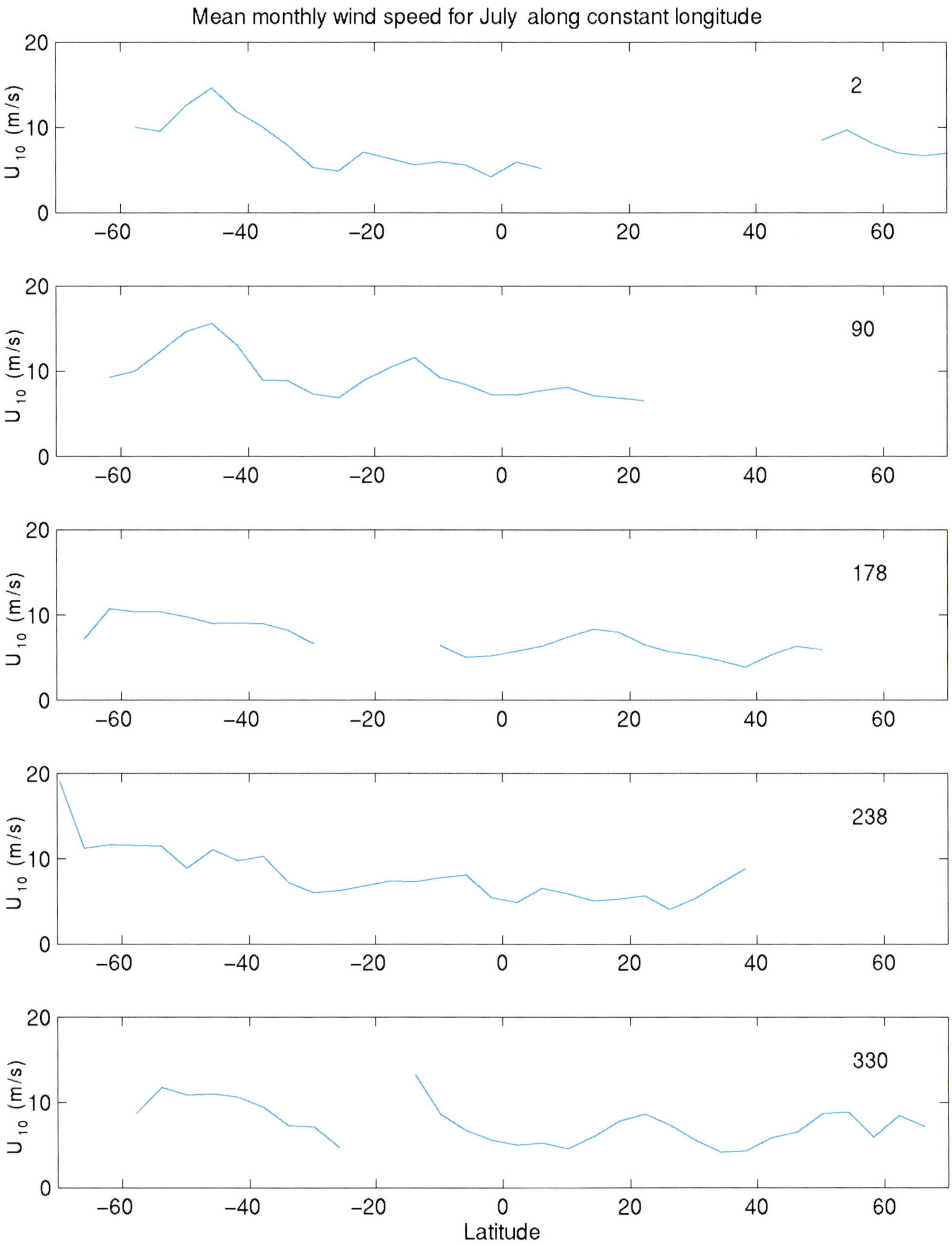

Fig. 7.25(e) Values of Mean Monthly Wind Speed for the month of July along the lines of constant longitude shown in Fig. 7.24. The values of longitude are shown in the top right corner of each panel.

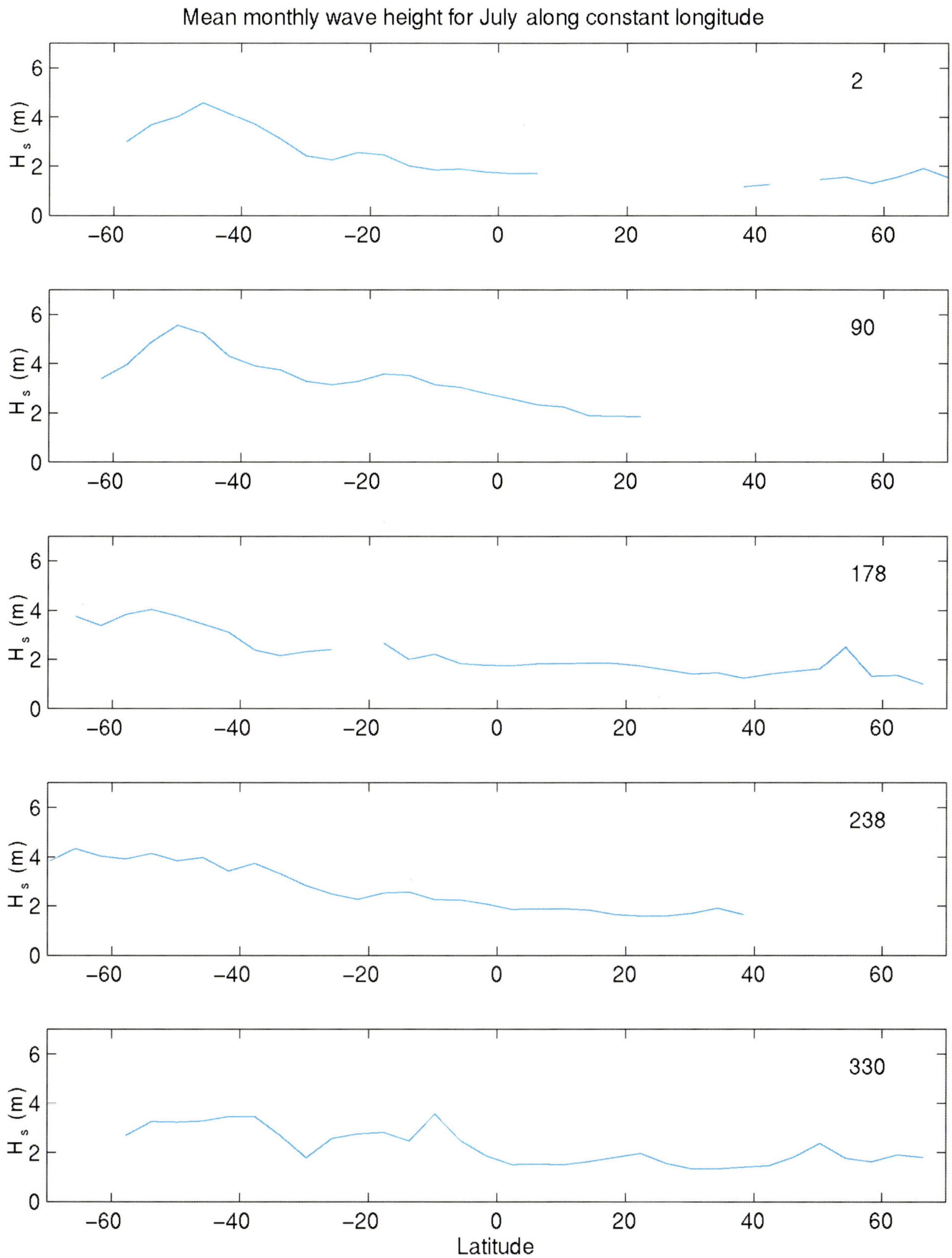

Fig. 7.25(f) Values of Mean Monthly Significant Wave Height for the month of July along the lines of constant longitude shown in Fig. 7.24. The values of longitude are shown in the top right corner of each panel.

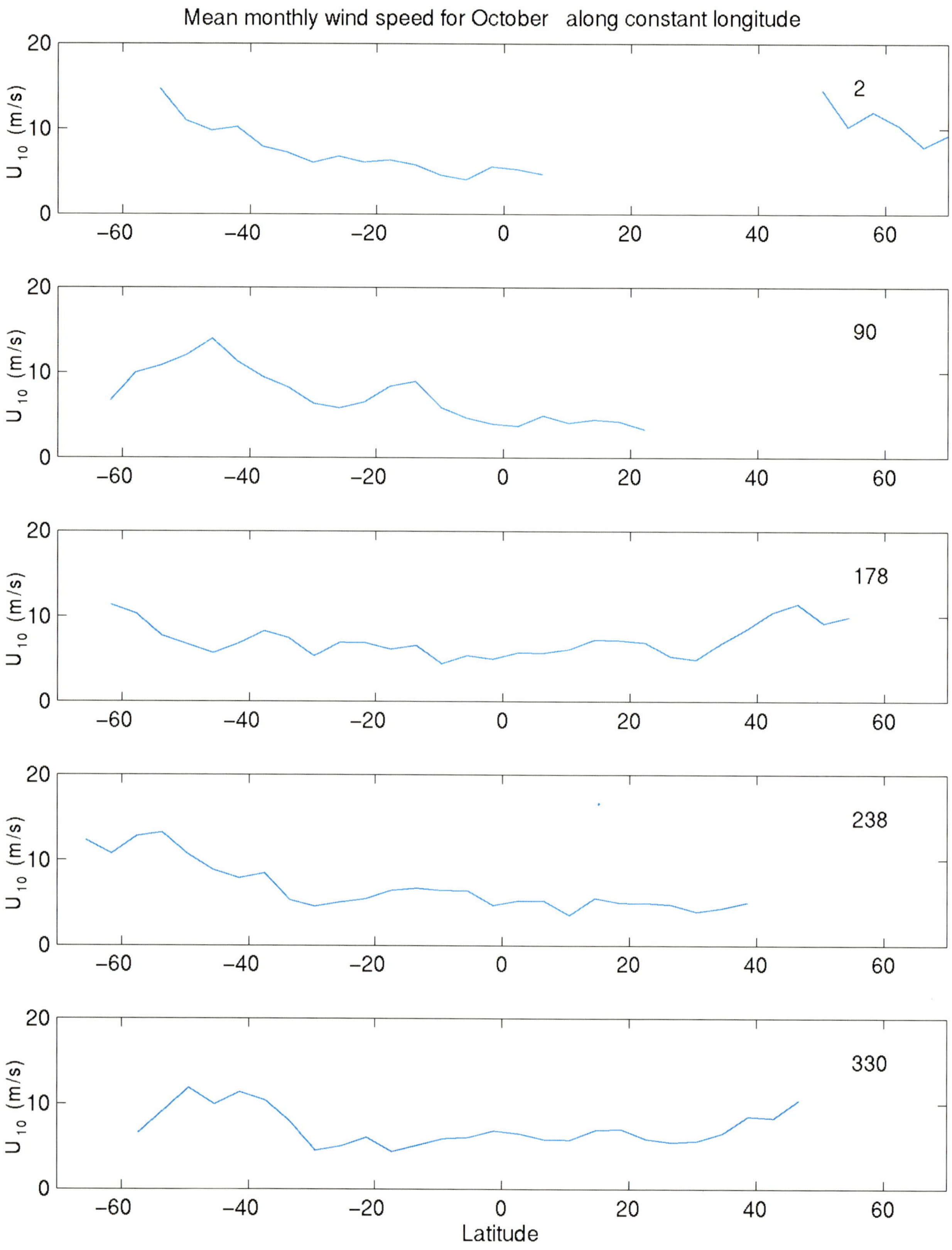

Fig. 7.25(g) Values of Mean Monthly Wind Speed for the month of October along the lines of constant longitude shown in Fig. 7.24. The values of longitude are shown in the top right corner of each panel.

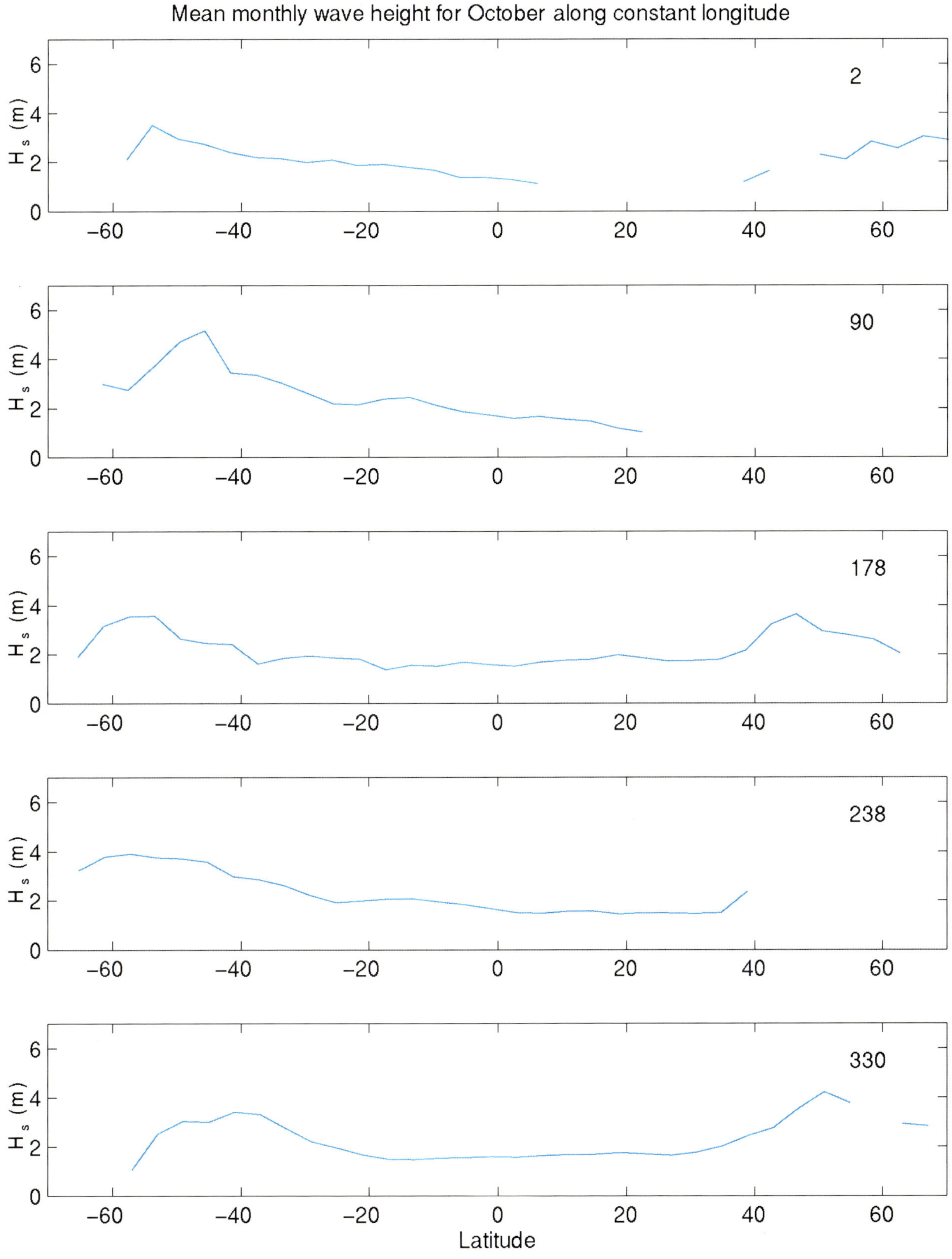

Fig. 7.25(h) Values of Mean Monthly Significant Wave Height for the month of October along the lines of constant longitude shown in Fig. 7.24. The values of longitude are shown in the top right corner of each panel.

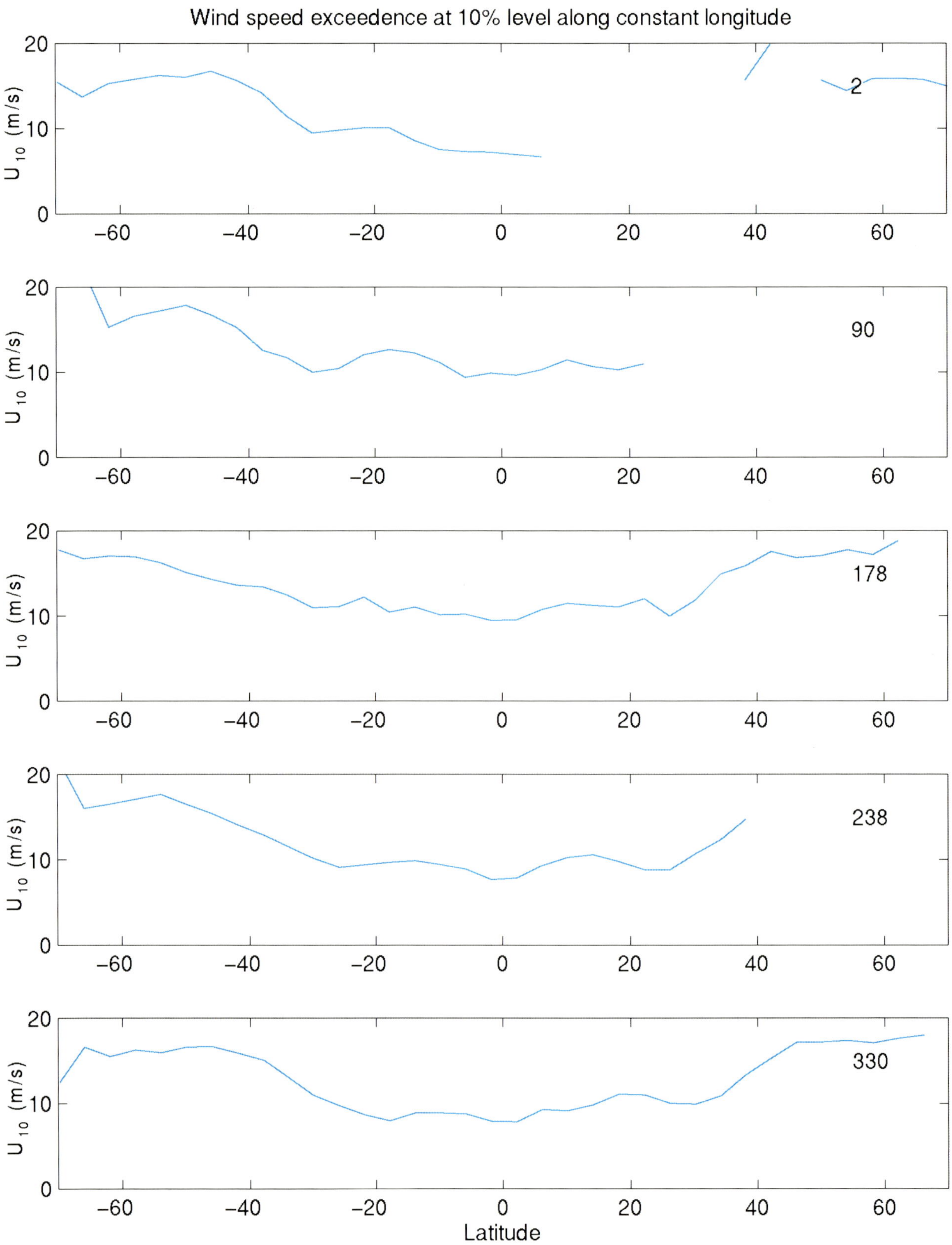

Fig. 7.26(a) Values of Wind Speed which will be exceeded 10% of the time along the lines of constant longitude shown in Fig. 7.24. The values of longitude are shown in the top right corner of each panel.

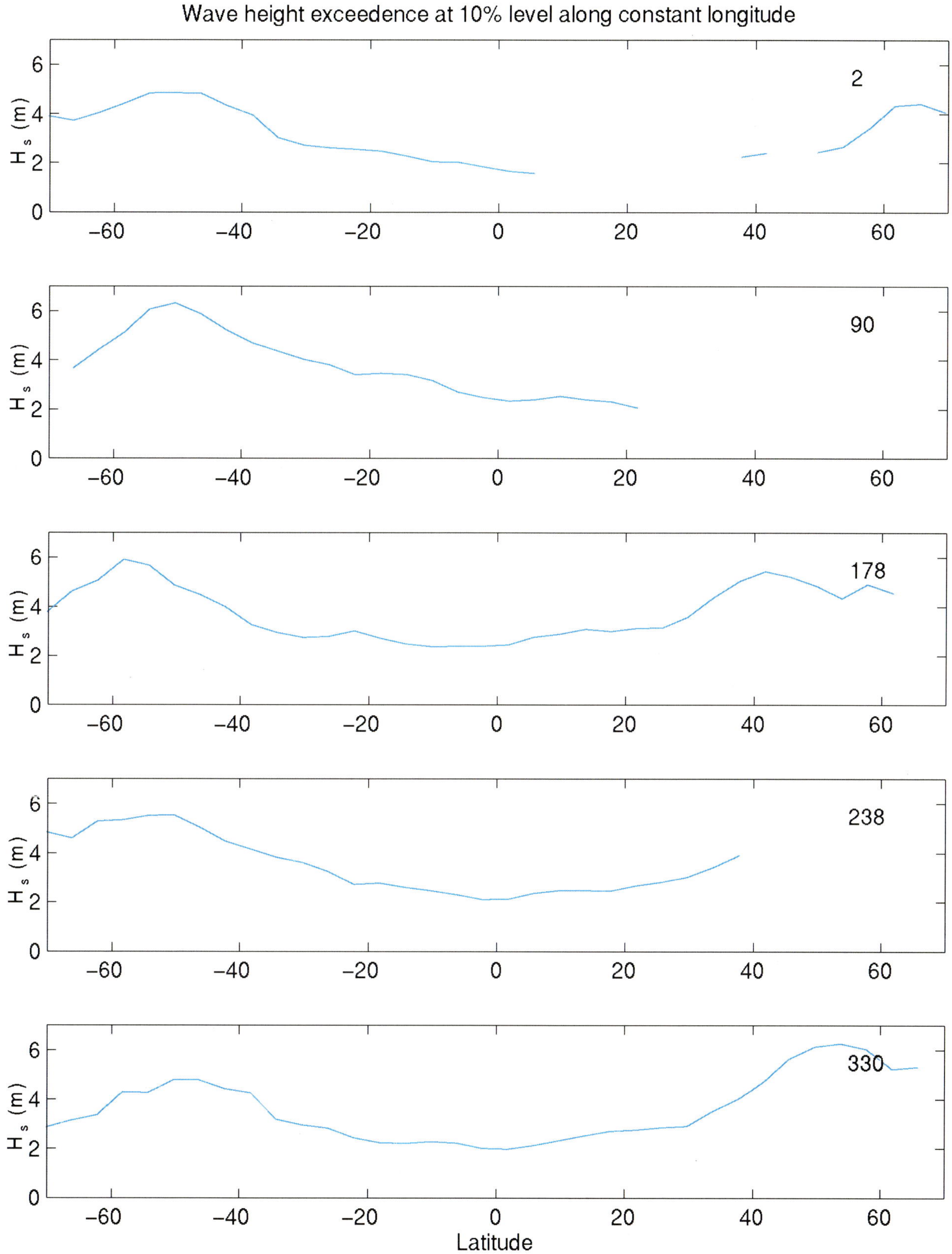

Fig. 7.26(b) Values of Significant Wave Height which will be exceeded 10% of the time along the lines of constant longitude shown in Fig. 7.24. The values of longitude are shown in the top right corner of each panel.

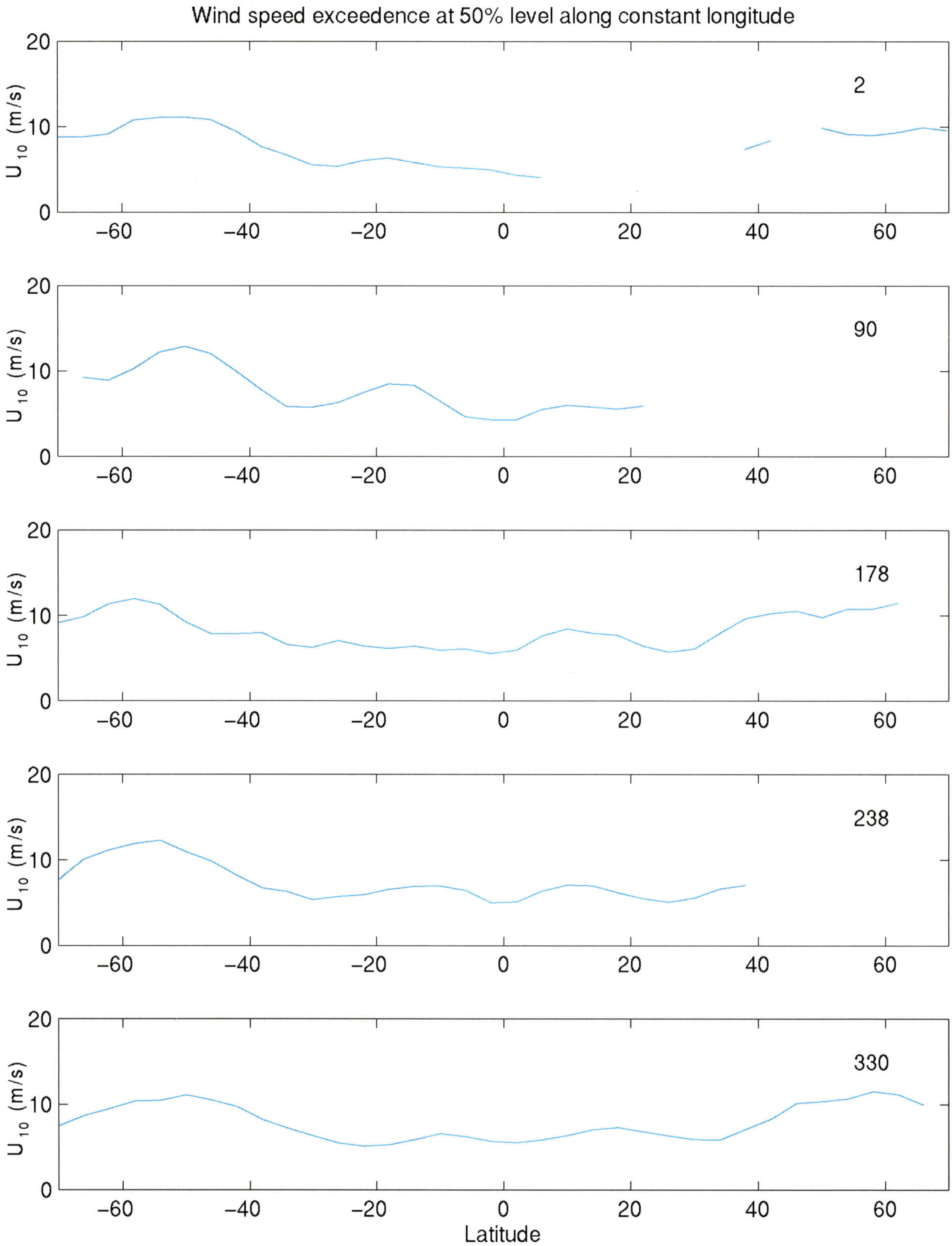

Fig. 7.26(c) Values of Wind Speed which will be exceeded 50% of the time along the lines of constant longitude shown in Fig. 7.24. The values of longitude are shown in the top right corner of each panel.

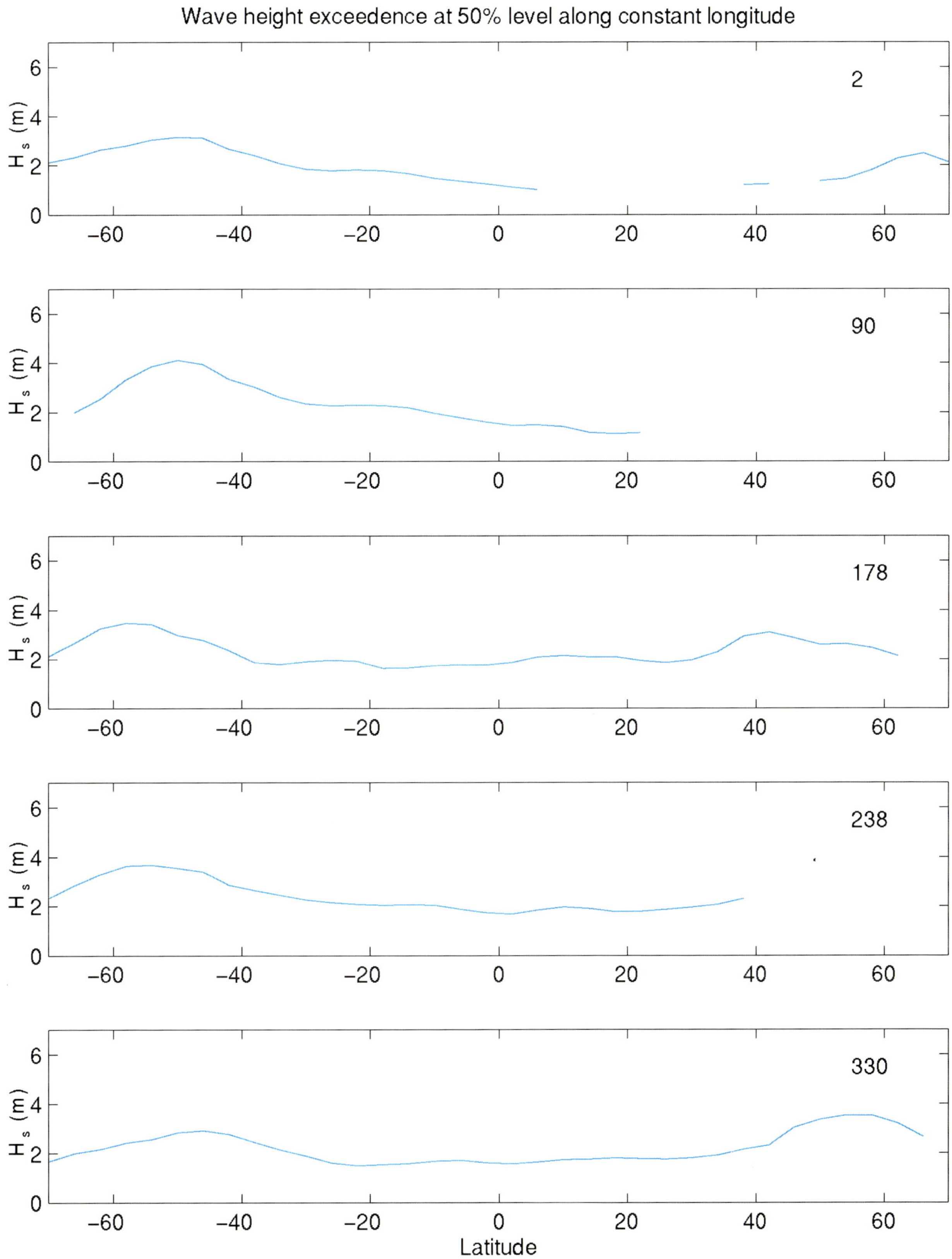

Fig. 7.26(d) Values of Significant Wave Height which will be exceeded 50% of the time along the lines of constant longitude shown in Fig. 7.24. The values of longitude are shown in the top right corner of each panel.

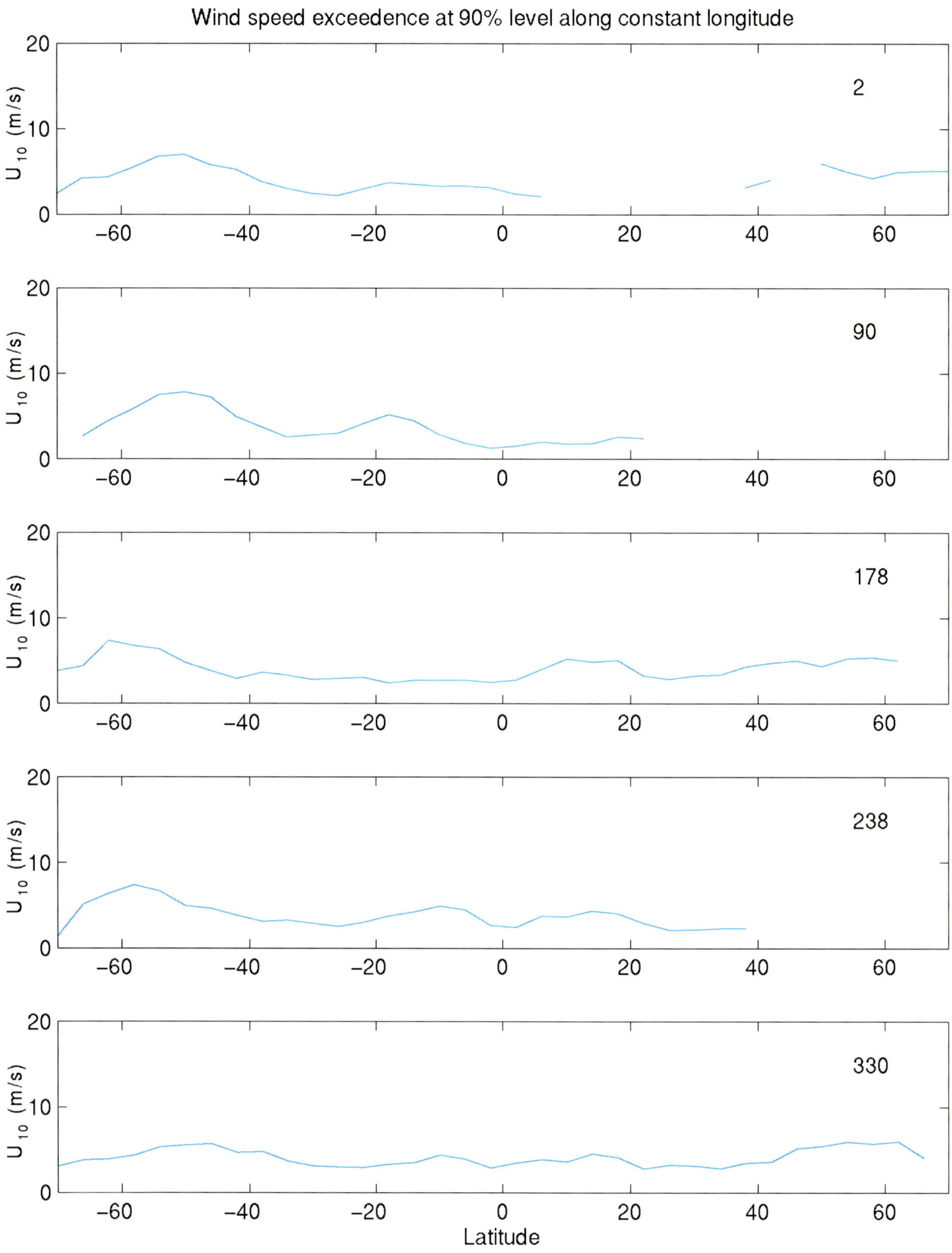

Fig. 7.26(e) Values of Wind Speed which will be exceeded 90% of the time along the lines of constant longitude shown in Fig. 7.24. The values of longitude are shown in the top right corner of each panel.

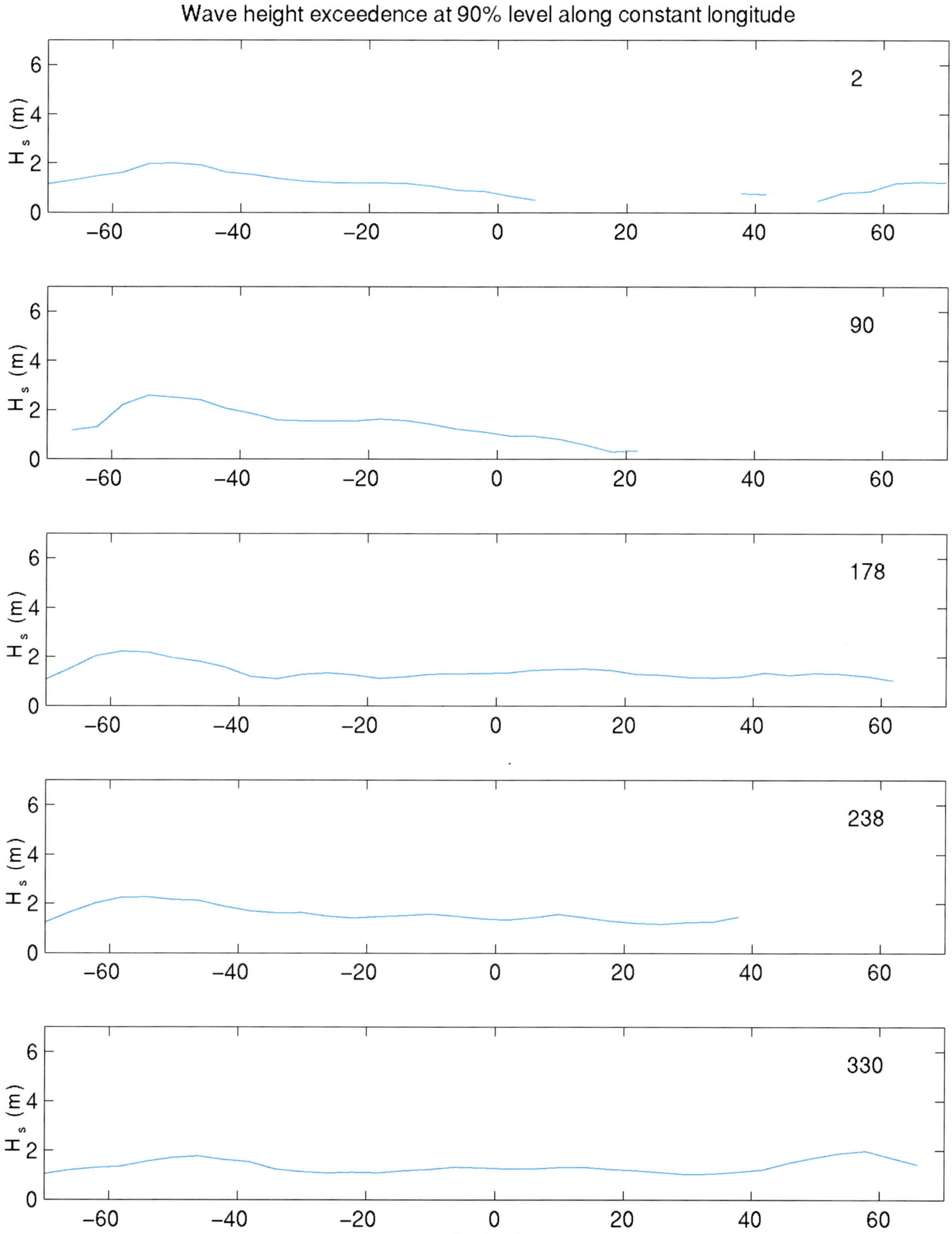

Fig. 7.26(f) Values of Significant Wave Height which will be exceeded 90% of the time along the lines of constant longitude shown in Fig. 7.24. The values of longitude are shown in the top right corner of each panel.

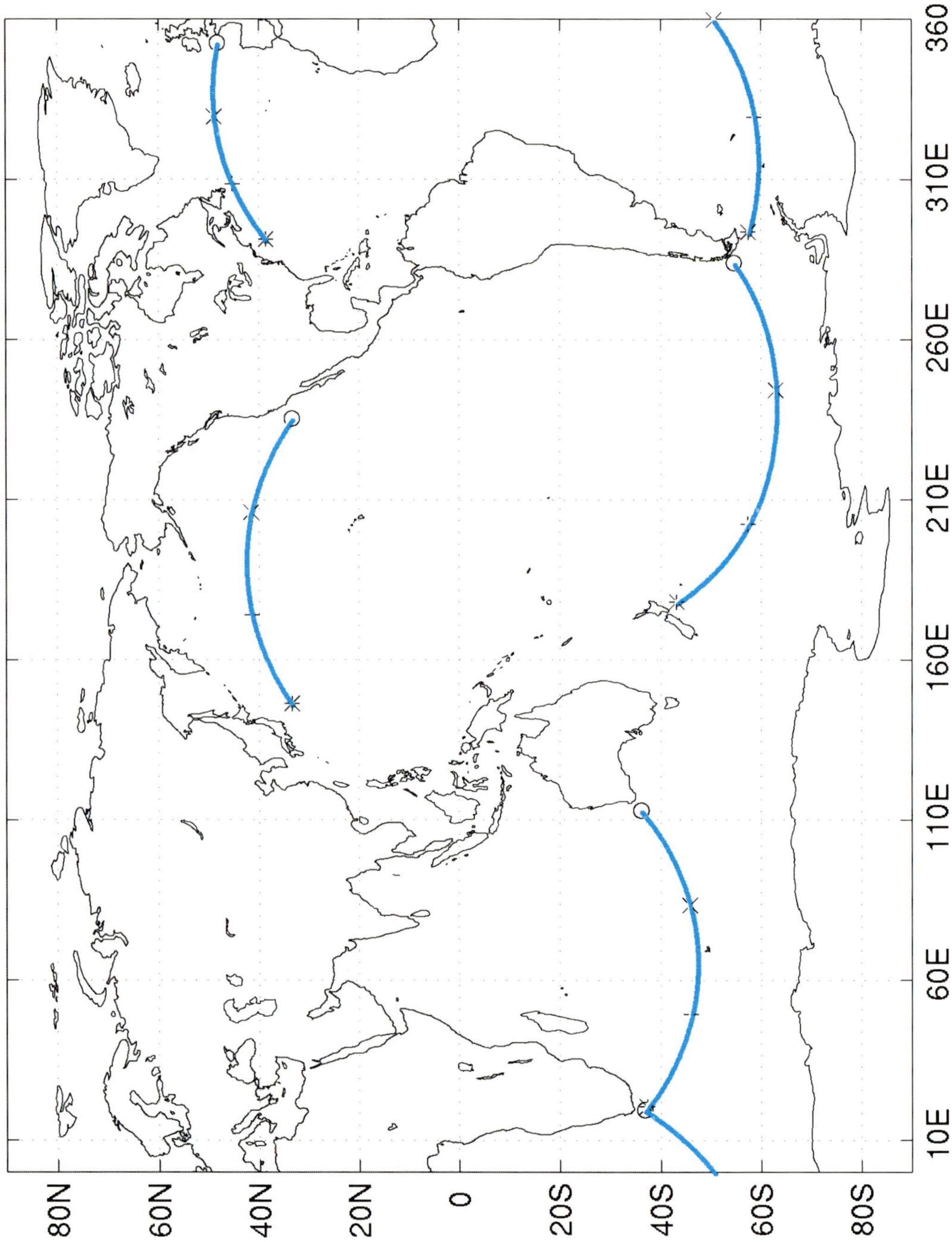

Fig. 7.27 Great Circle paths along which results are presented in Fig. 7.28.

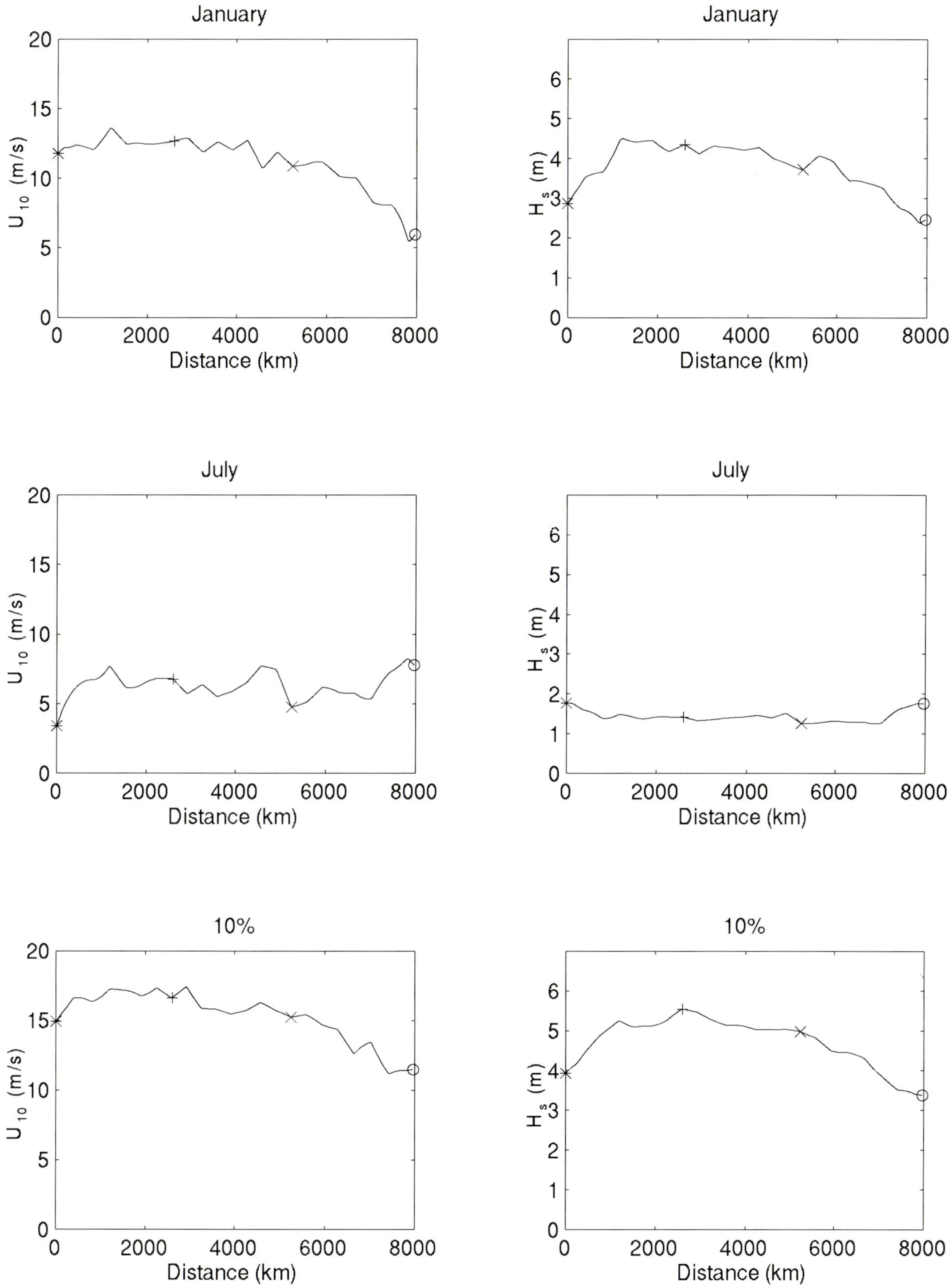

Fig. 7.28(a) Values of parameters along the Great Circle path between Japan and the west coast of North America (see Fig. 7.27). Values shown are: Mean Monthly Wind Speed for January (top left), Mean Monthly Wind Speed for July (middle left), Wind Speed exceeded 10% of the time (bottom left), Mean Monthly Significant Wave Height for January (top right), Mean Monthly Significant Wave Height for July (middle right) and Significant Wave Height exceeded 10% of the time (bottom right). The four markers in each panel correspond to the similarly marked positions shown on the corresponding Great Circle path in Fig. 7.27.

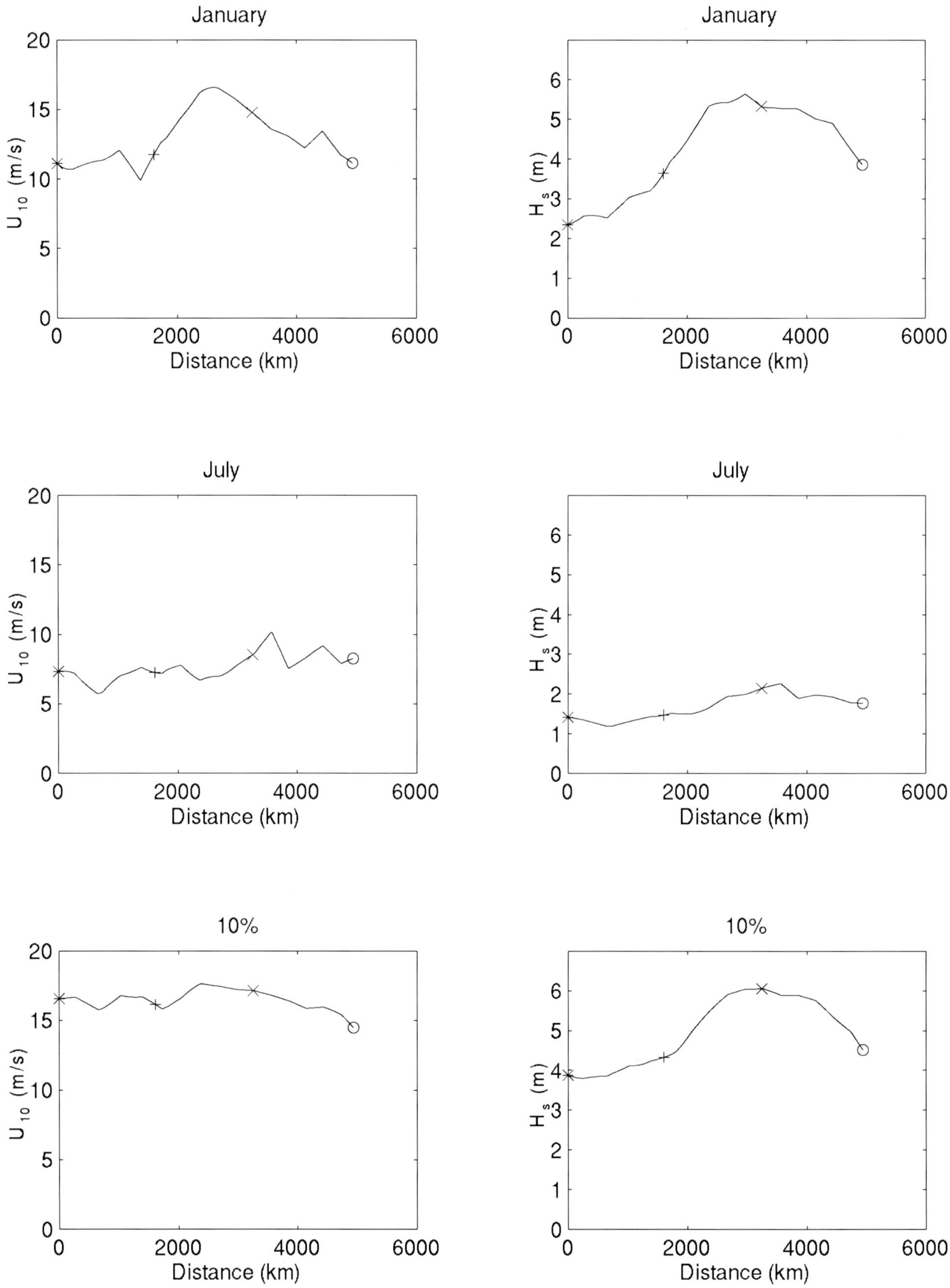

Fig. 7.28(b) Values of parameters along the Great Circle path between the east coast of North America and Europe (see Fig. 7.27). Values shown are: Mean Monthly Wind Speed for January (top left), Mean Monthly Wind Speed for July (middle left), Wind Speed exceeded 10% of the time (bottom left), Mean Monthly Significant Wave Height for January (top right), Mean Monthly Significant Wave Height for July (middle right) and Significant Wave Height exceeded 10% of the time (bottom right). The four markers in each panel correspond to the similarly marked positions shown on the corresponding Great Circle path in Fig. 7.27.

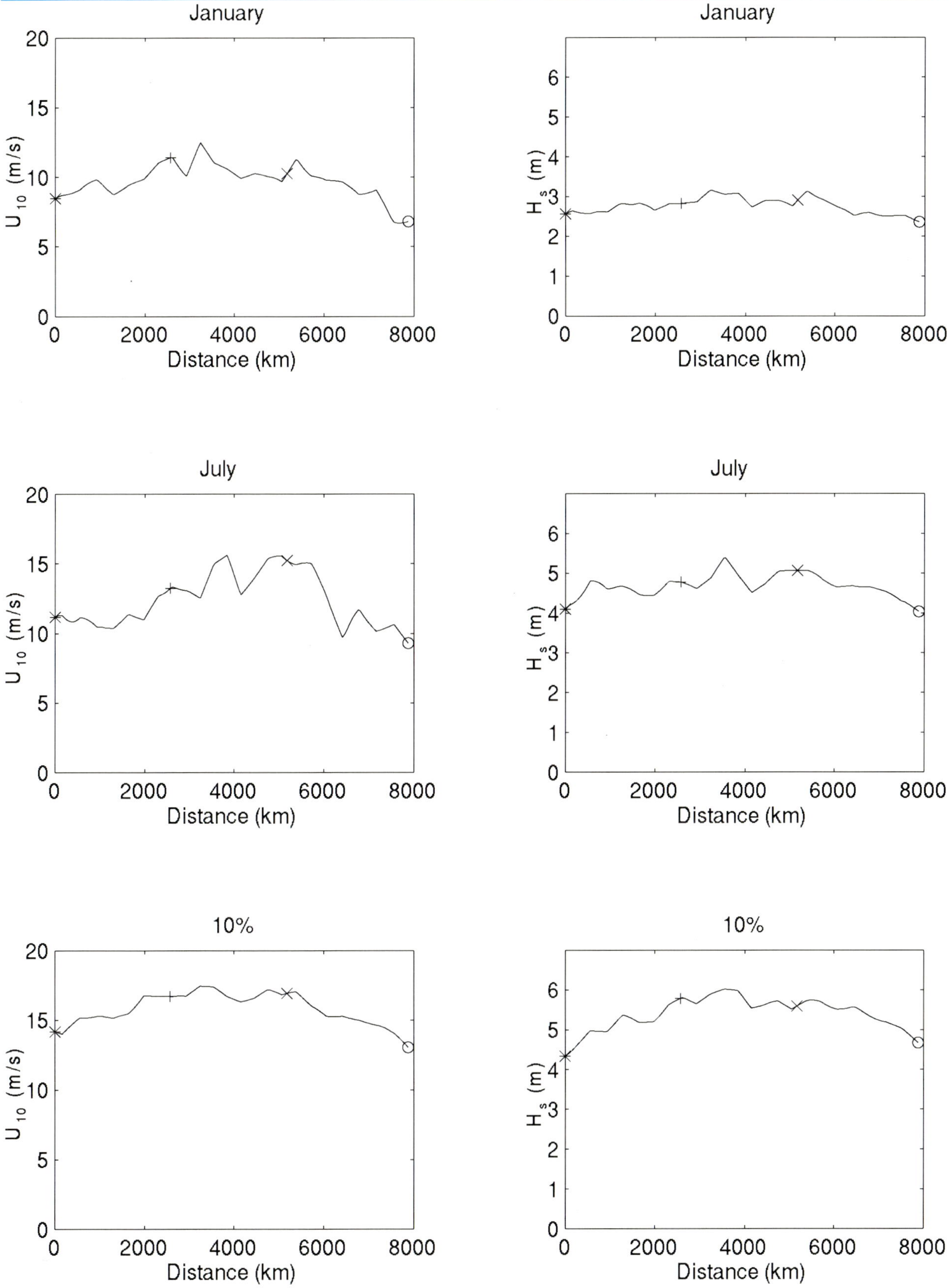

Fig. 7.28(c) Values of parameters along the Great Circle path between South Africa and the south-west tip of Australia (see Fig. 7.27). Values shown are: Mean Monthly Wind Speed for January (top left), Mean Monthly Wind Speed for July (middle left), Wind Speed exceeded 10% of the time (bottom left), Mean Monthly Significant Wave Height for January (top right), Mean Monthly Significant Wave Height for July (middle right) and Significant Wave Height exceeded 10% of the time (bottom right). The four markers in each panel correspond to the similarly marked positions shown on the corresponding Great Circle path in Fig. 7.27.

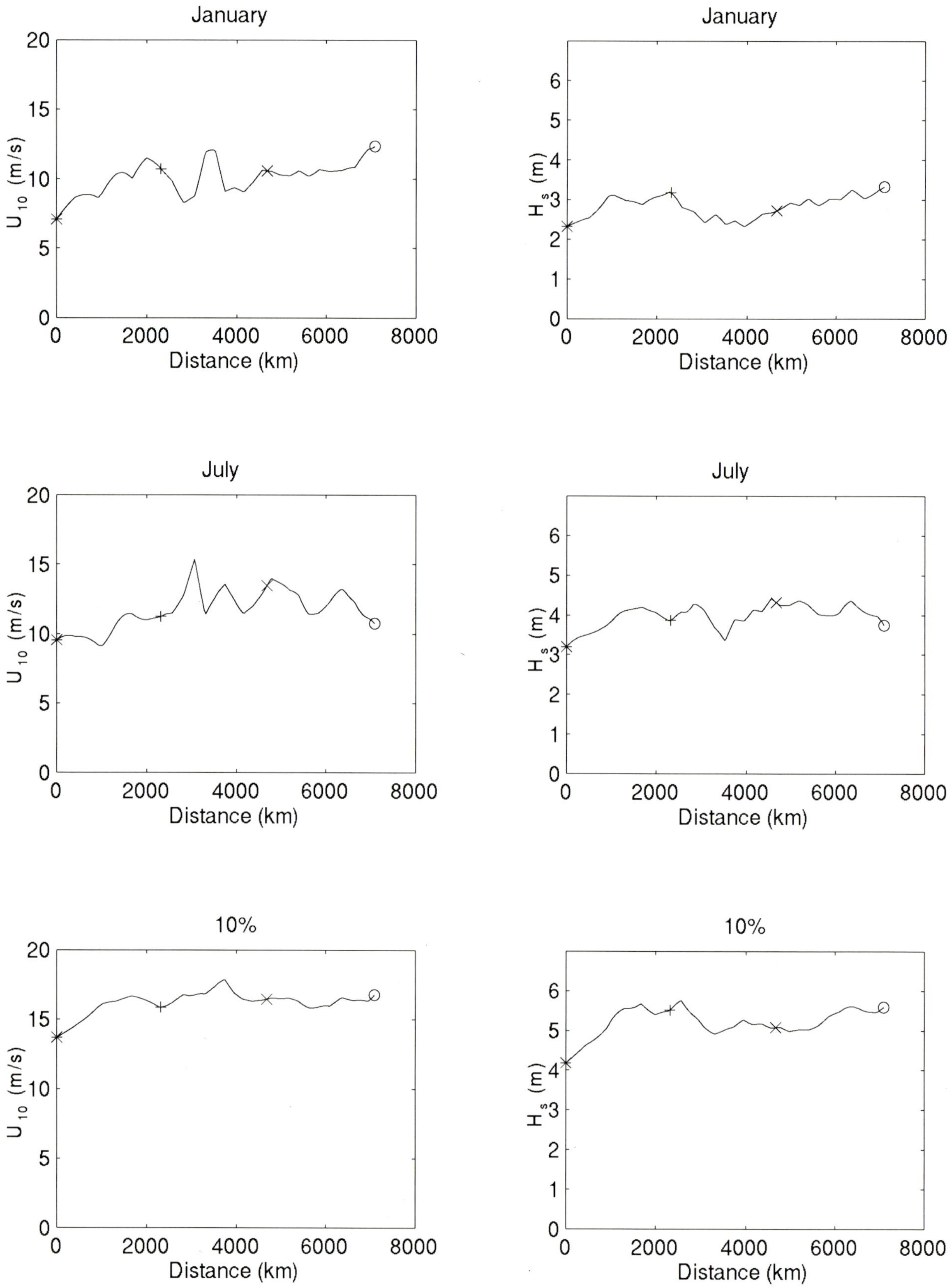

Fig. 7.28(d) Values of parameters along the Great Circle path between New Zealand and the southern tip of South America (see Fig. 7.27). Values shown are: Mean Monthly Wind Speed for January (top left), Mean Monthly Wind Speed for July (middle left), Wind Speed exceeded 10% of the time (bottom left), Mean Monthly Significant Wave Height for January (top right), Mean Monthly Significant Wave Height for July (middle right) and Significant Wave Height exceeded 10% of the time (bottom right). The four markers in each panel correspond to the similarly marked positions shown on the corresponding Great Circle path in Fig. 7.27.

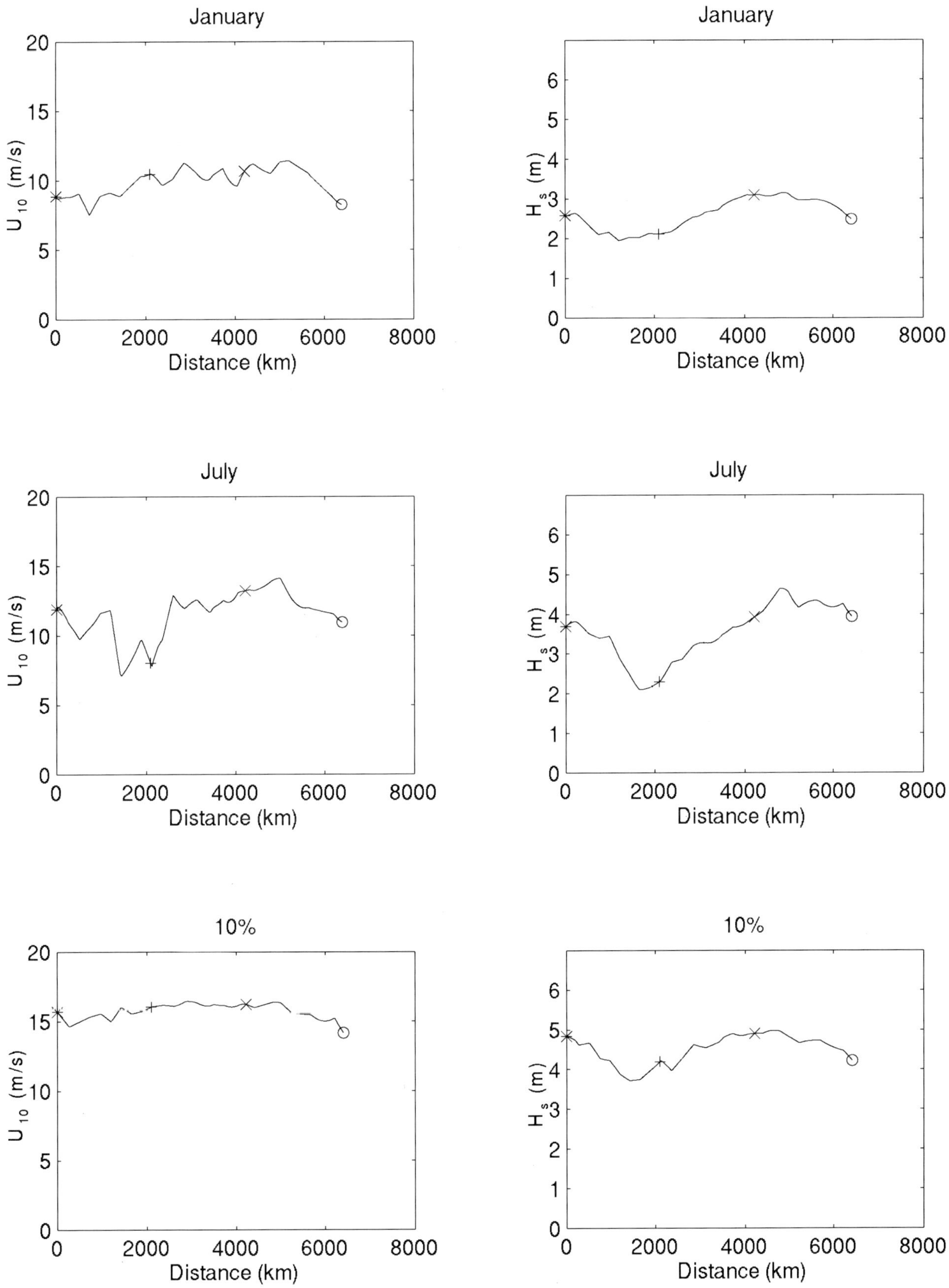

Fig. 7.28(e) Values of parameters along the Great Circle path between the southern tip of South America and South Africa (see Fig. 7.27). Values shown are: Mean Monthly Wind Speed for January (top left), Mean Monthly Wind Speed for July (middle left), Wind Speed exceeded 10% of the time (bottom left), Mean Monthly Significant Wave Height for January (top right), Mean Monthly Significant Wave Height for July (middle right) and Significant Wave Height exceeded 10% of the time (bottom right). The four markers in each panel correspond to the similarly marked positions shown on the corresponding Great Circle path in Fig. 7.27.

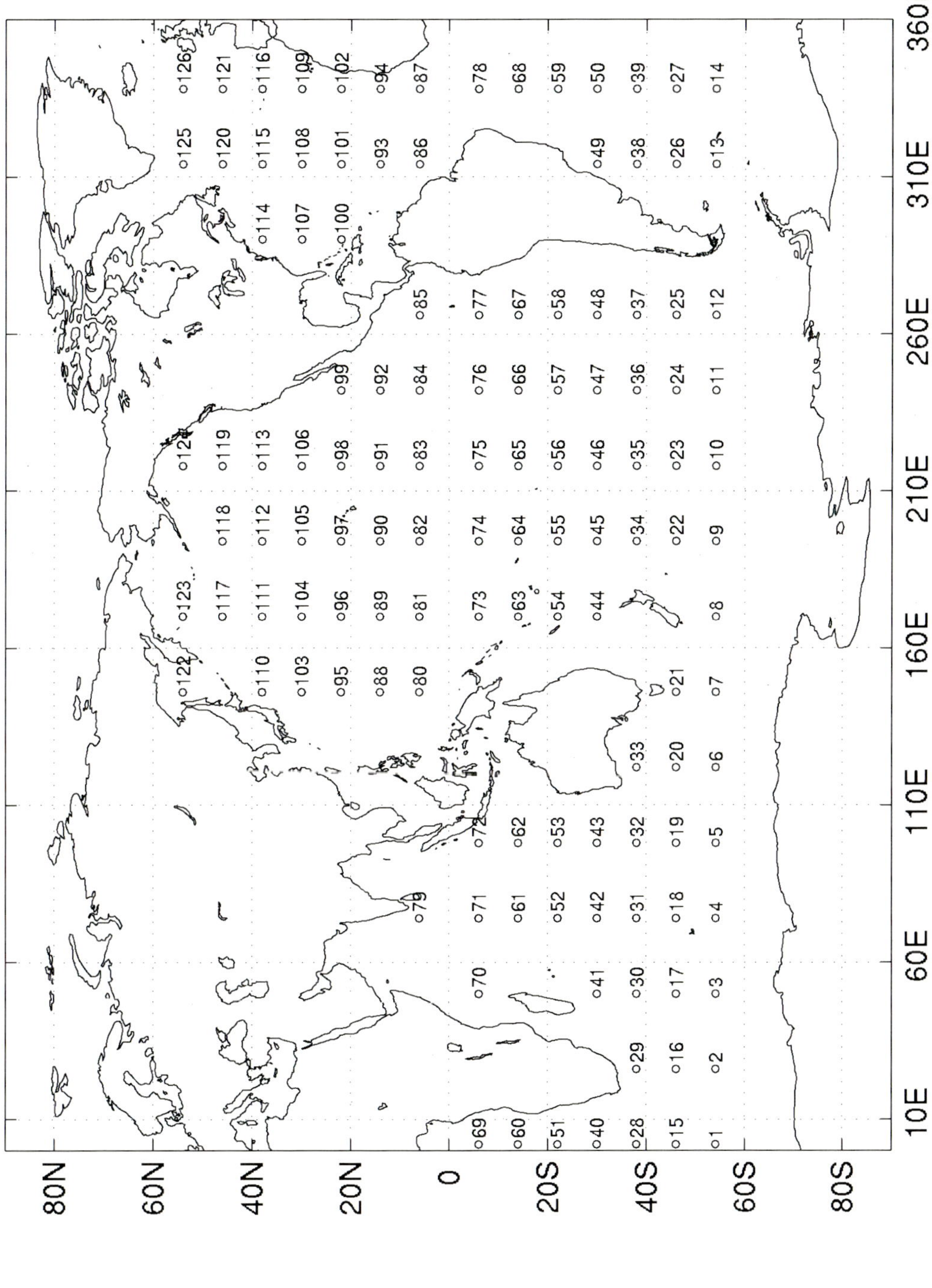

Fig. 7.29 Positions at which detailed output is presented in Figs 7.30 and 7.31.

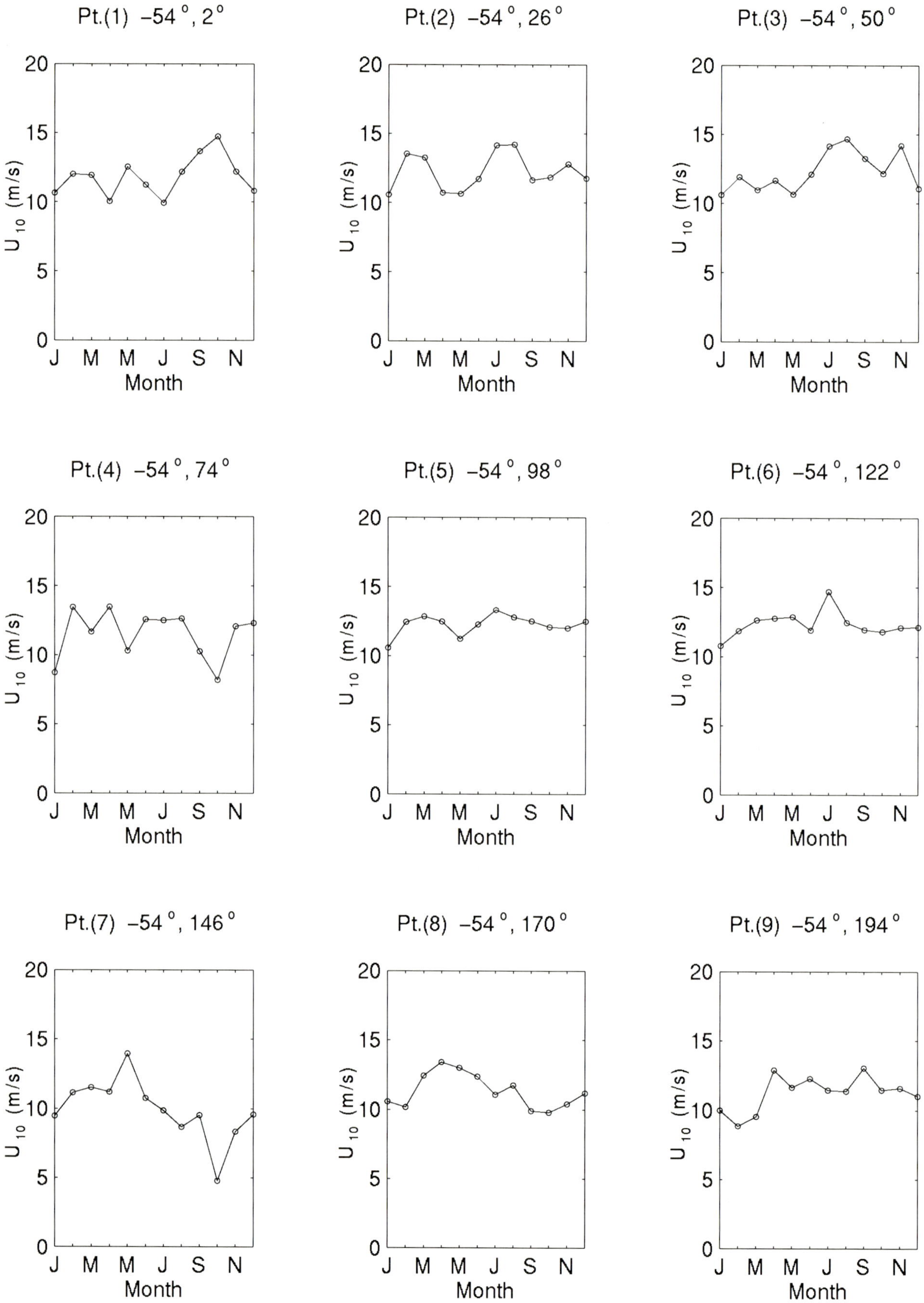

Fig. 7.30(a) Seasonal variability in values of Mean Monthly Wind Speed at points 1–9 of Fig. 7.29.

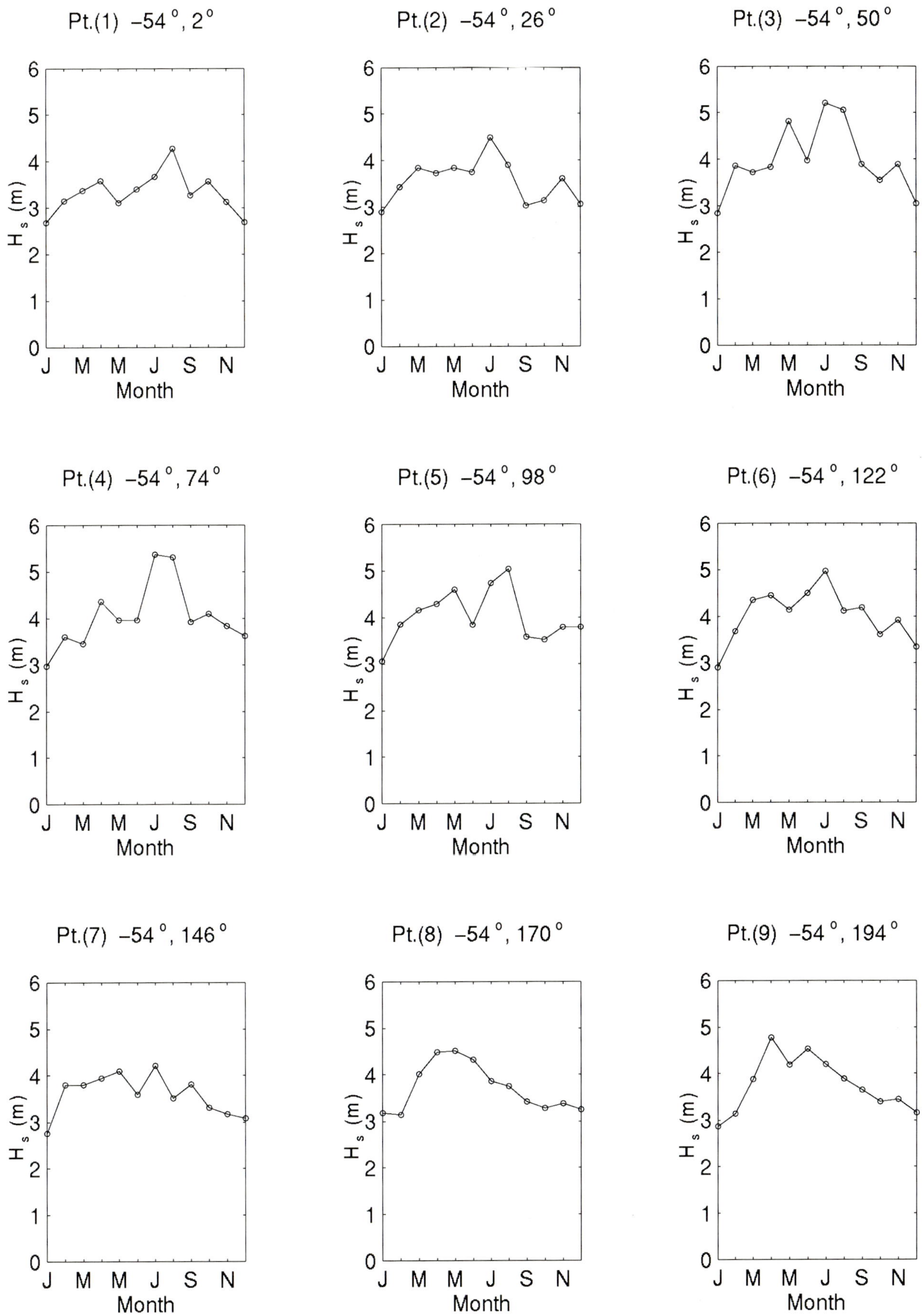

Fig. 7.30(b) Seasonal variability in values of Mean Monthly Significant Wave Height at points 1–9 of Fig. 7.29.

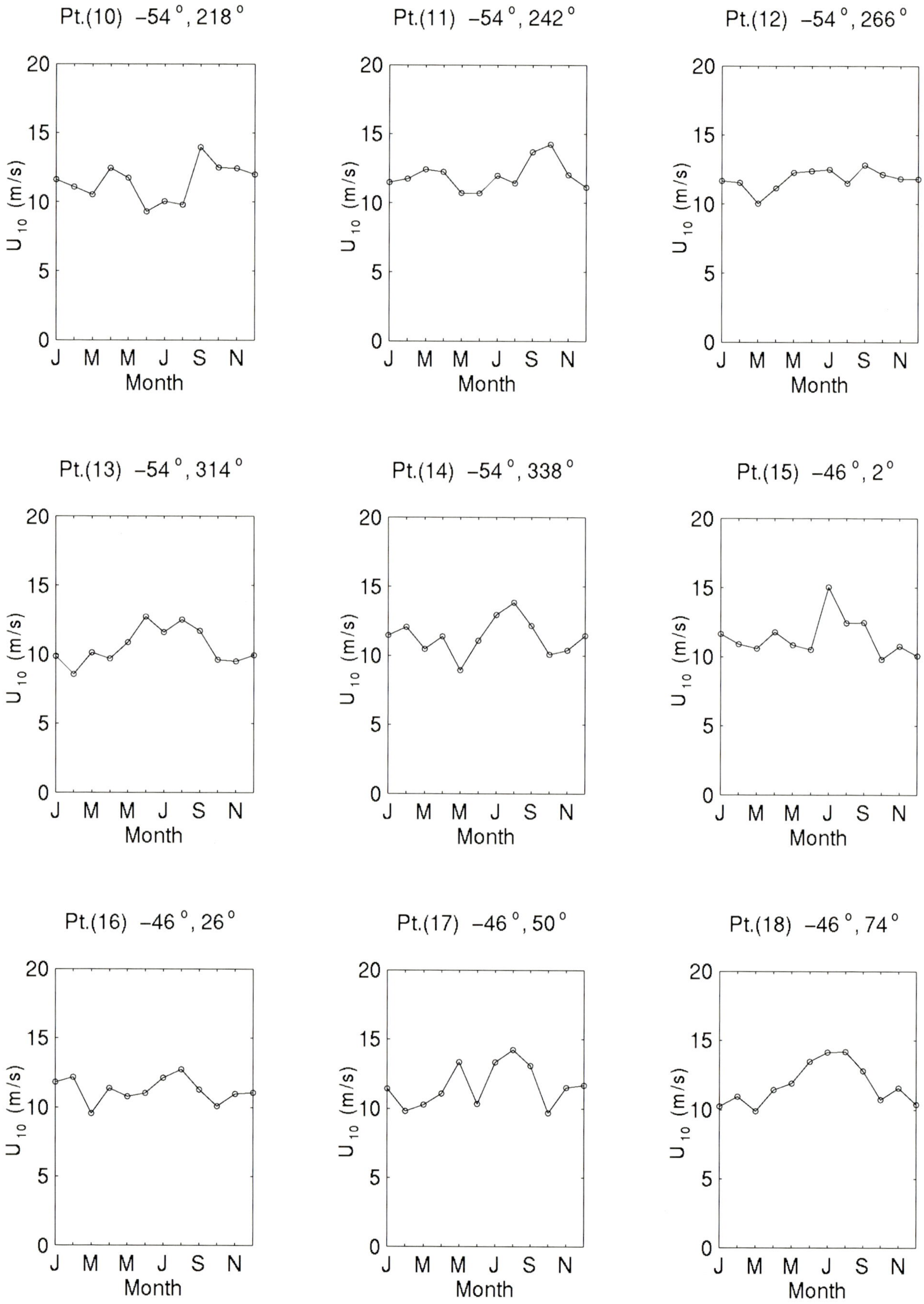

Fig. 7.30(c) Seasonal variability in values of Mean Monthly Wind Speed at points 10–18 of Fig. 7.29.

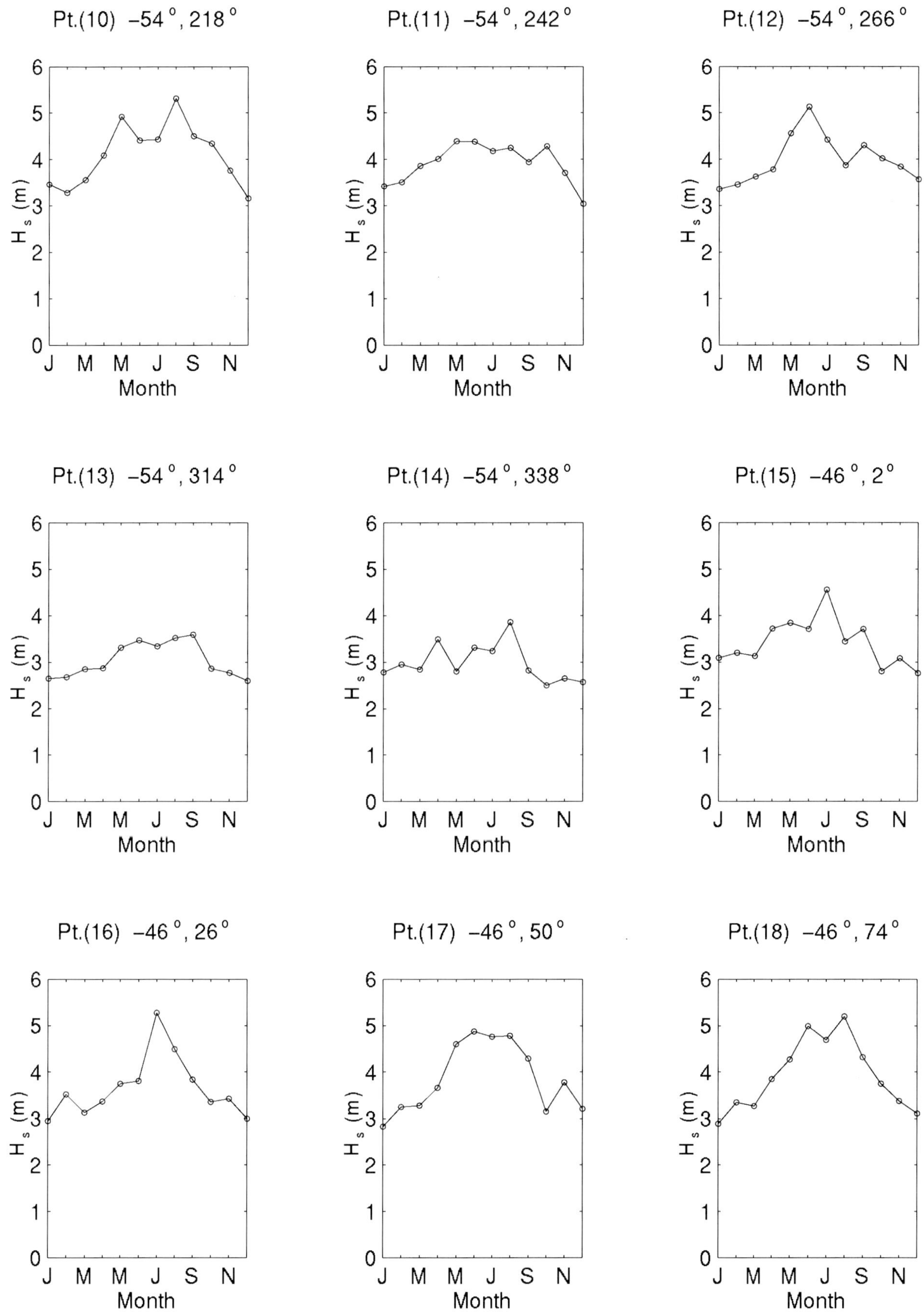

Fig. 7.30(d) Seasonal variability in values of Mean Monthly Significant Wave Height at points 10–18 of Fig. 7.29.

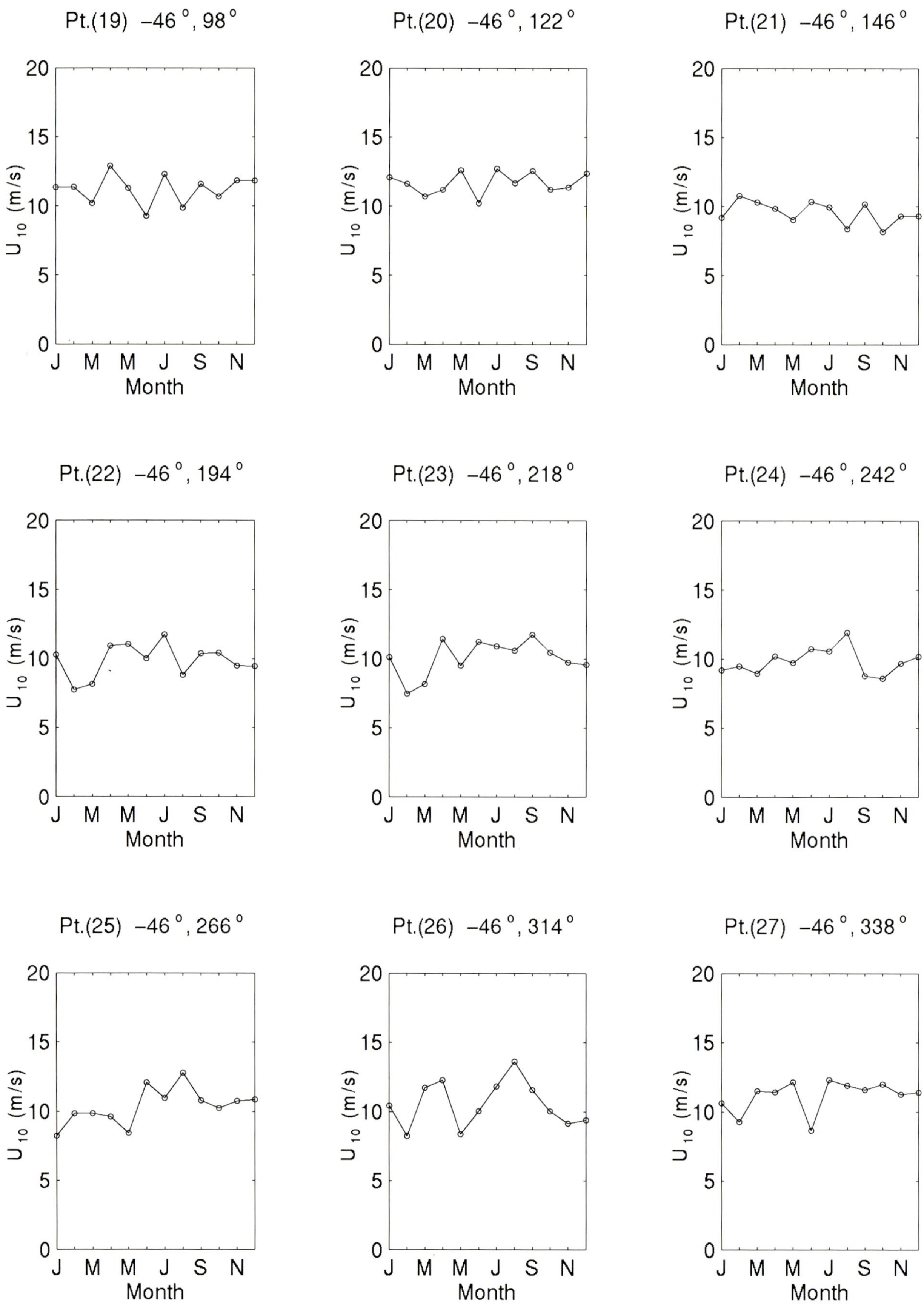

Fig. 7.30(e) Seasonal variability in values of Mean Monthly Wind Speed at points 19–27 of Fig. 7.29.

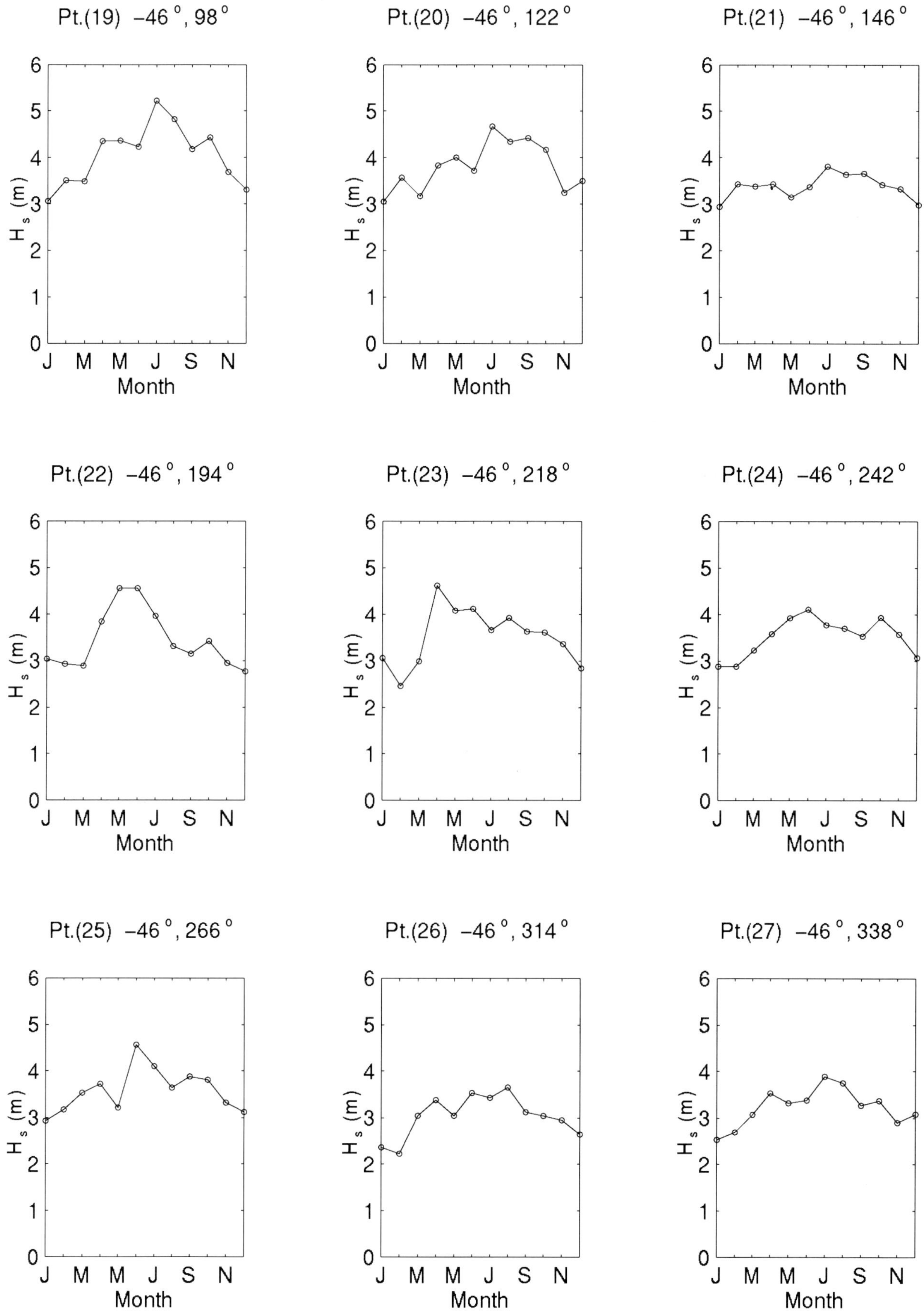

Fig. 7.30(f) Seasonal variability in values of Mean Monthly Significant Wave Height at points 19–27 of Fig. 7.29.

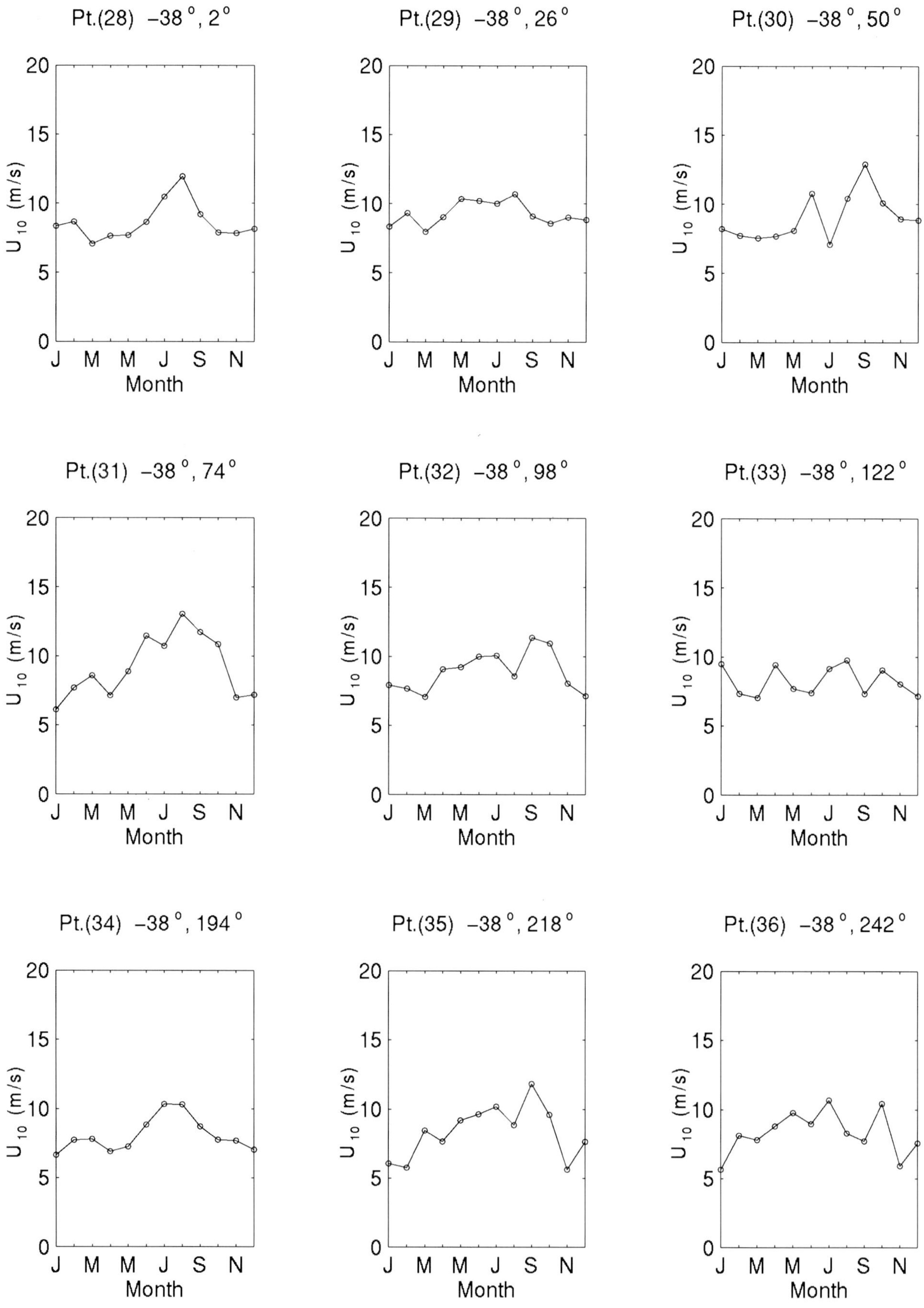

Fig. 7.30(g) Seasonal variability in values of Mean Monthly Wind Speed at points 28–36 of Fig. 7.29.

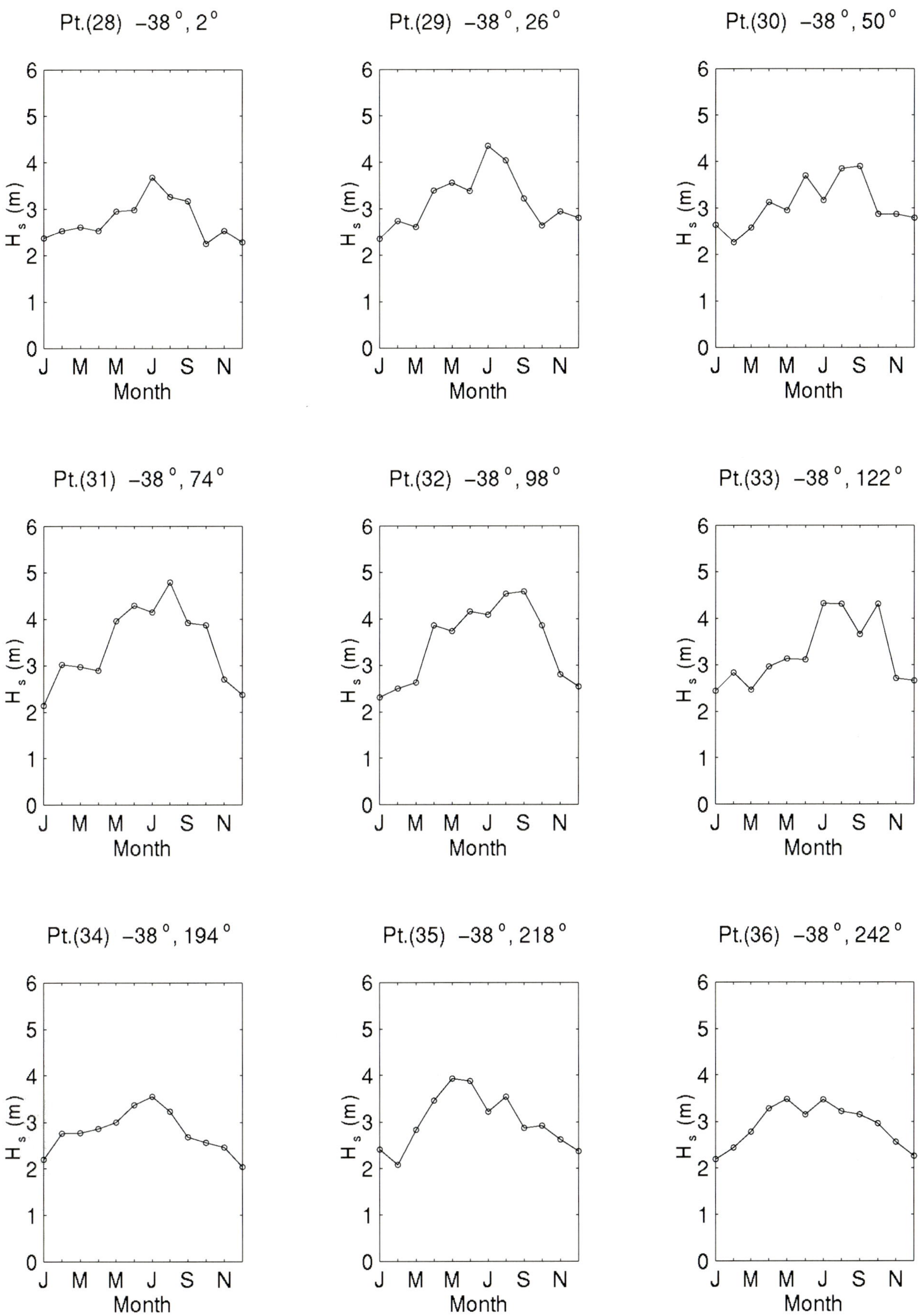

Fig. 7.30(h) Seasonal variability in values of Mean Monthly Significant Wave Height at points 28–36 of Fig. 7.29.

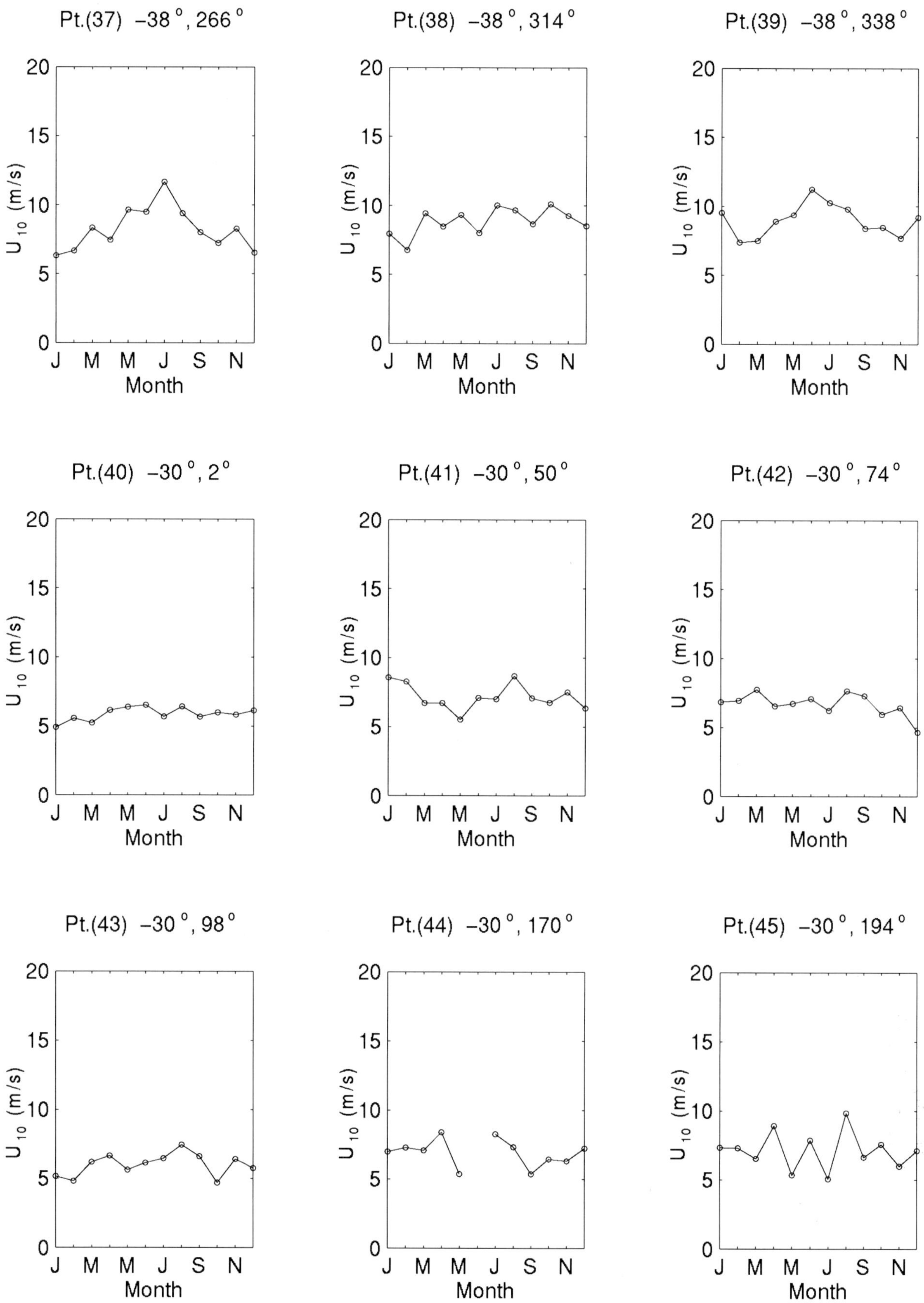

Fig. 7.30(i) Seasonal variability in values of Mean Monthly Wind Speed at points 37–45 of Fig. 7.29.

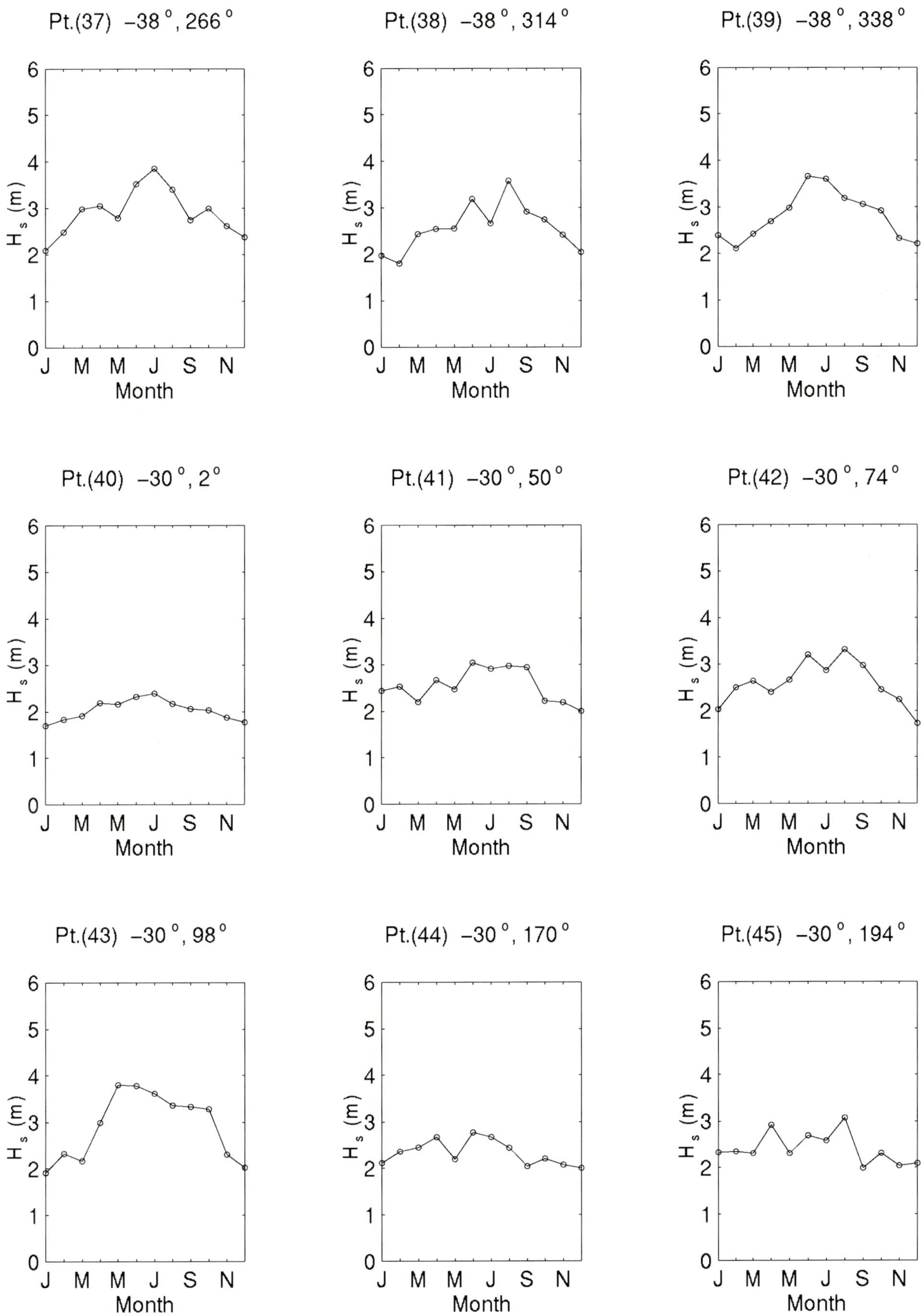

Fig. 7.30(j) Seasonal variability in values of Mean Monthly Significant Wave Height at points 37–45 of Fig. 7.29.

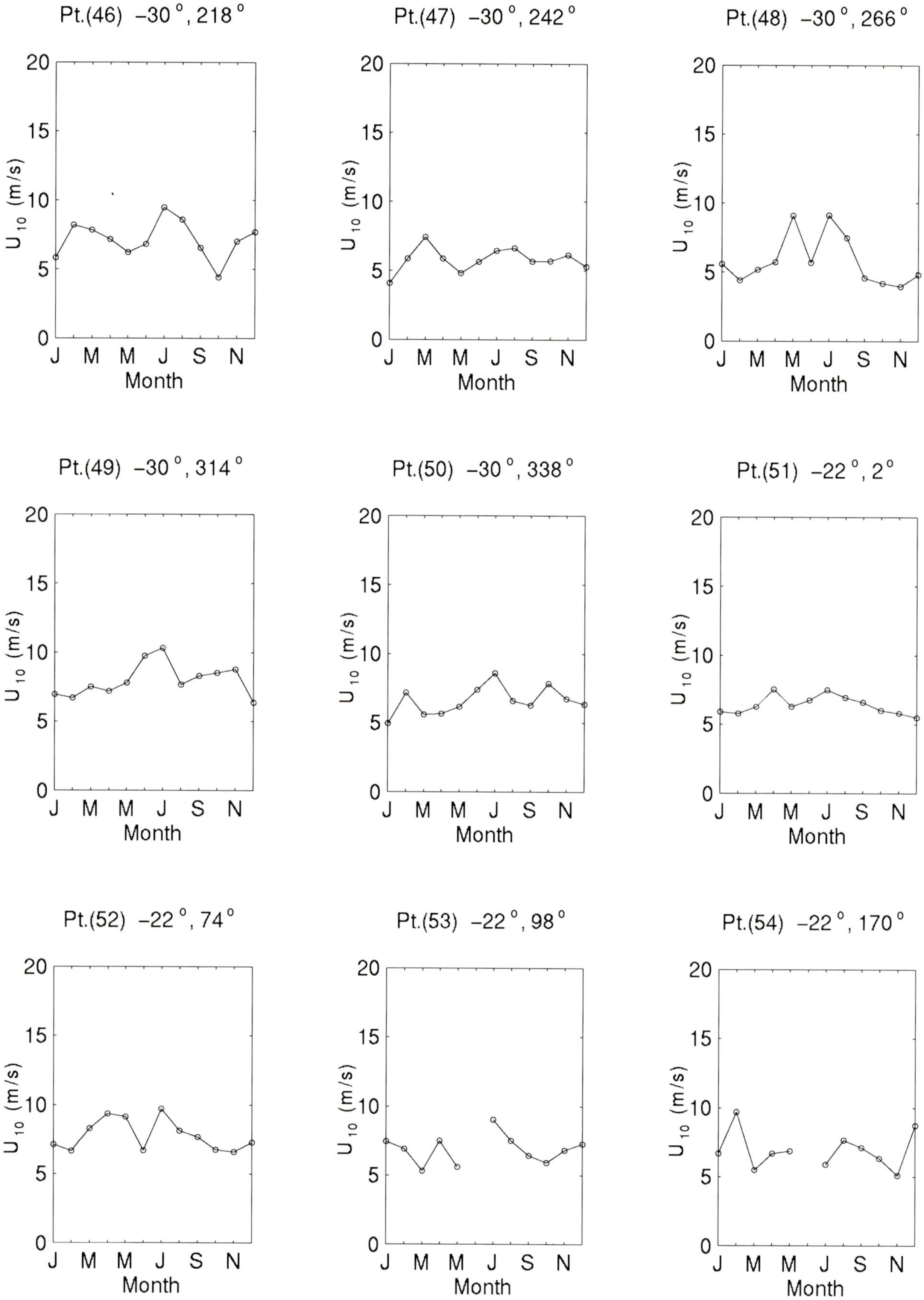

Fig. 7.30(k) Seasonal variability in values of Mean Monthly Wind Speed at points 46–54 of Fig. 7.29.

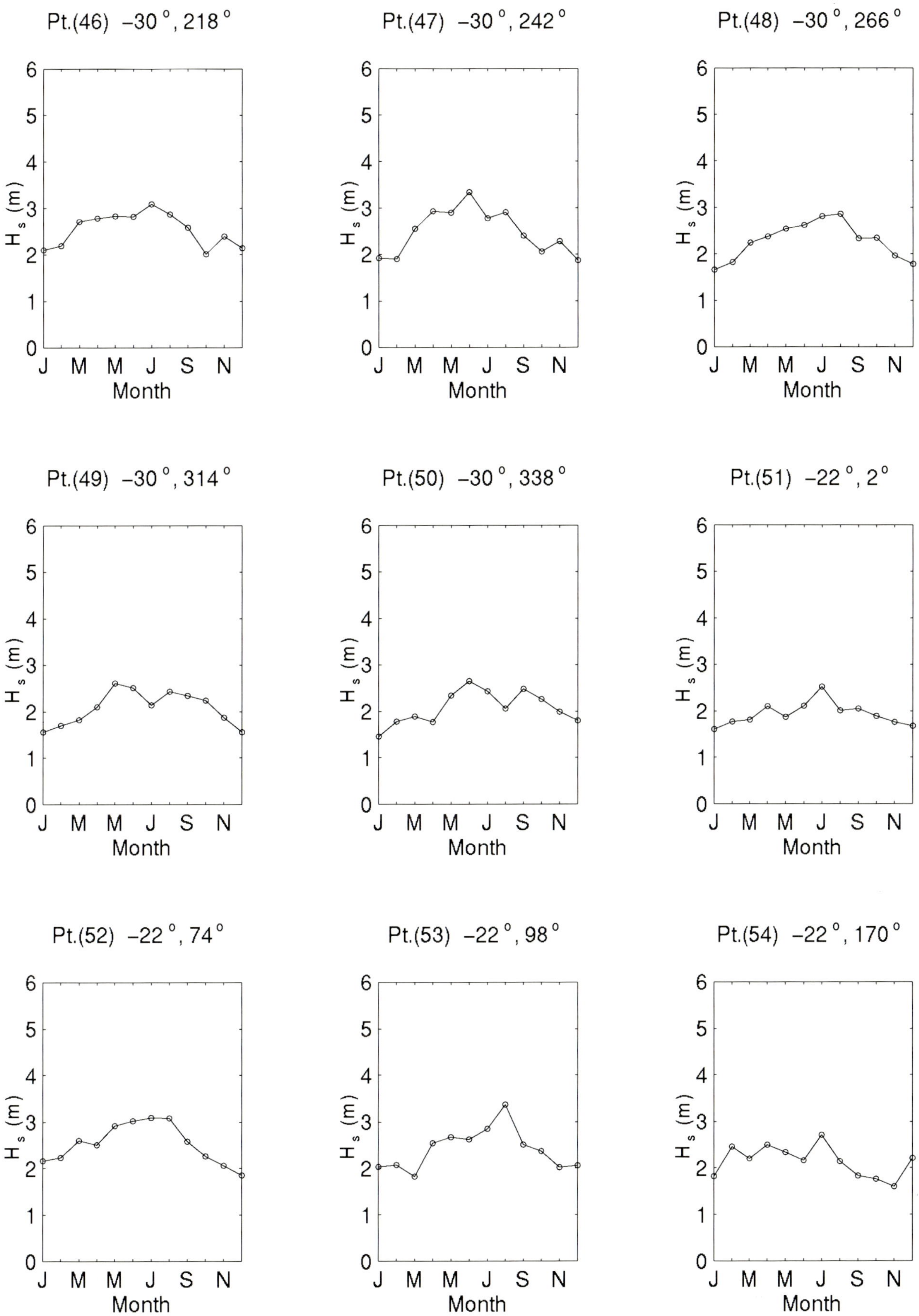

Fig. 7.30(l) Seasonal variability in values of Mean Monthly Significant Wave Height at points 46–54 of Fig. 7.29.

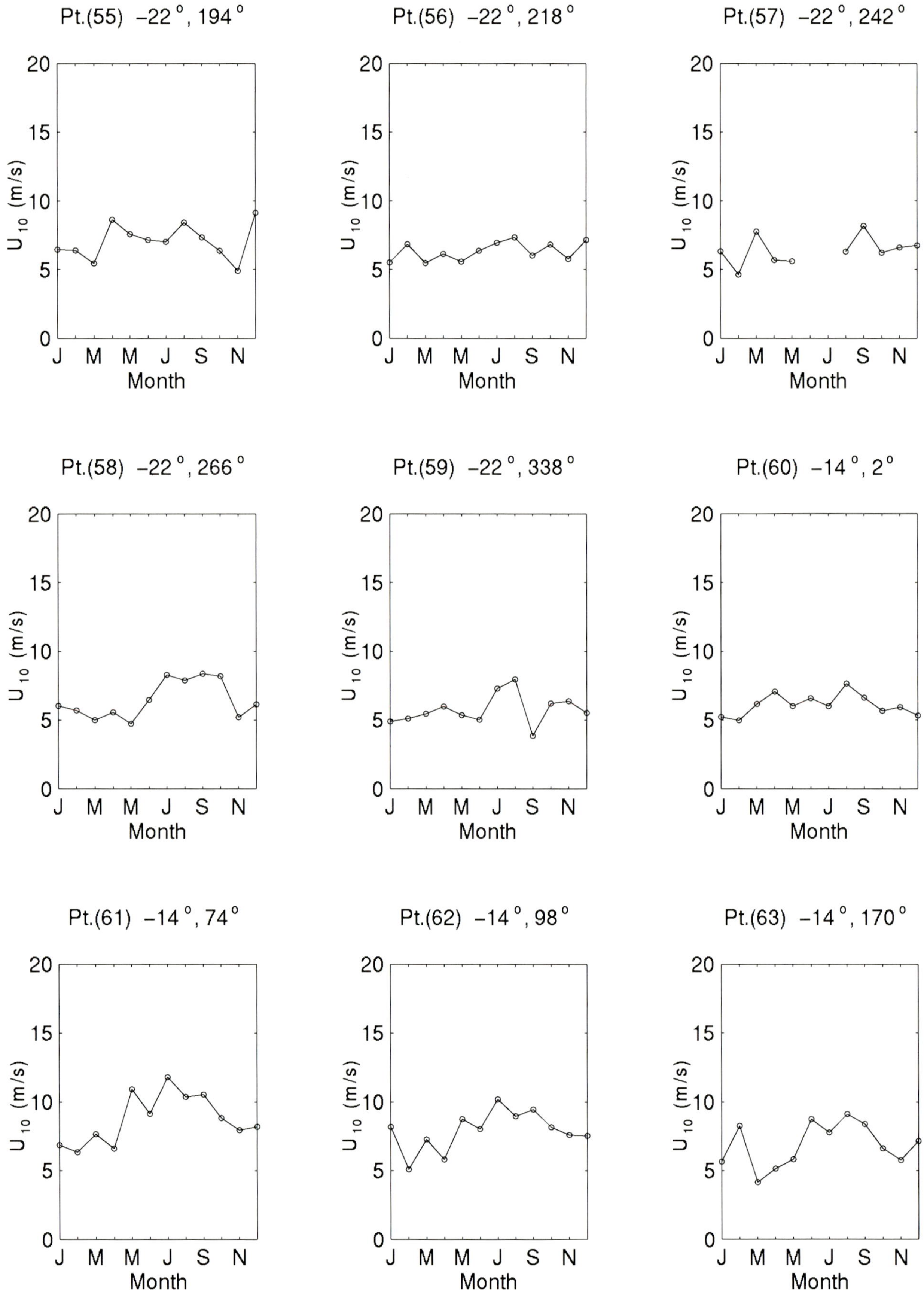

Fig. 7.30(m) Seasonal variability in values of Mean Monthly Wind Speed at points 55–63 of Fig. 7.29.

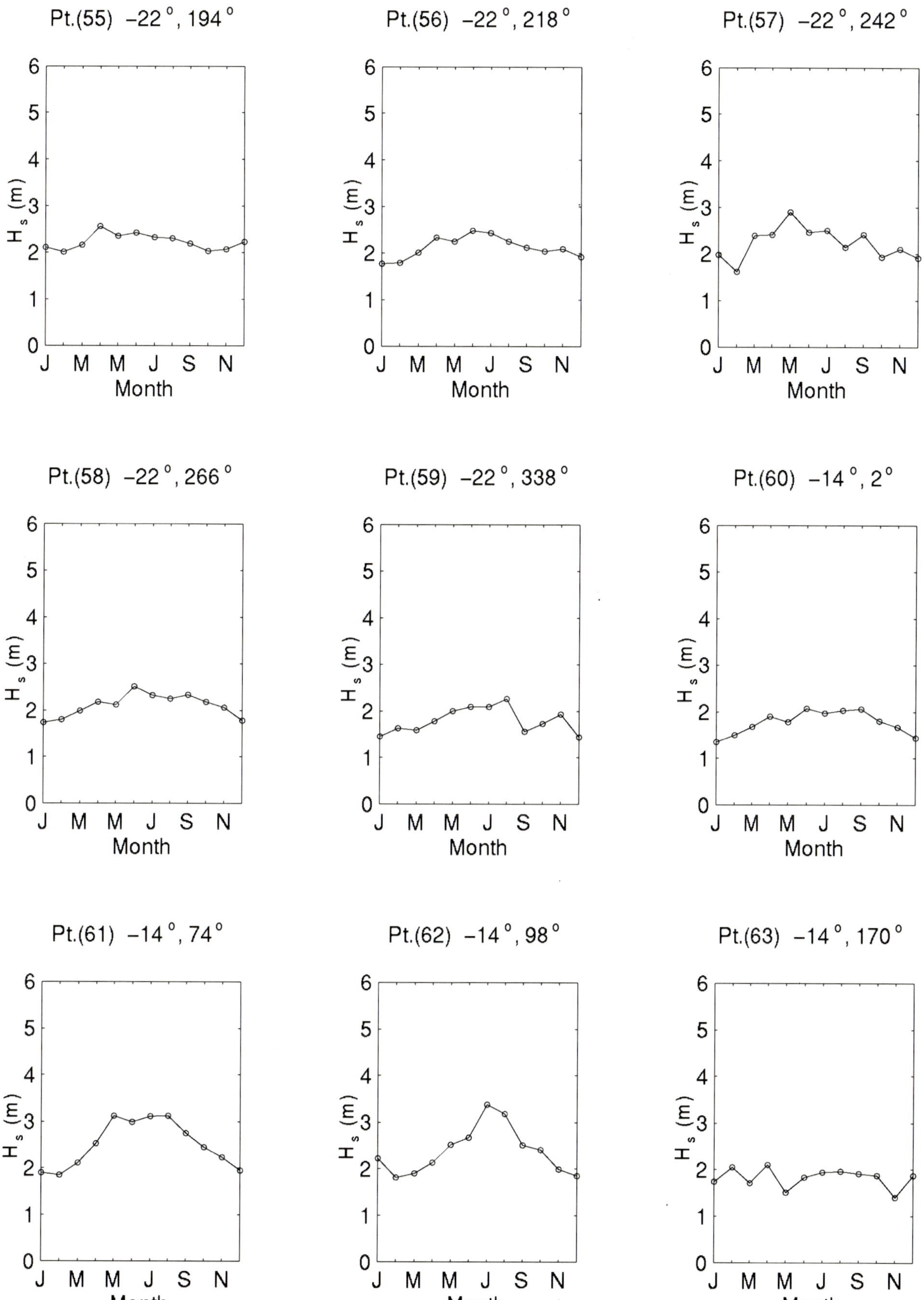

Fig. 7.30(n) Seasonal variability in values of Mean Monthly Significant Wave Height at points 55–63 of Fig. 7.29.

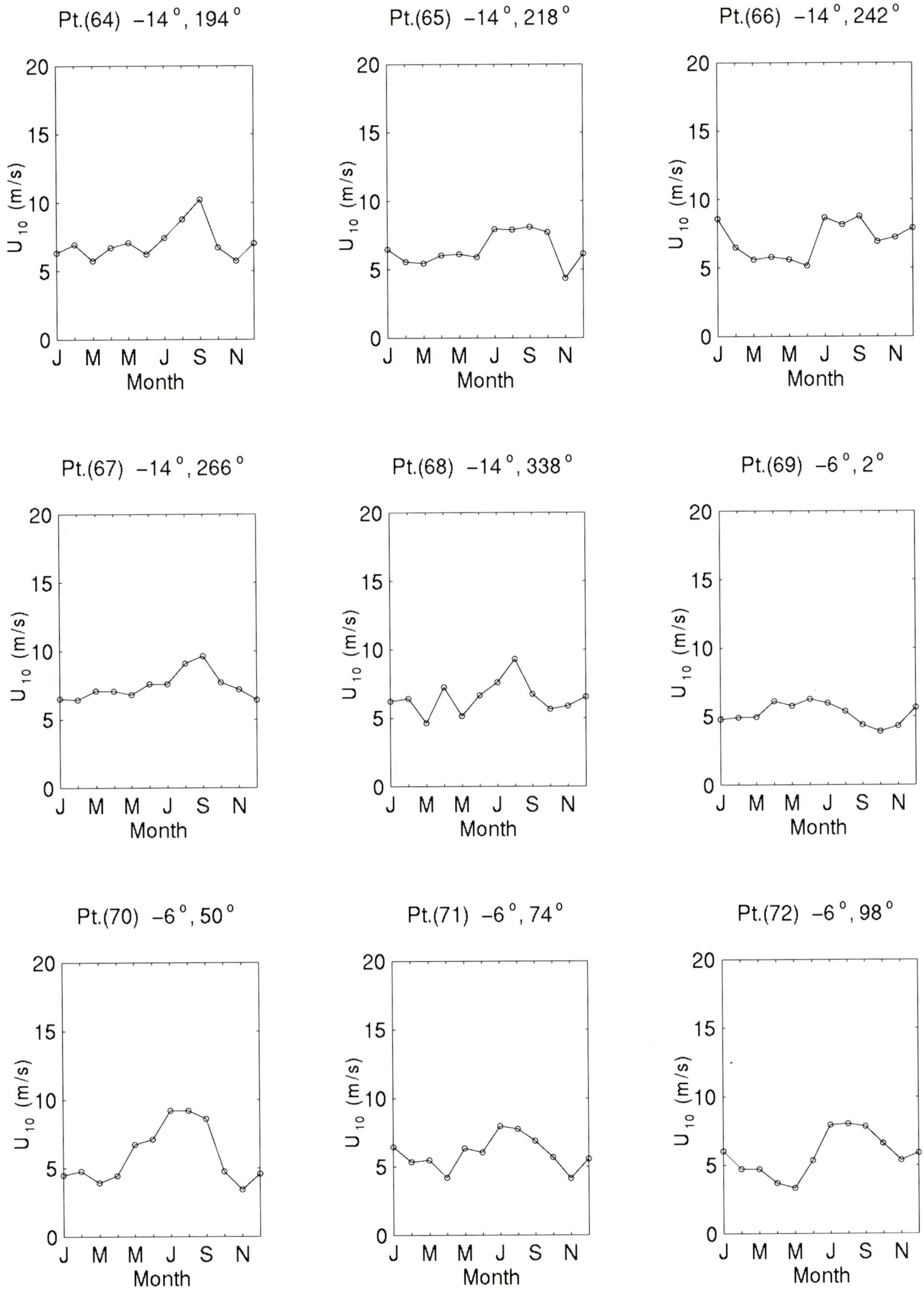

Fig. 7.30(o) Seasonal variability in values of Mean Monthly Wind Speed at points 64–72 of Fig. 7.29.

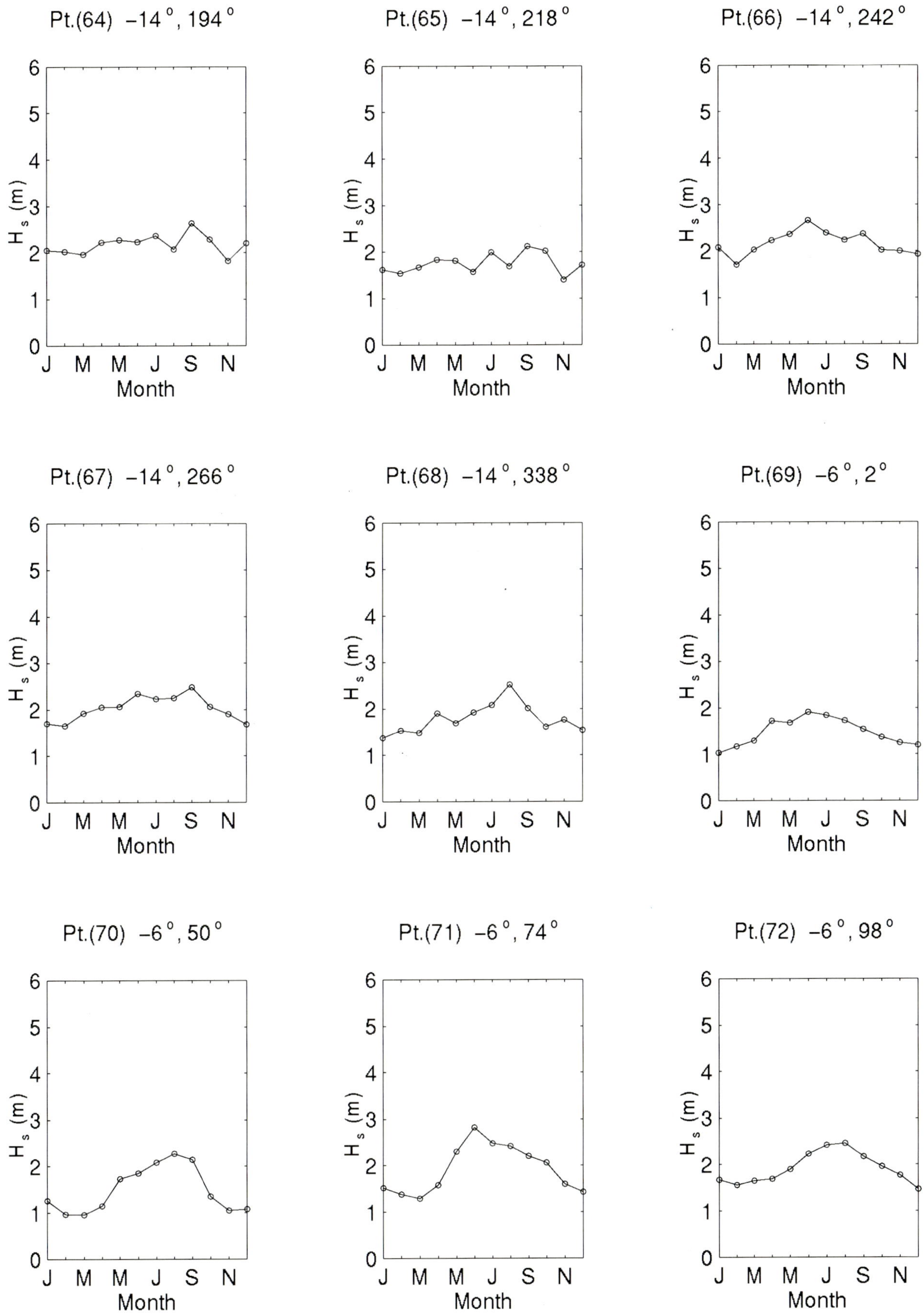

Fig. 7.30(p) Seasonal variability in values of Mean Monthly Significant Wave Height at points 64–72 of Fig. 7.29.

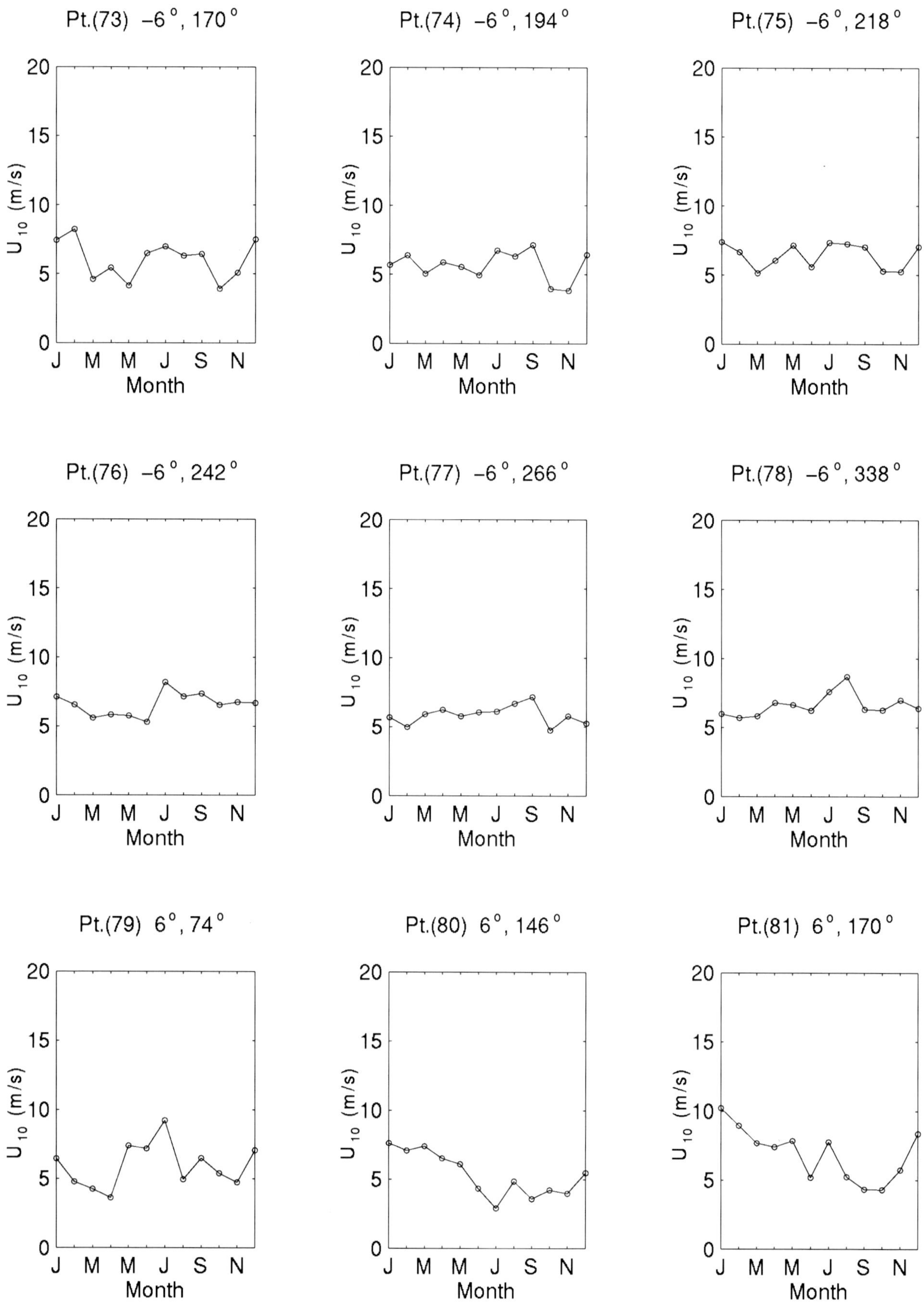

Fig. 7.30(q) Seasonal variability in values of Mean Monthly Wind Speed at points 73–81 of Fig. 7.29.

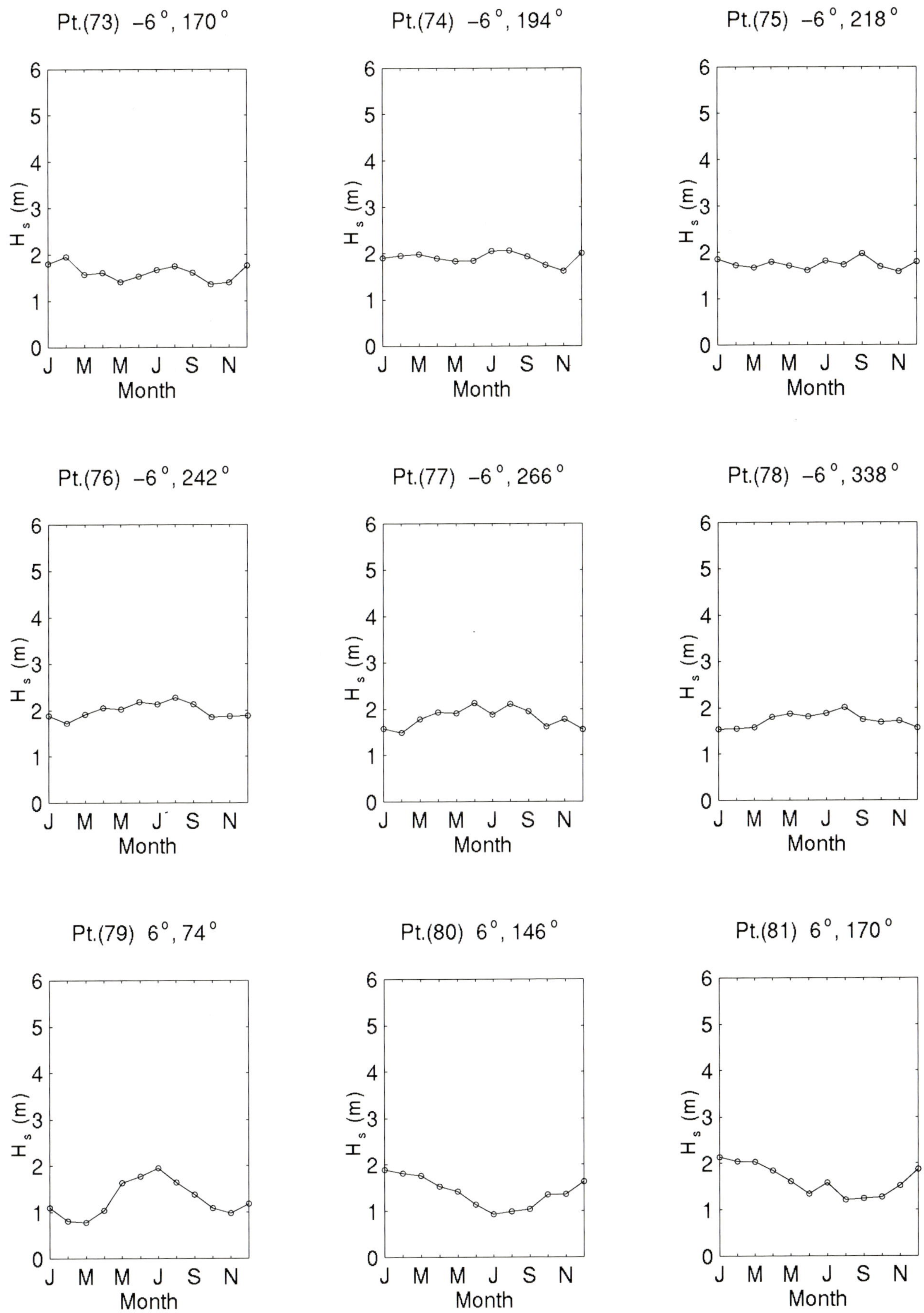

Fig. 7.30(r) Seasonal variability in values of Mean Monthly Significant Wave Height at points 73–81 of Fig. 7.29.

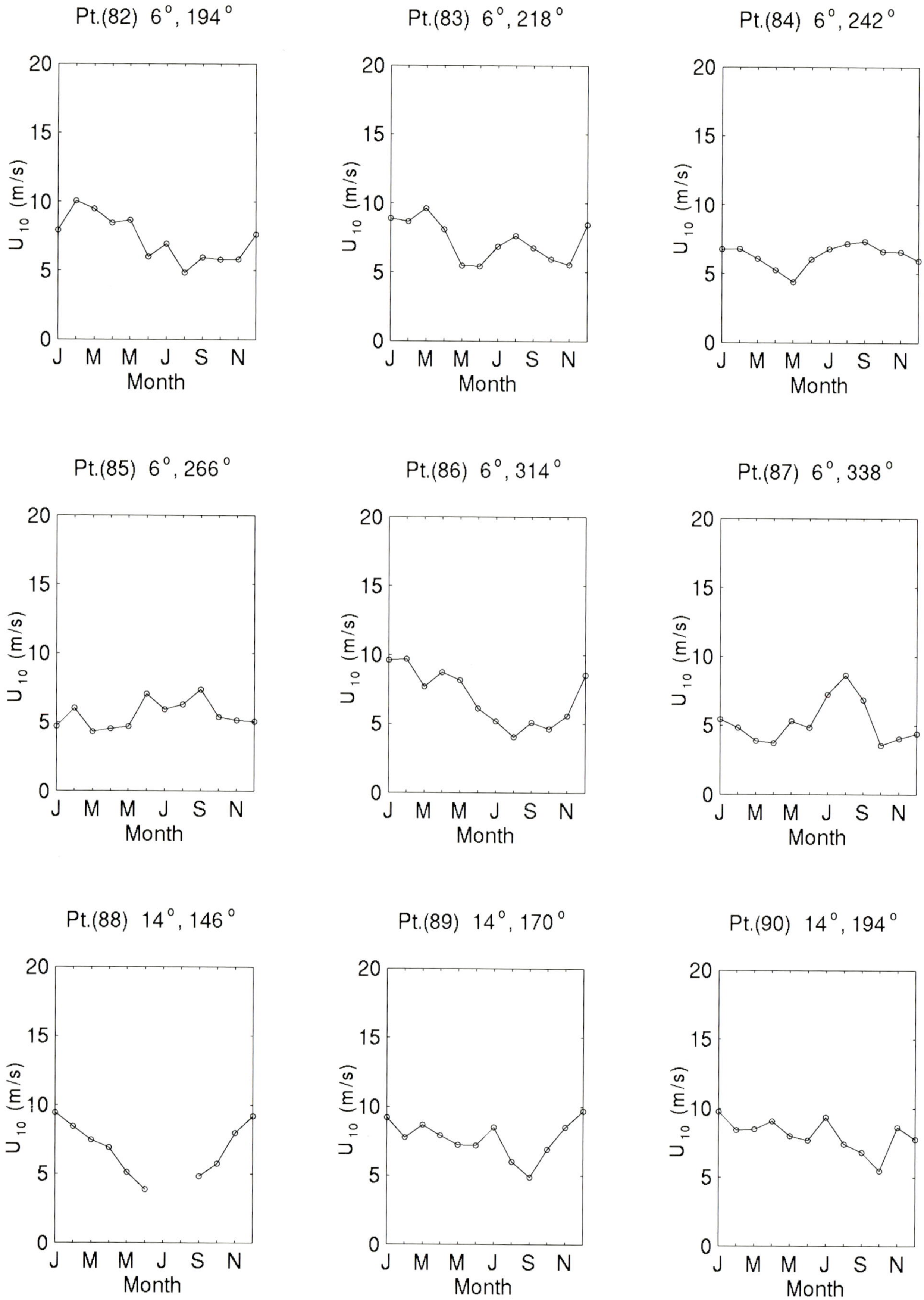

Fig. 7.30(s) Seasonal variability in values of Mean Monthly Wind Speed at points 82–90 of Fig. 7.29.

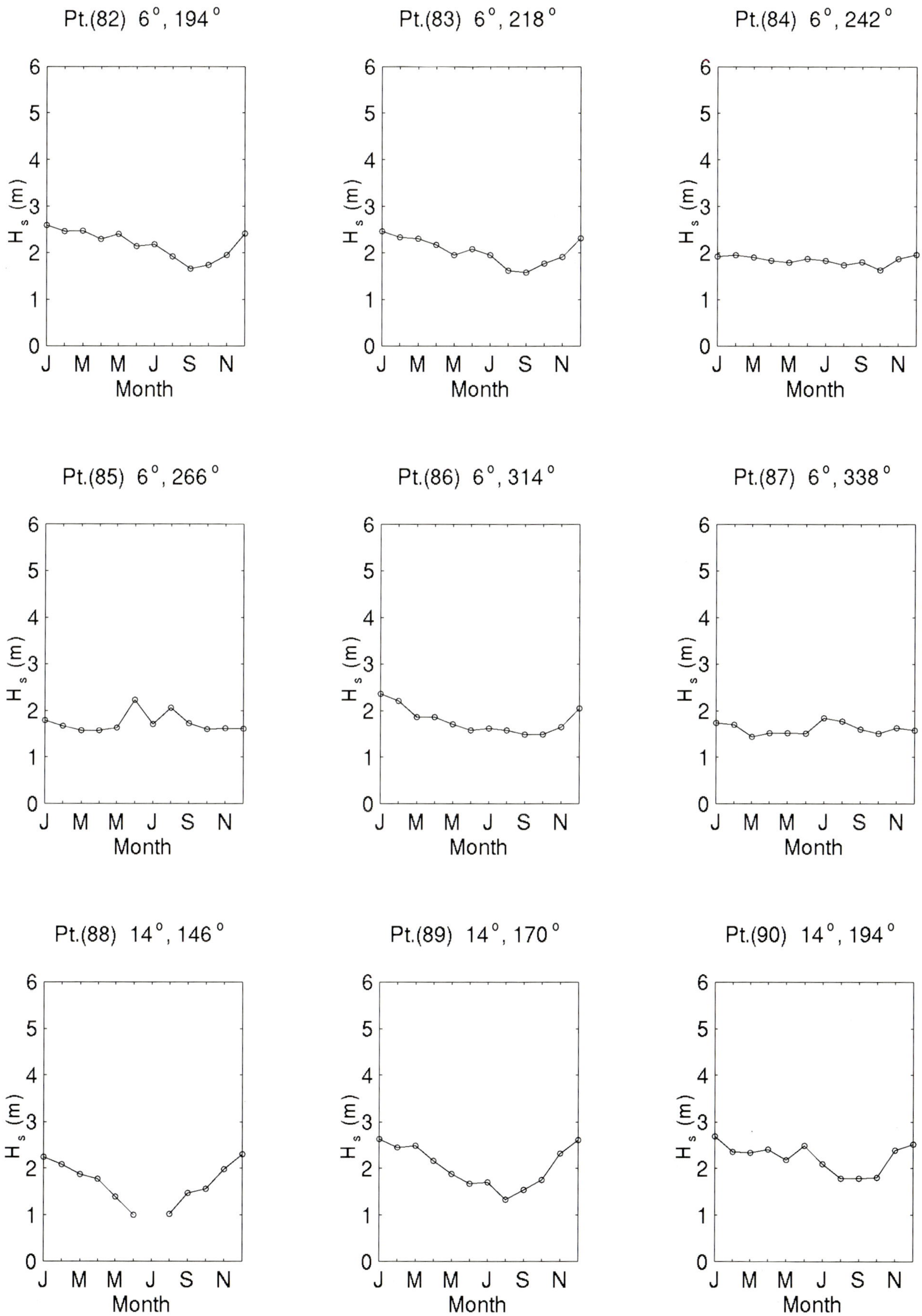

Fig. 7.30(t) Seasonal variability in values of Mean Monthly Significant Wave Height at points 82–90 of Fig. 7.29.

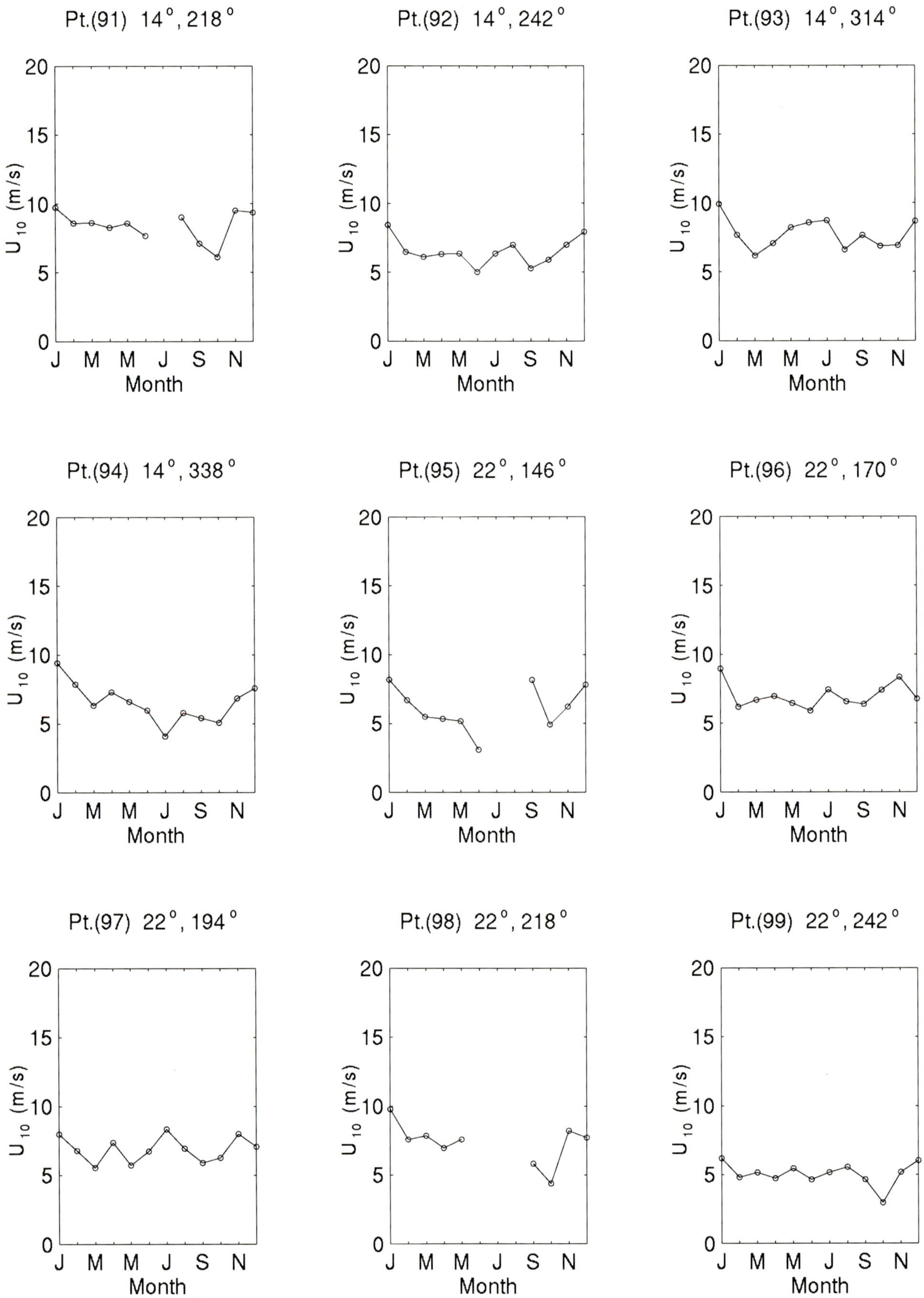

Fig. 7.30(u) Seasonal variability in values of Mean Monthly Wind Speed at points 91–99 of Fig. 7.29.

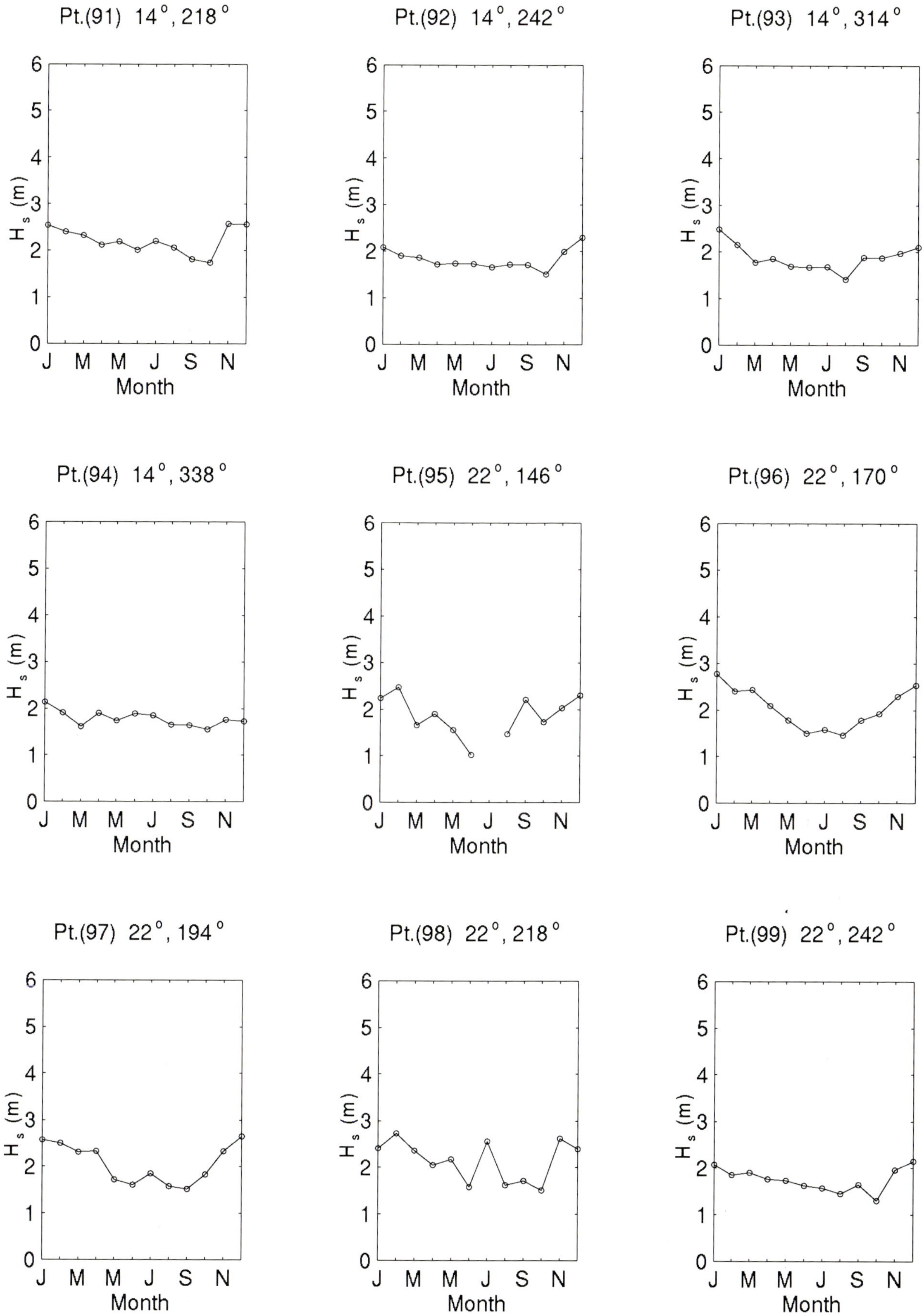

Fig. 7.30(v) Seasonal variability in values of Mean Monthly Significant Wave Height at points 91–99 of Fig. 7.29.

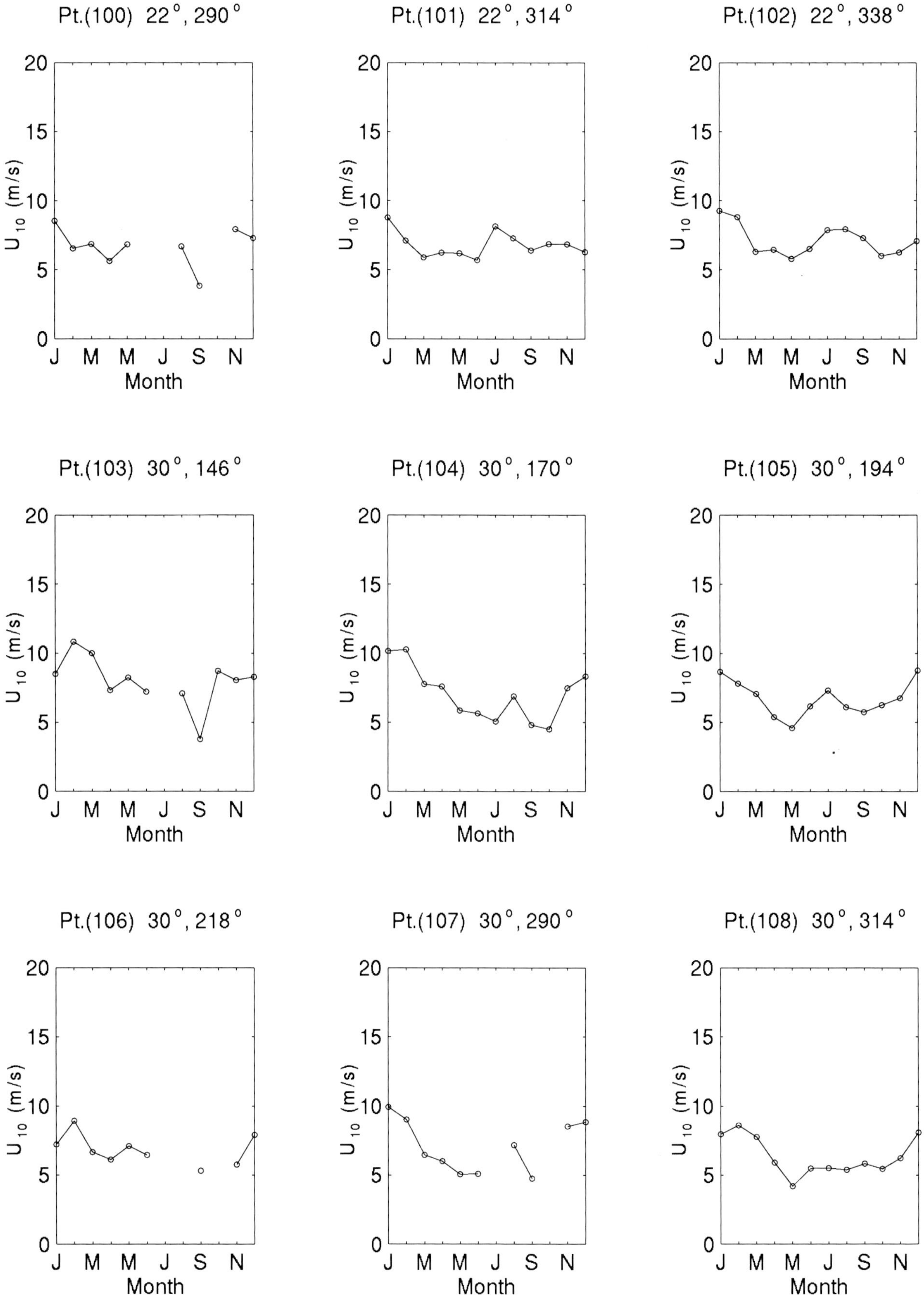

Fig. 7.30(w) Seasonal variability in values of Mean Monthly Wind Speed at points 100–108 of Fig. 7.29.

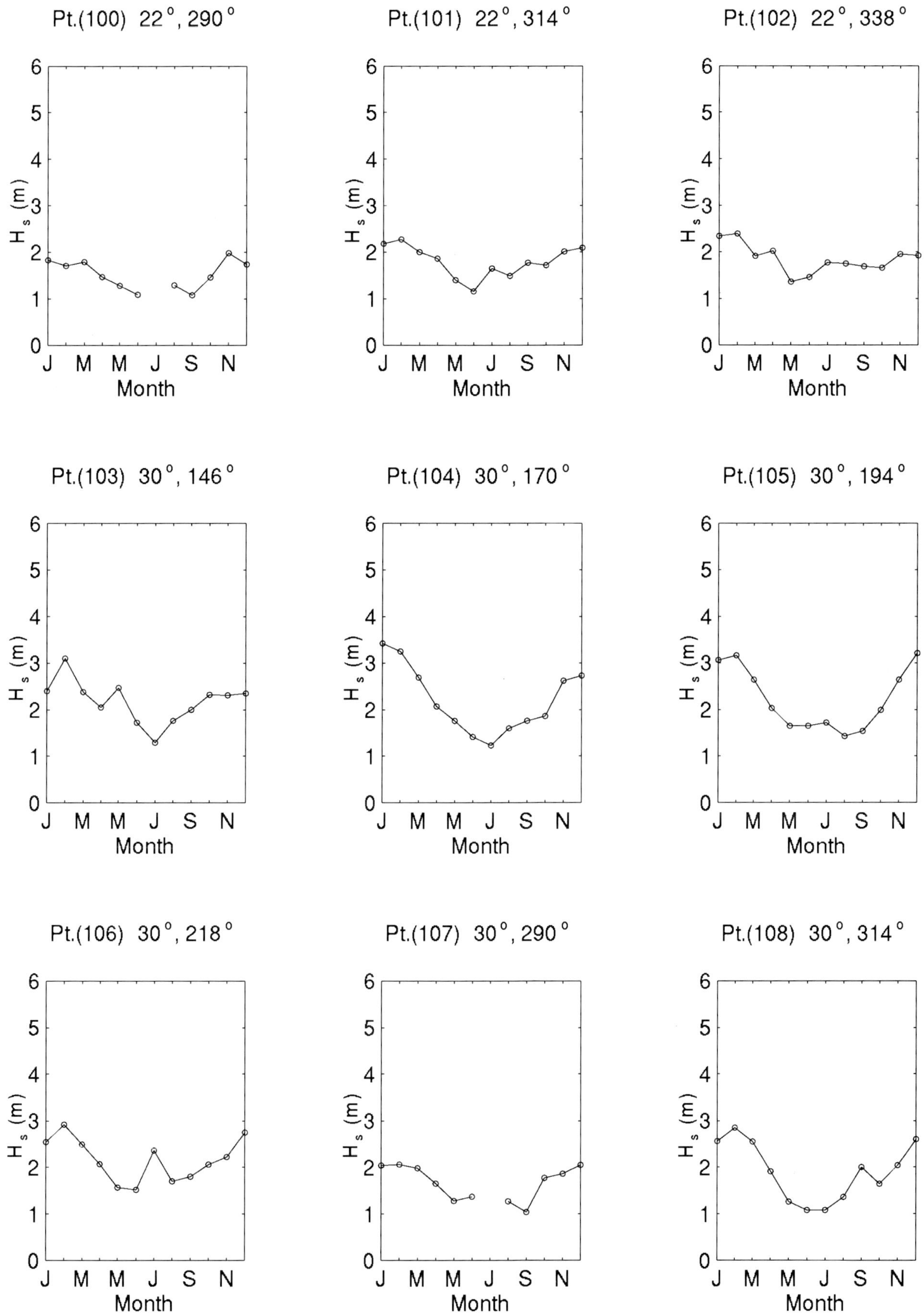

Fig. 7.30(x) Seasonal variability in values of Mean Monthly Significant Wave Height at points 100–108 of Fig. 7.29.

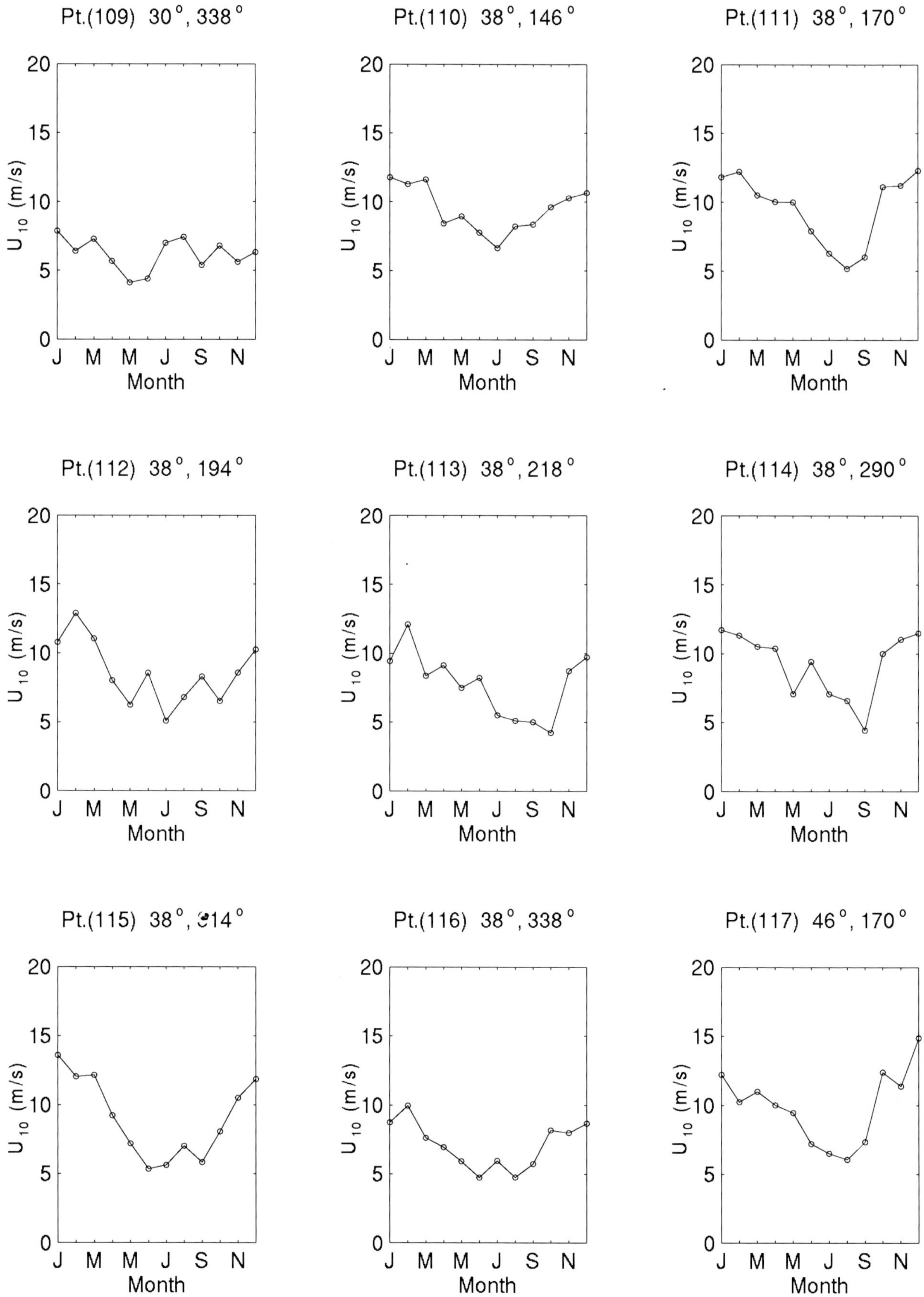

Fig. 7.30(y) Seasonal variability in values of Mean Monthly Wind Speed at points 109–117 of Fig. 7.29.

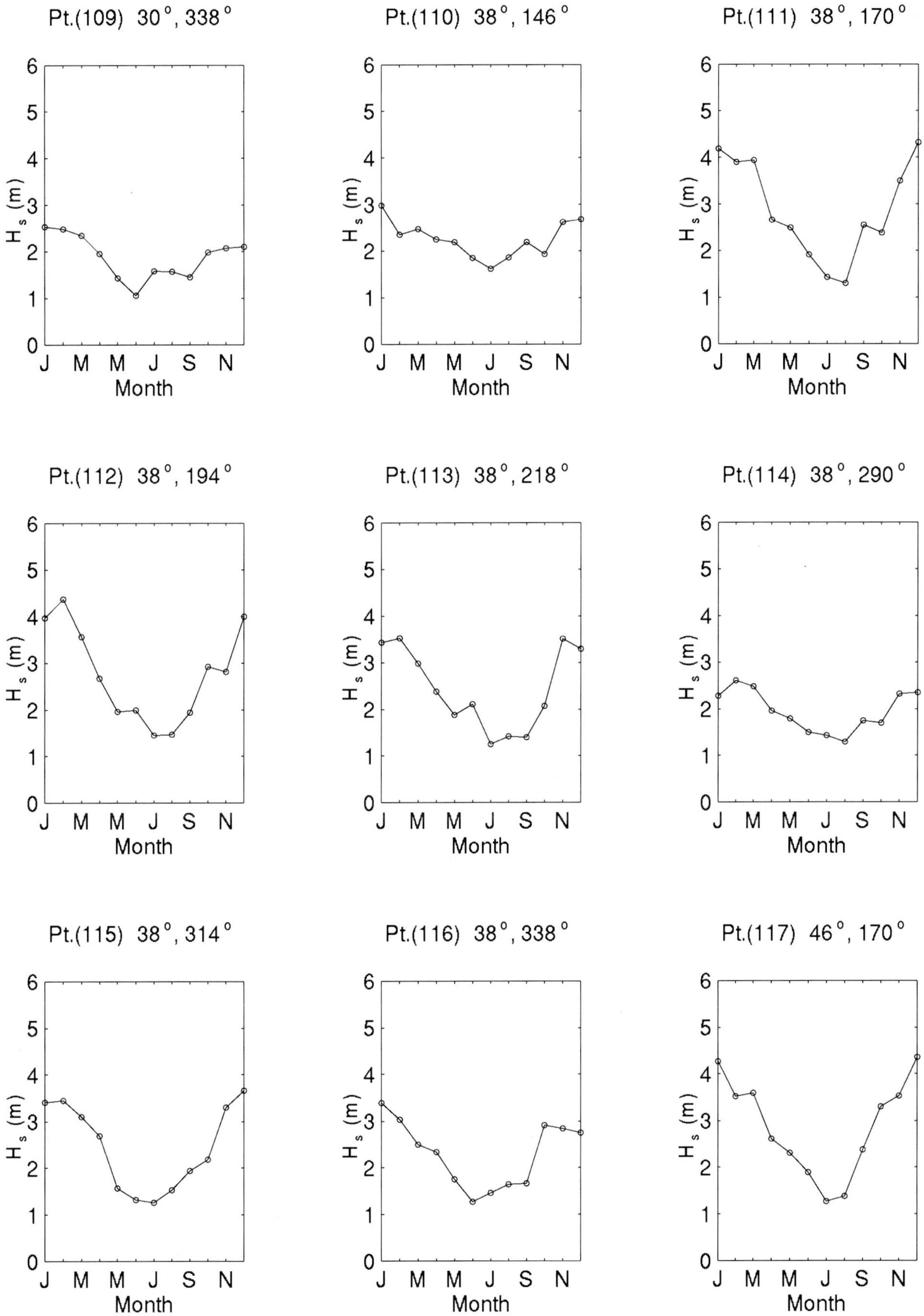

Fig. 7.30(z) Seasonal variability in values of Mean Monthly Significant Wave Height at points 109–117 of Fig. 7.29.

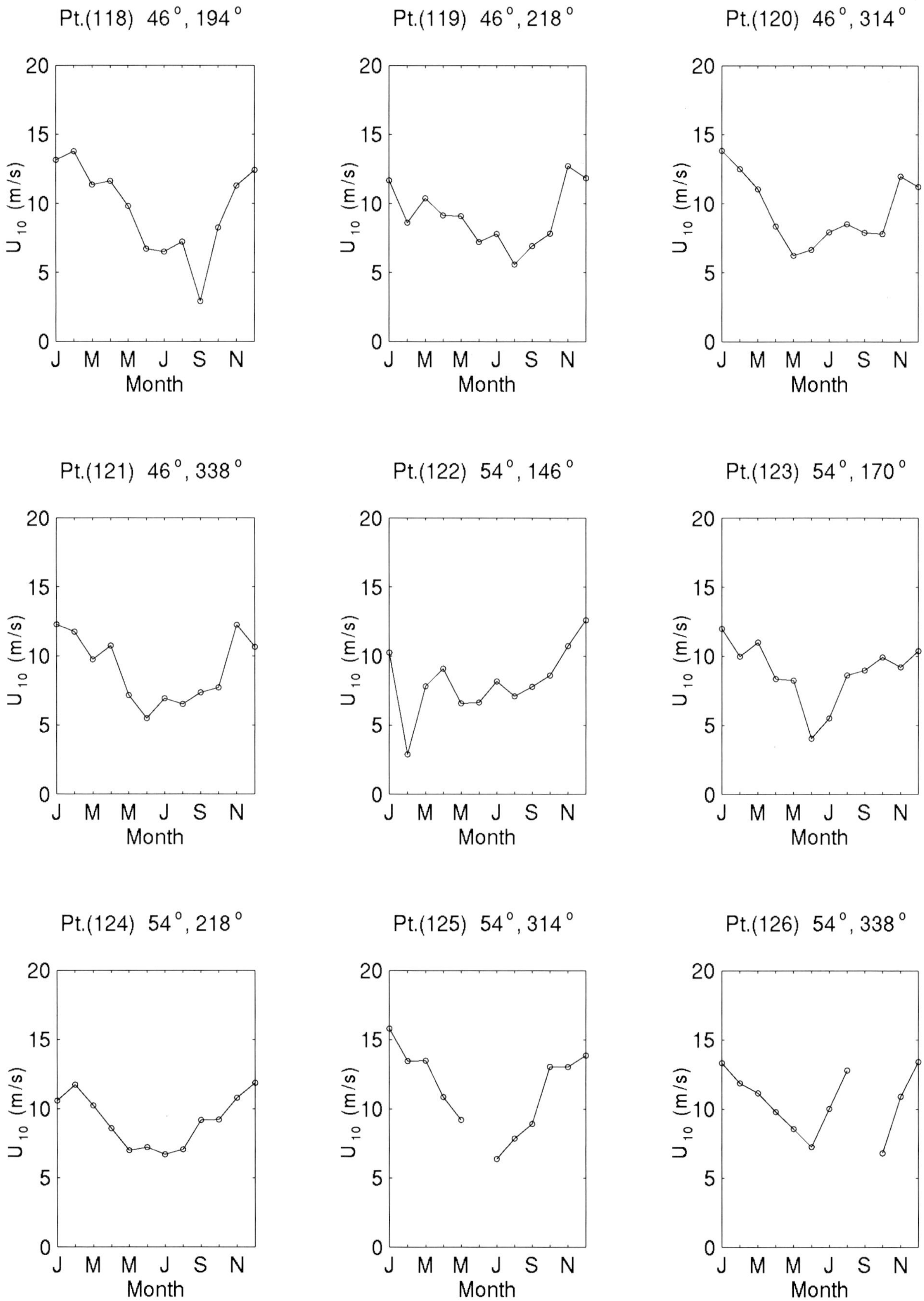

Fig. 7.30(aa) Seasonal variability in values of Mean Monthly Wind Speed at points 118–126 of Fig. 7.29.

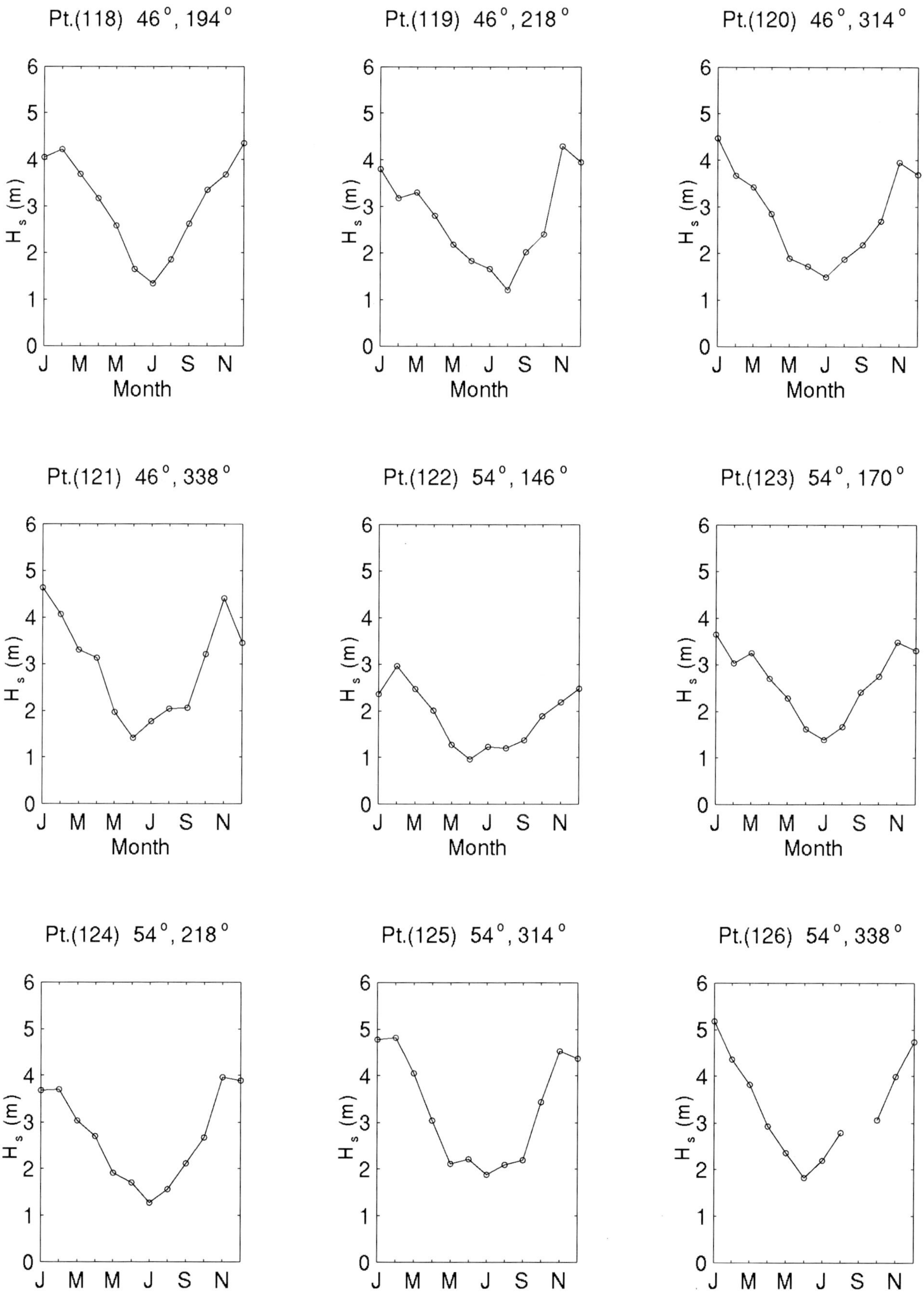

Fig. 7.30(bb) Seasonal variability in values of Mean Monthly Significant Wave Height at points 118–126 of Fig. 7.29.

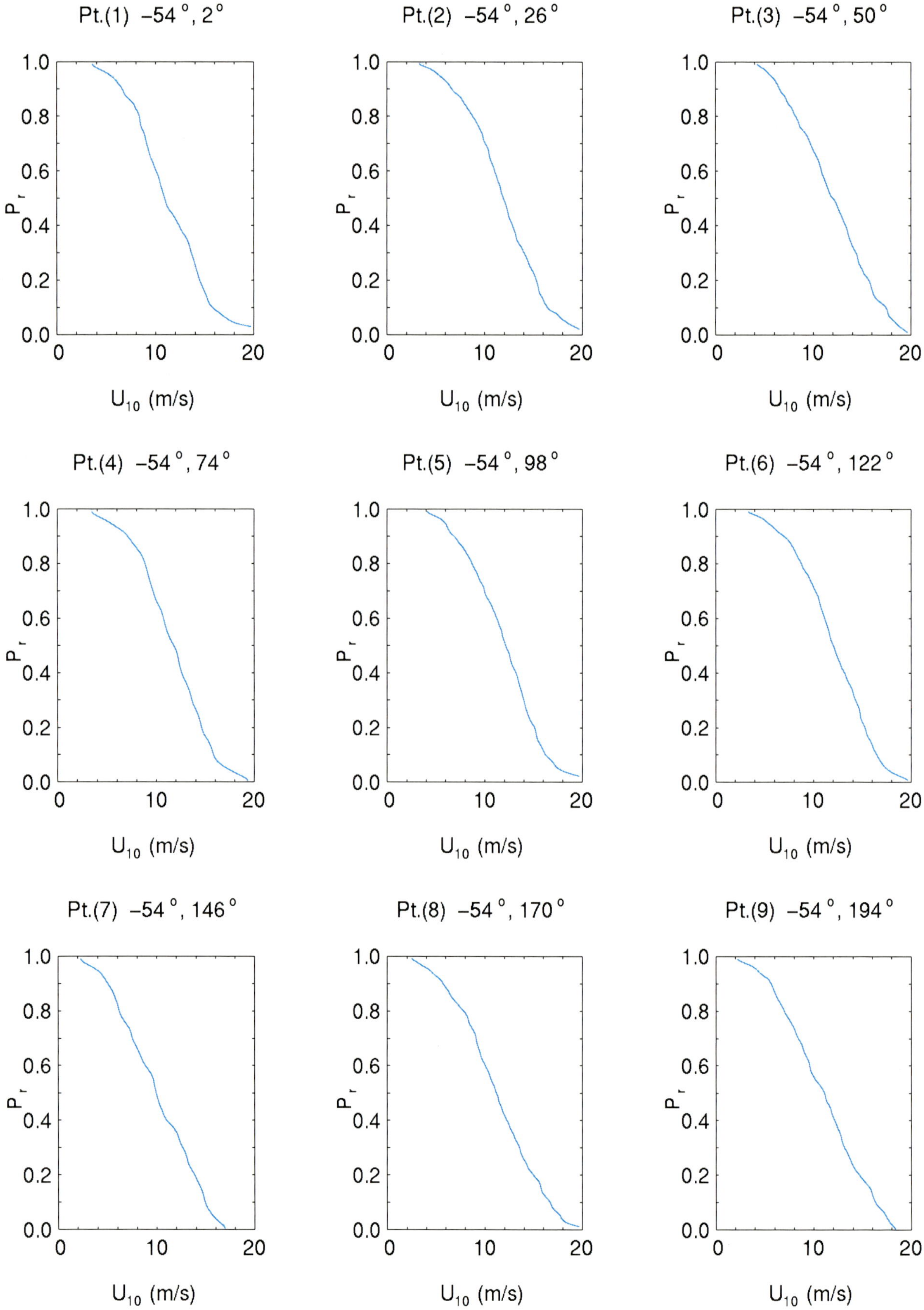

Fig. 7.31(a) Probability of Exceedence for Wind Speed at points 1–9 of Fig. 7.29.

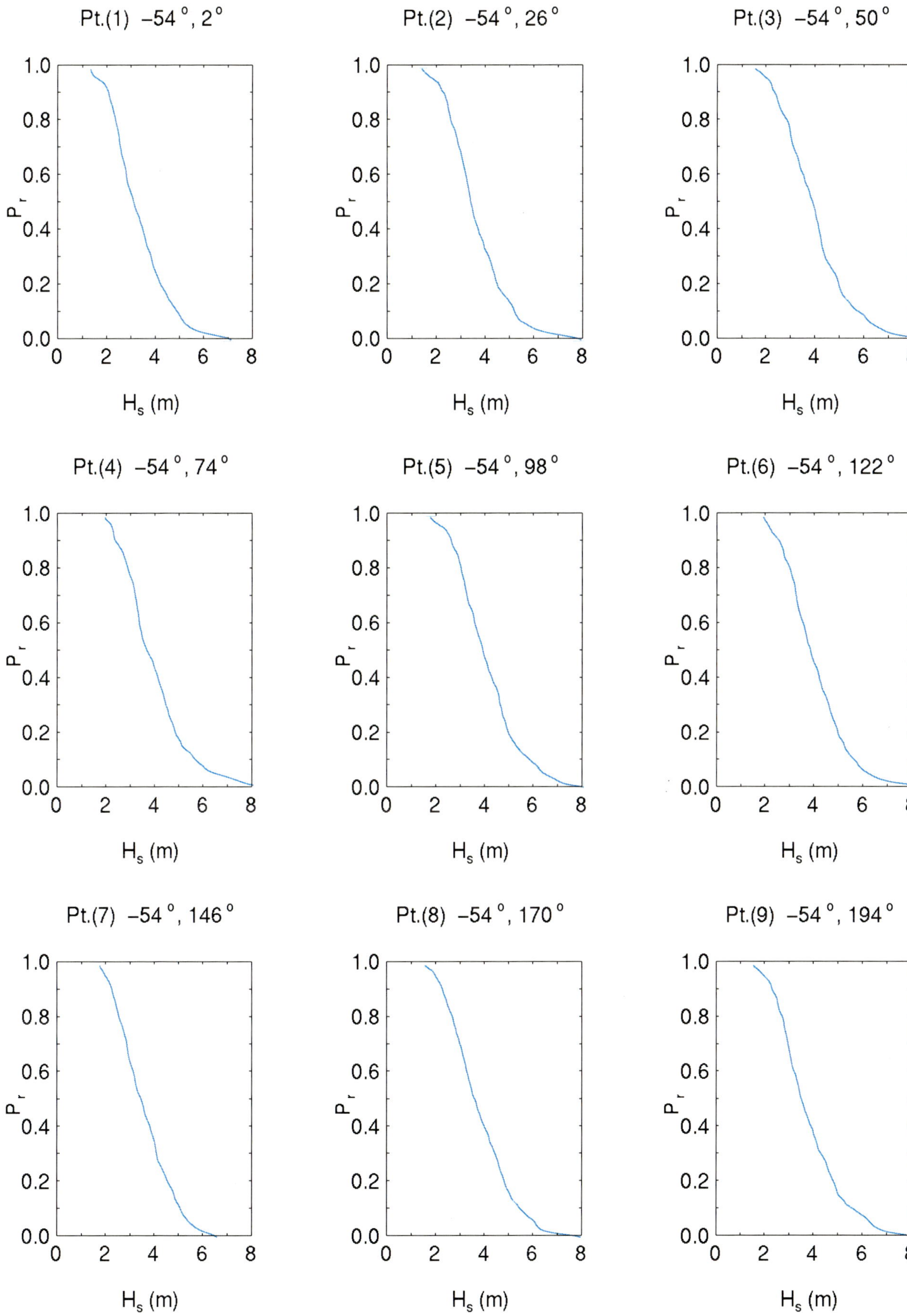

Fig. 7.31(b) Probability of Exceedence for Significant Wave Height at points 1–9 of Fig. 7.29.

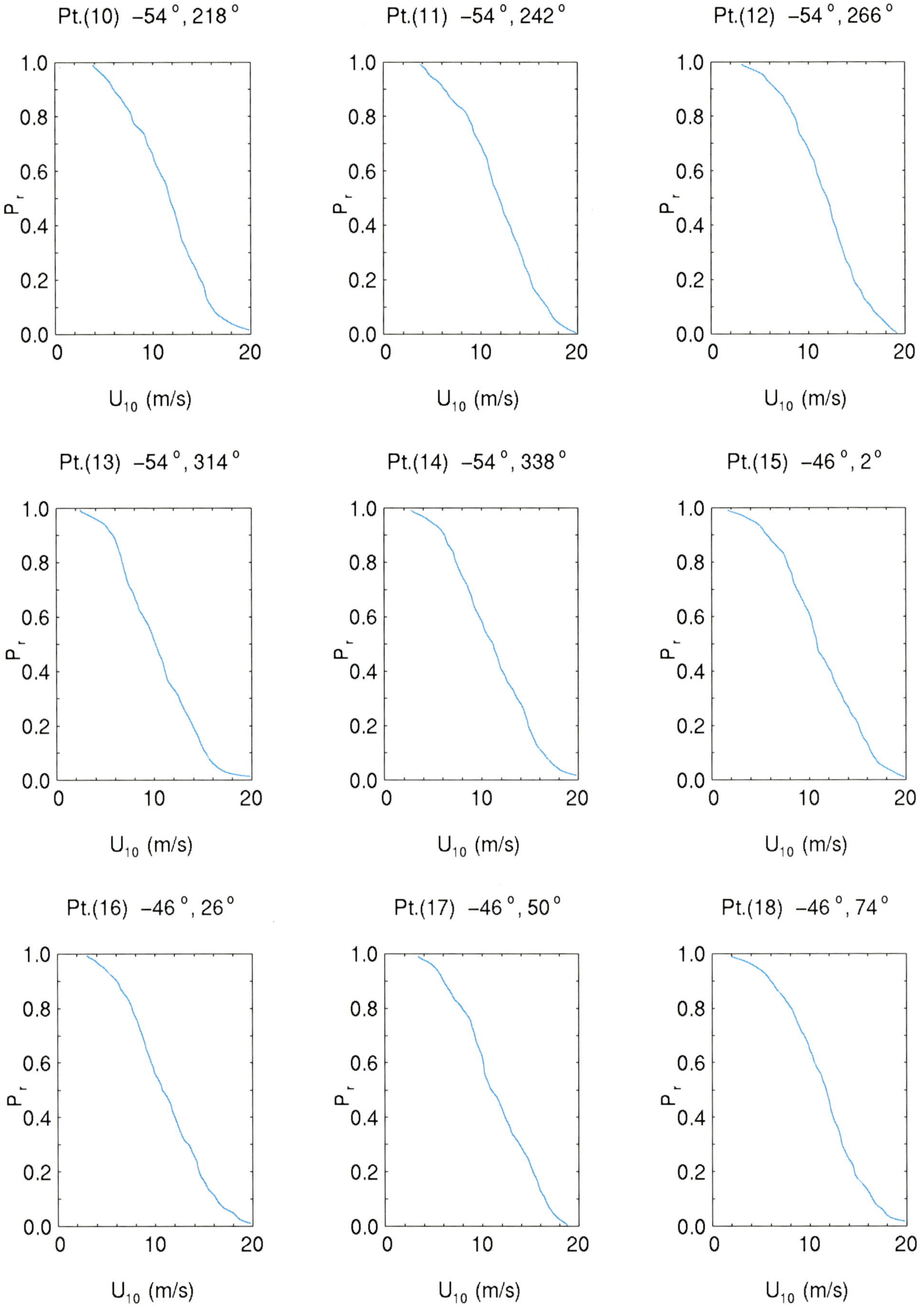

Fig. 7.31(c) Probability of Exceedence for Wind Speed at points 10–18 of Fig. 7.29.

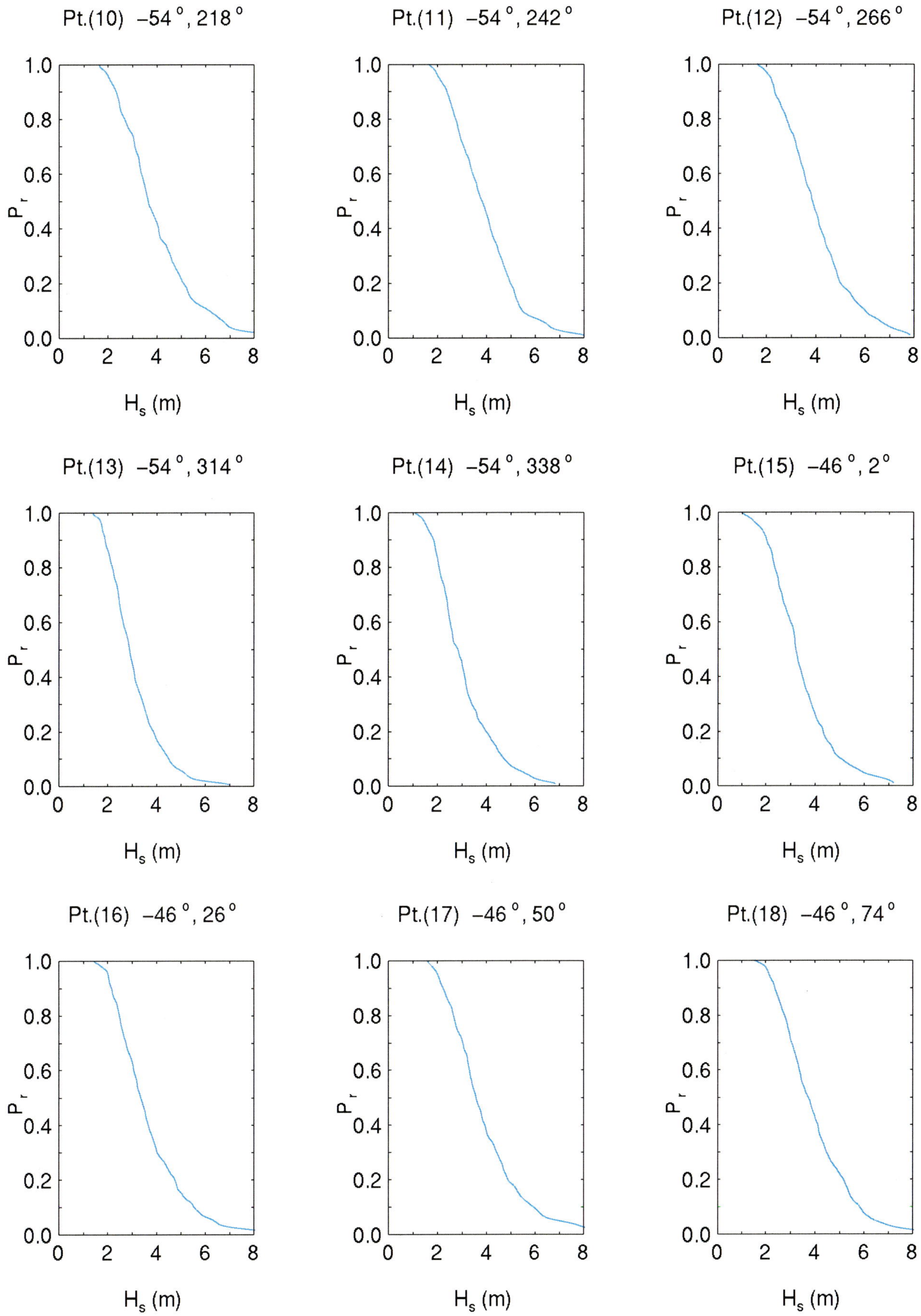

Fig. 7.31(d) Probability of Exceedence for Significant Wave Height at points 10–18 of Fig. 7.29.

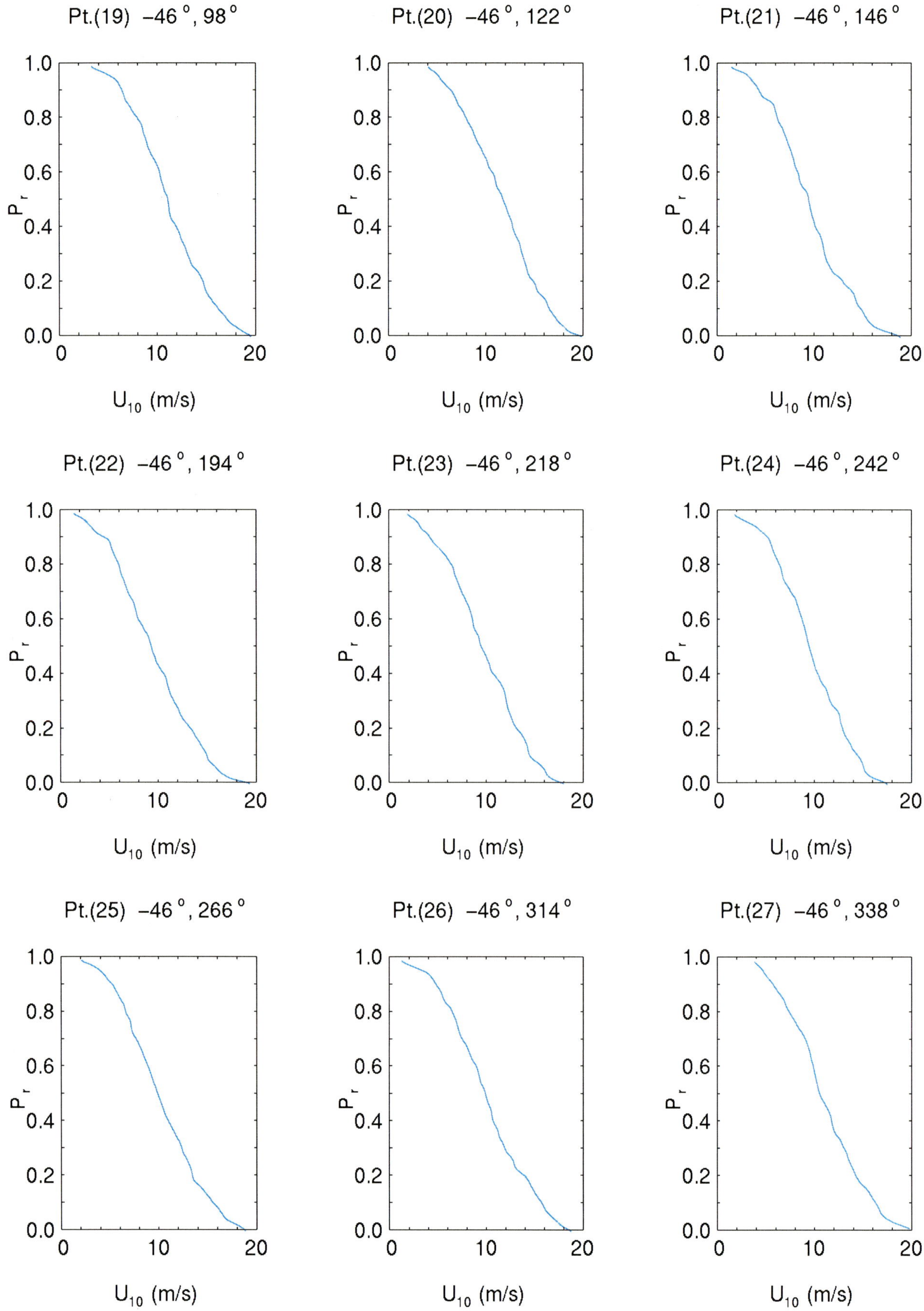

Fig. 7.31(e) Probability of Exceedence for Wind Speed at points 19–27 of Fig. 7.29.

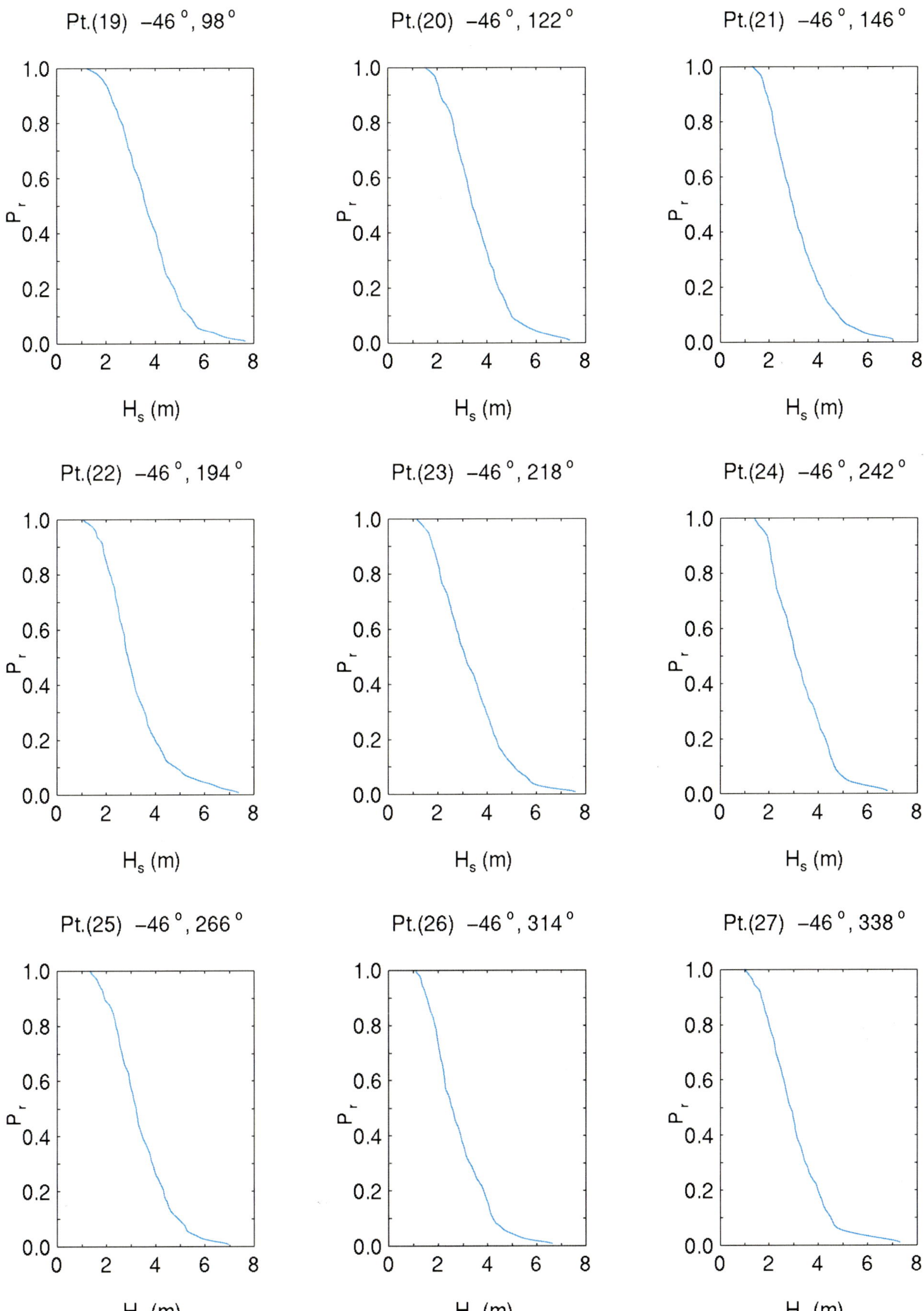

Fig. 7.31(f) Probability of Exceedence for Significant Wave Height at points 19–27 of Fig. 7.29.

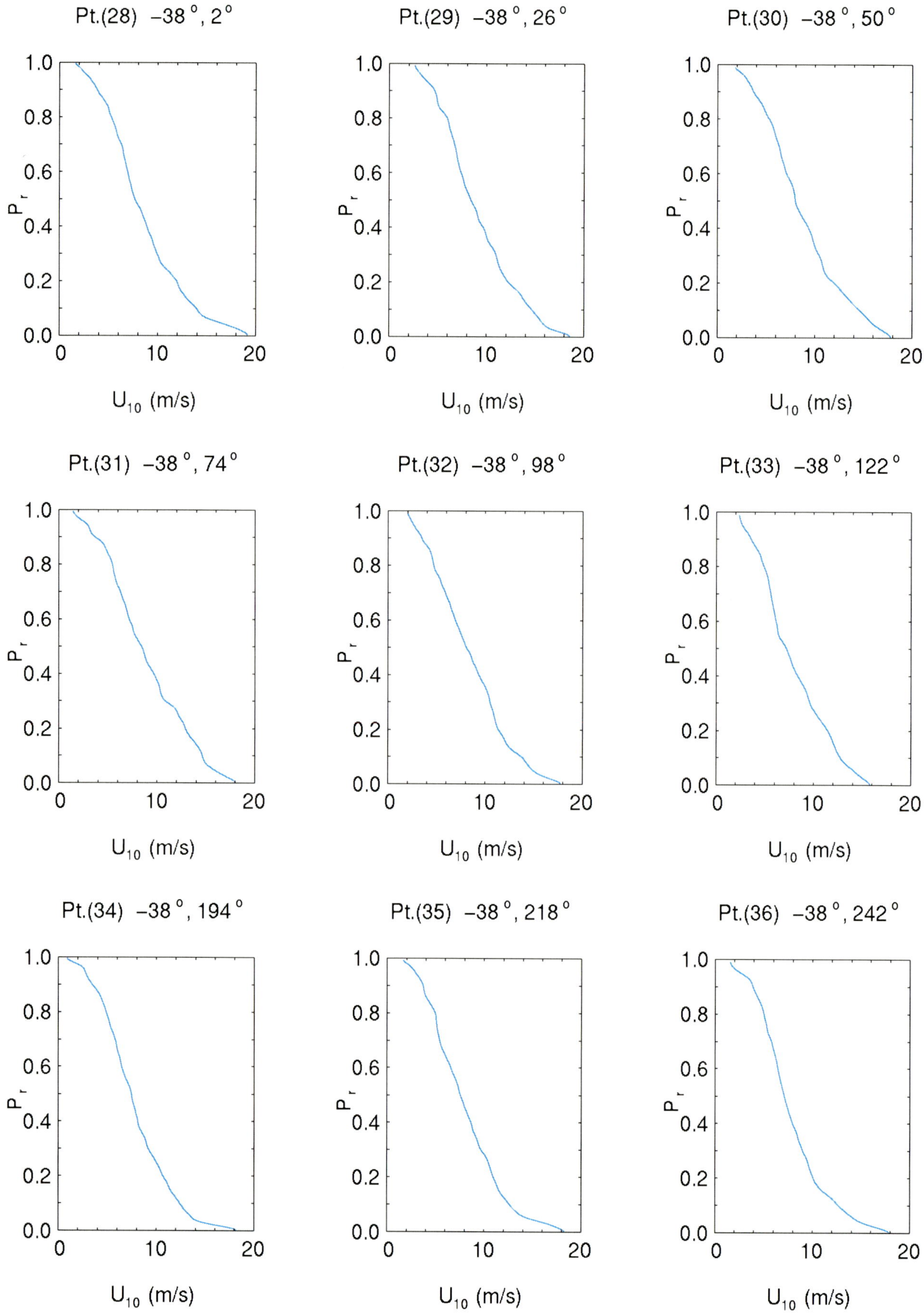

Fig. 7.31(g) Probability of Exceedence for Wind Speed at points 28–36 of Fig. 7.29.

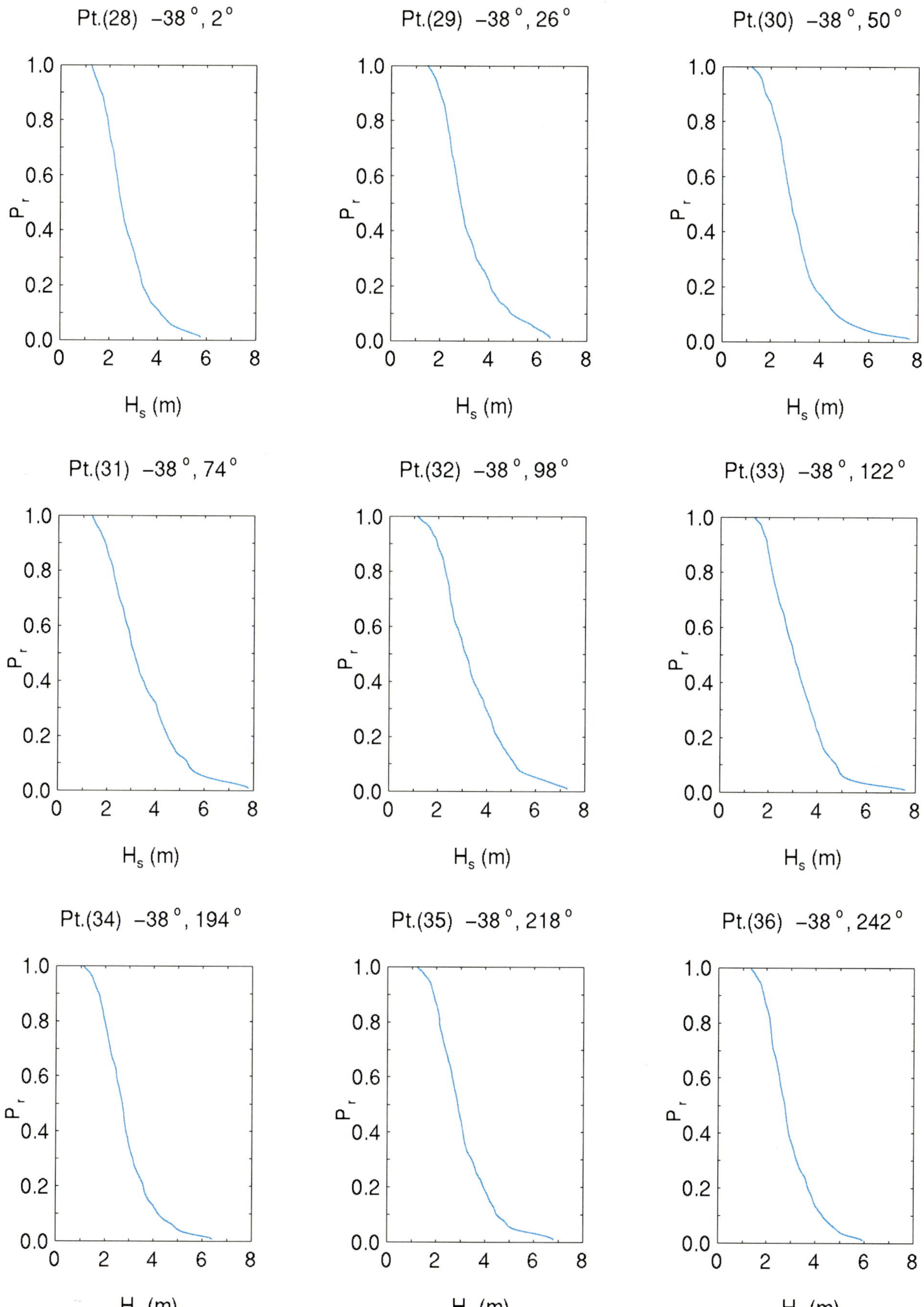

Fig. 7.31(h) Probability of Exceedence for Significant Wave Height at points 28–36 of Fig. 7.29.

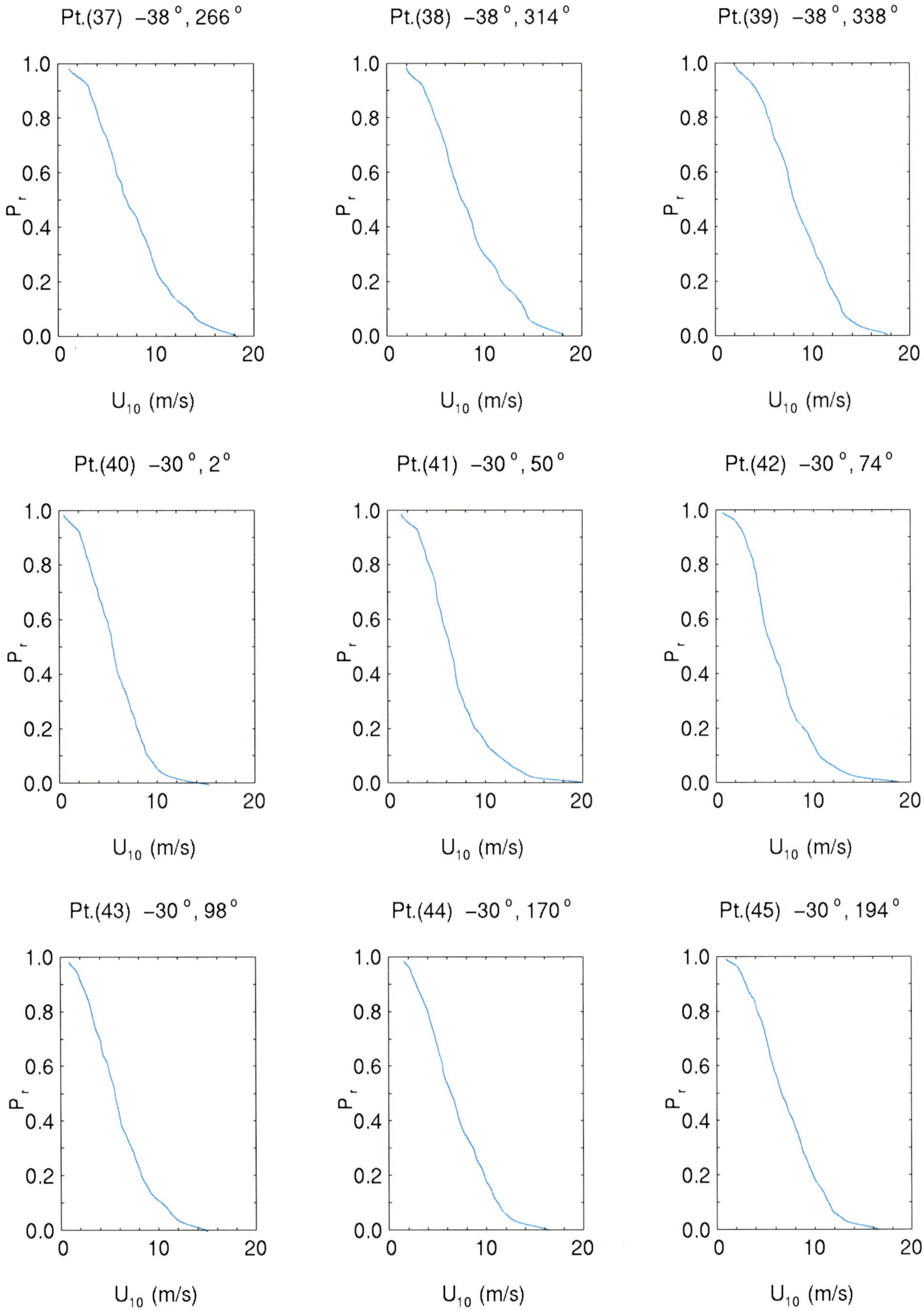

Fig. 7.31(i) Probability of Exceedence for Wind Speed at points 37–45 of Fig. 7.29.

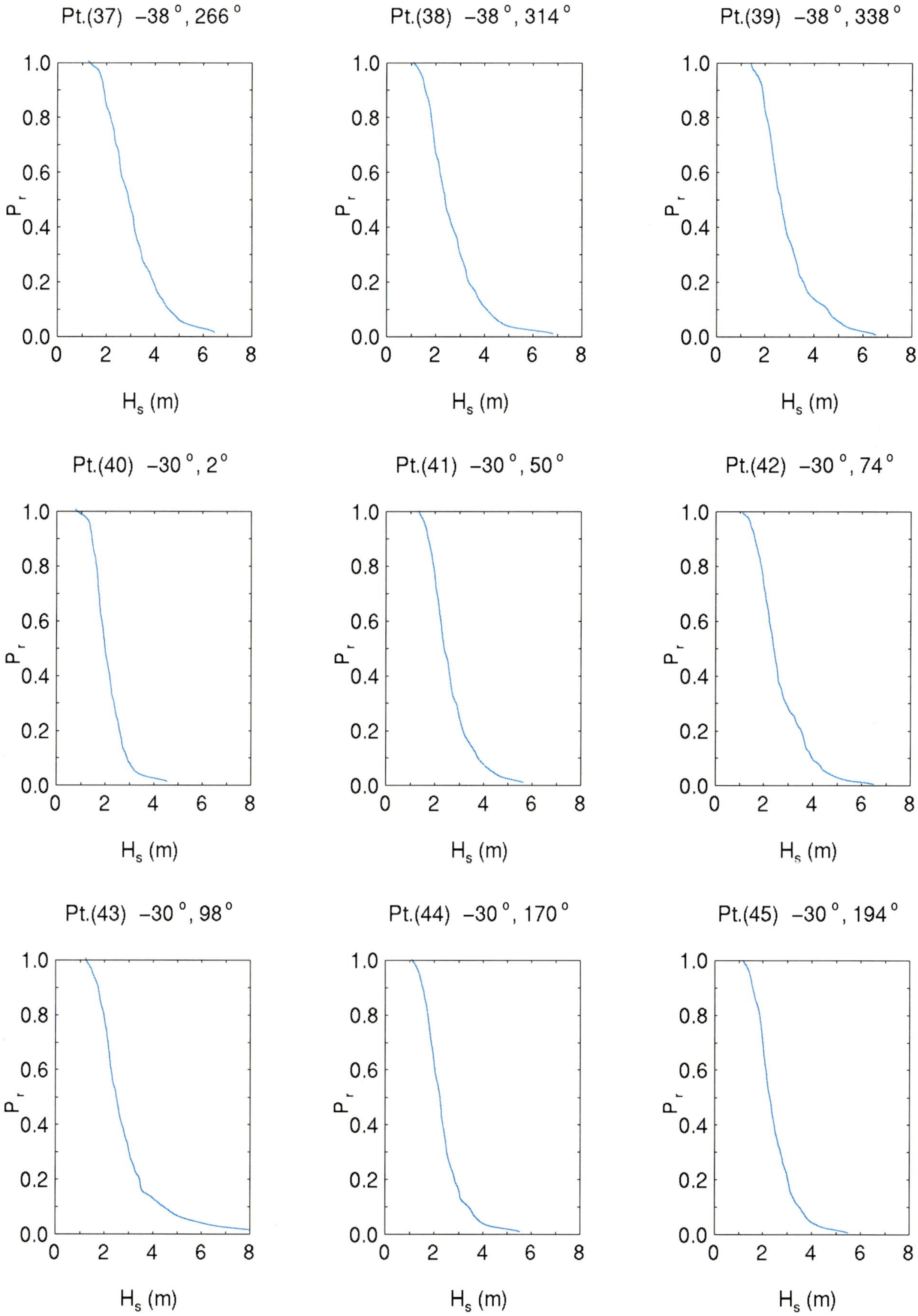

Fig. 7.31(j) Probability of Exceedence for Significant Wave Height at points 37–45 of Fig. 7.29.

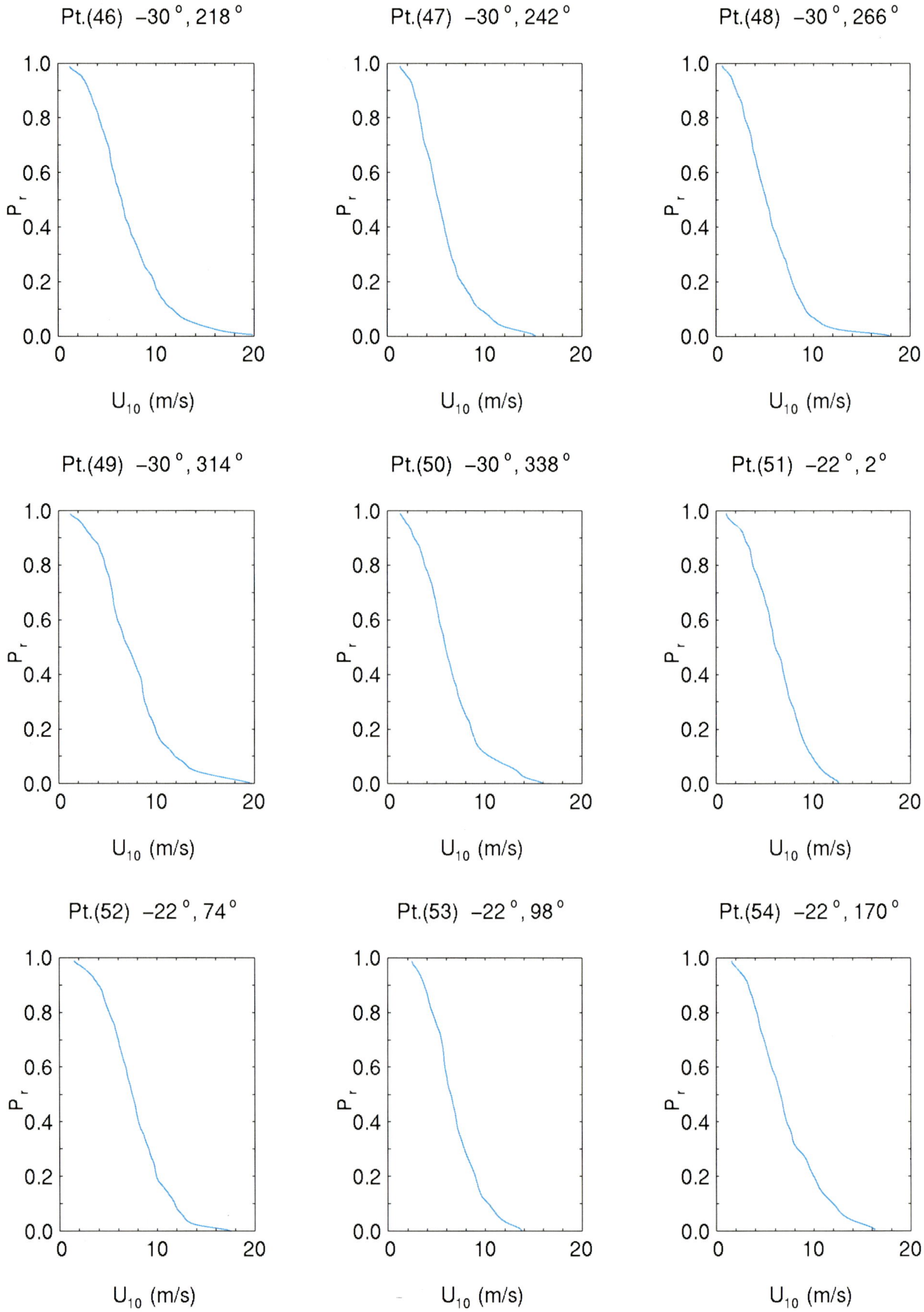

Fig. 7.31(k) Probability of Exceedence for Wind Speed at points 46–54 of Fig. 7.29.

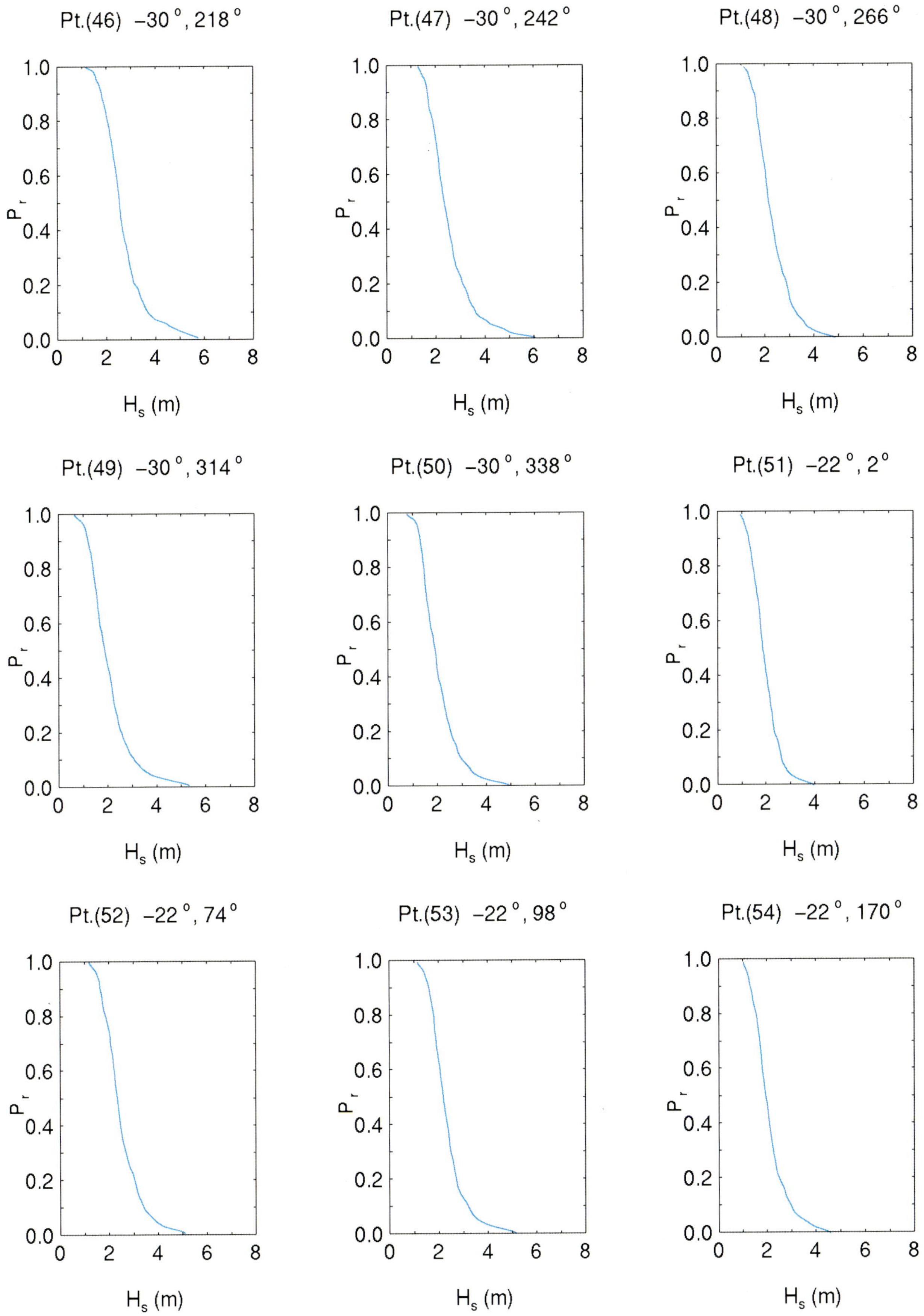

Fig. 7.31(l) Probability of Exceedence for Significant Wave Height at points 46–54 of Fig. 7.29.

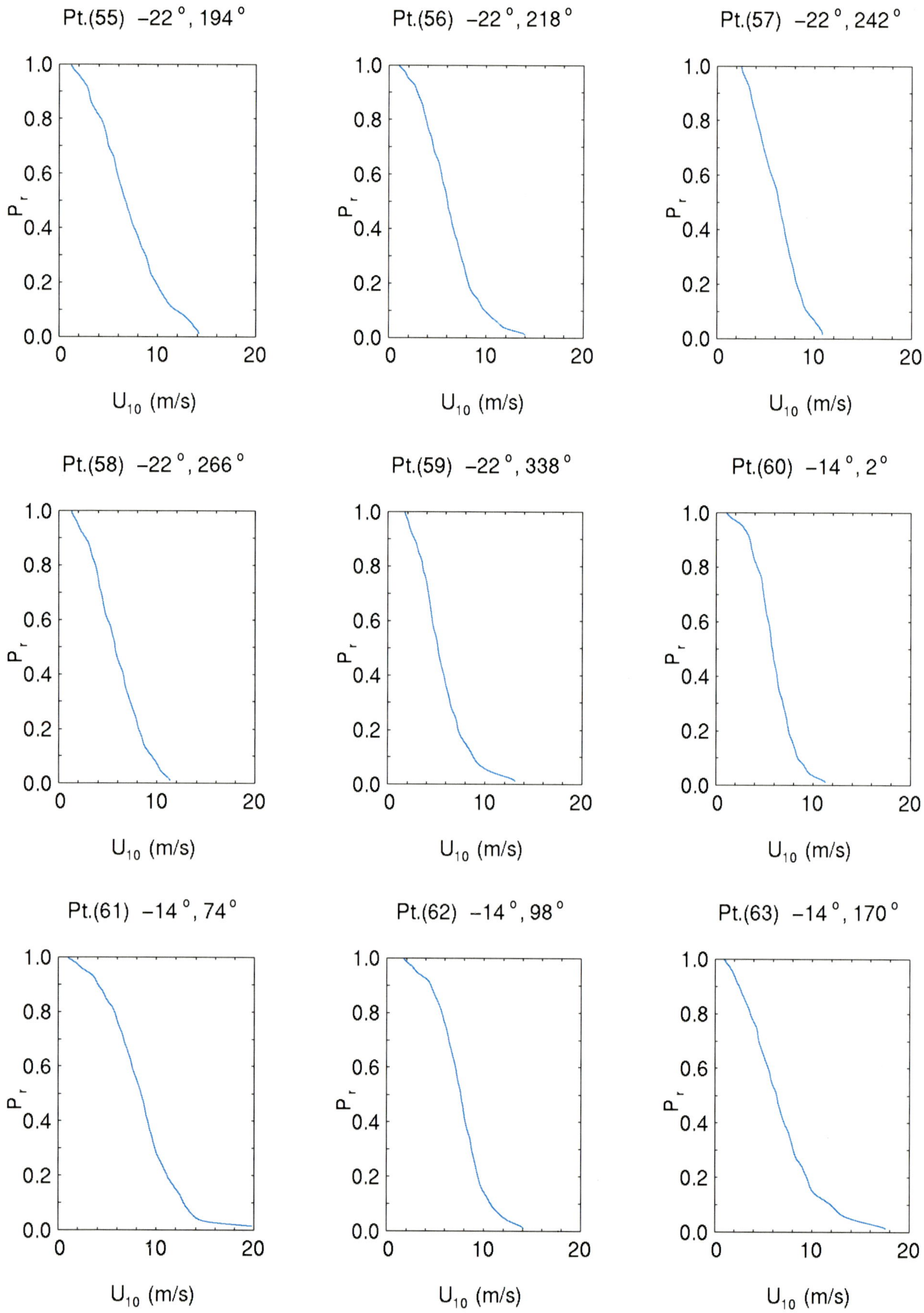

Fig. 7.31(m) Probability of Exceedence for Wind Speed at points 55–63 of Fig. 7.29.

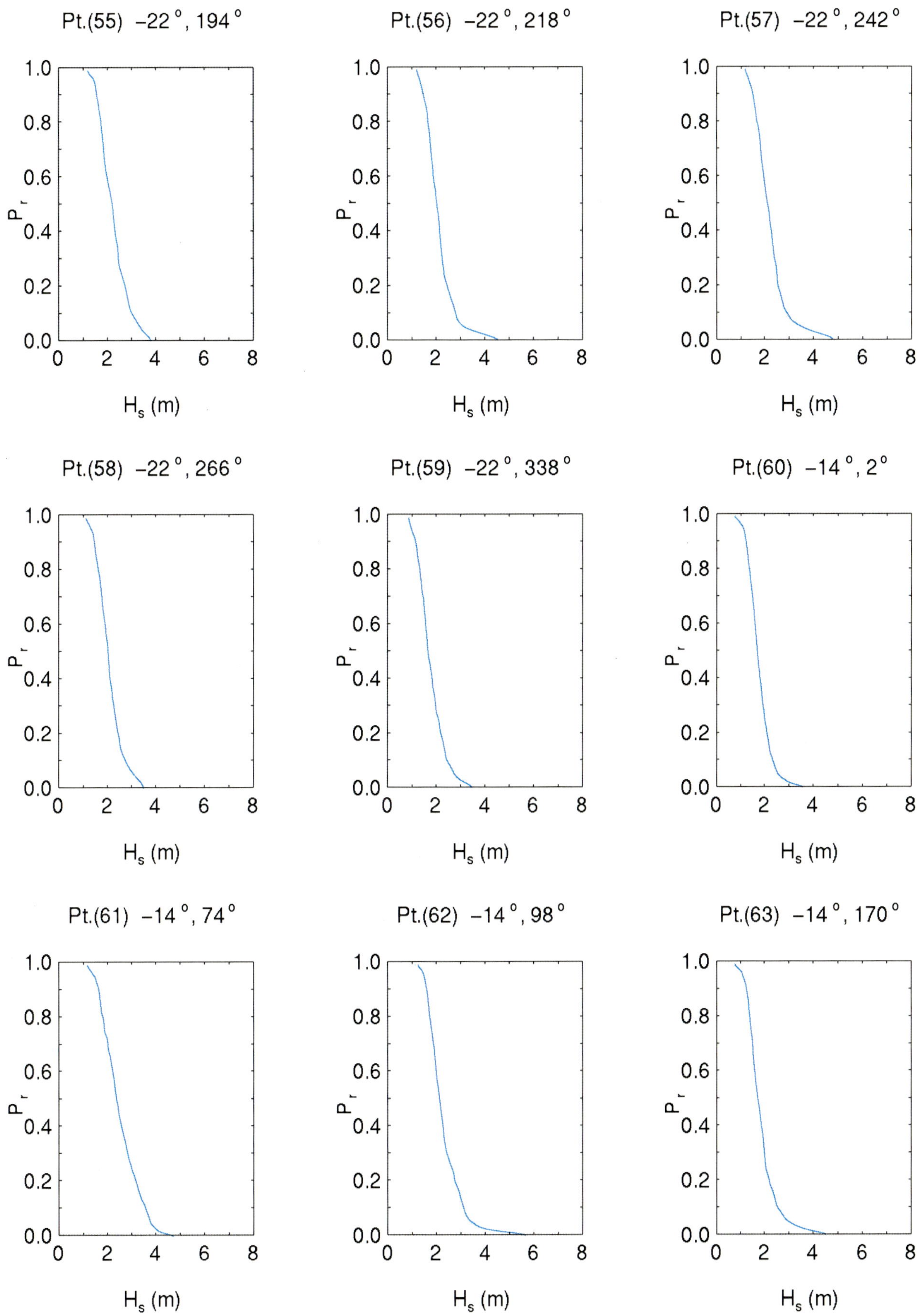

Fig. 7.31(n) Probability of Exceedence for Significant Wave Height at points 55–63 of Fig. 7.29.

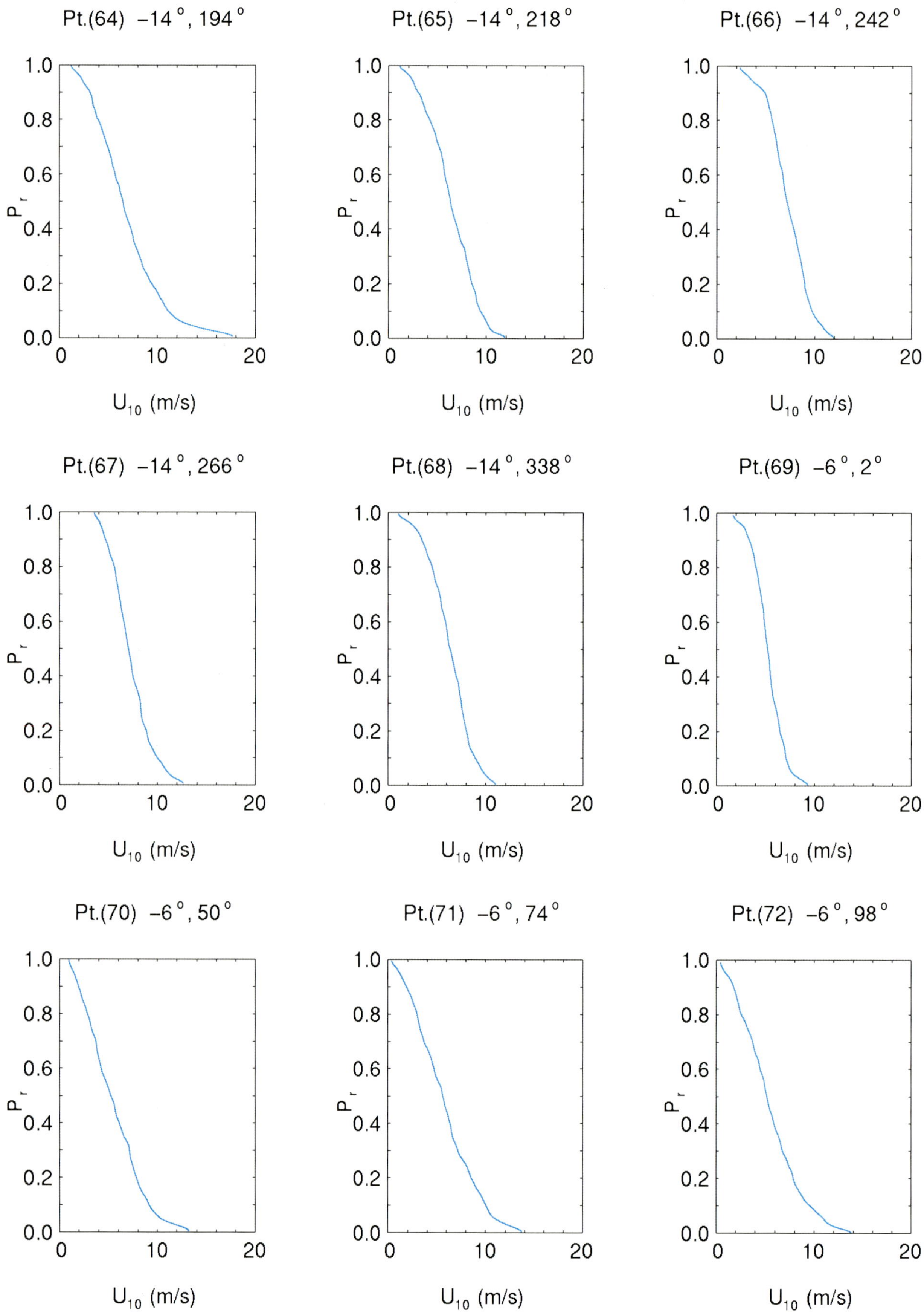

Fig. 7.31(o) Probability of Exceedence for Wind Speed at points 64–72 of Fig. 7.29.

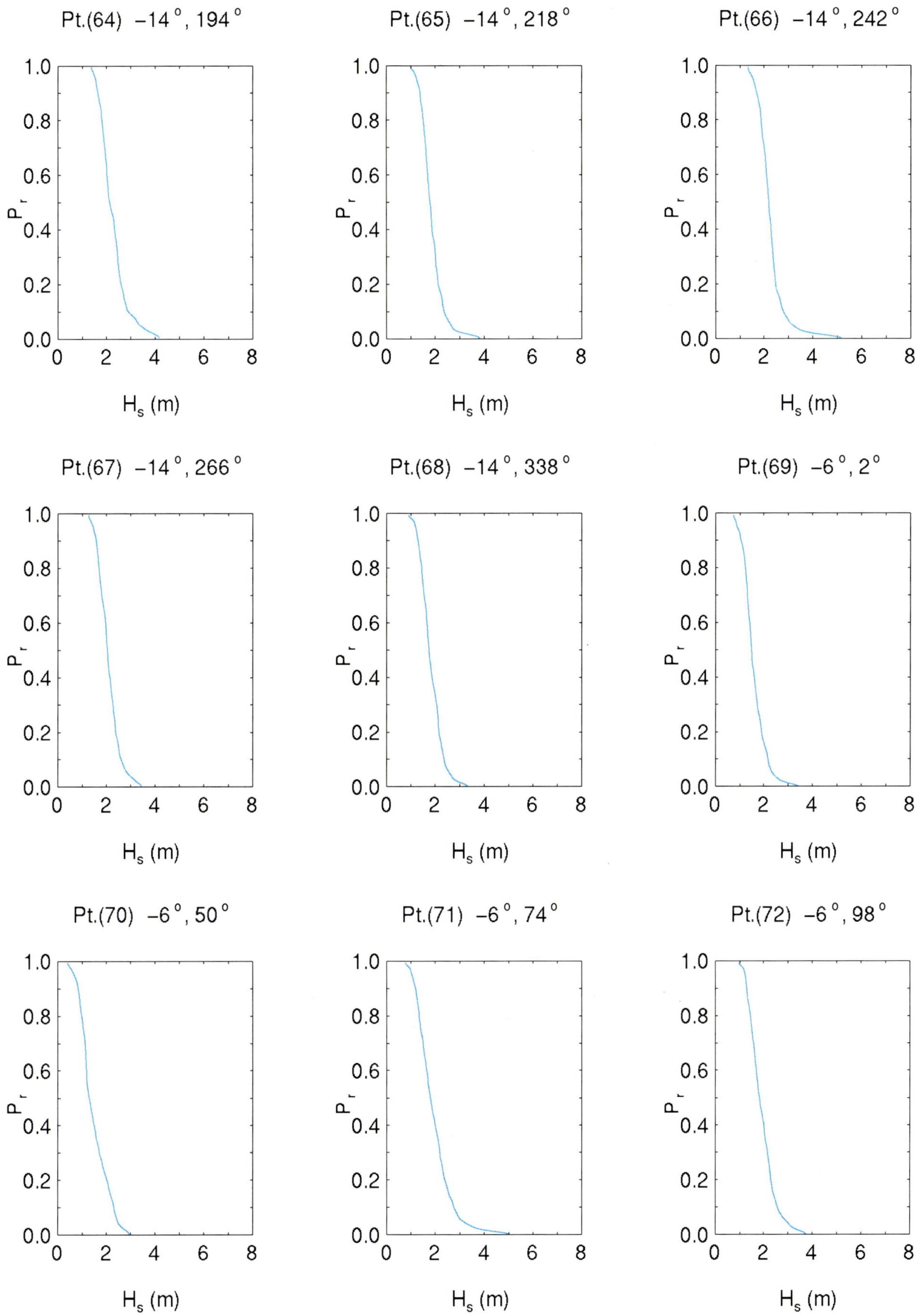

Fig. 7.31(p) Probability of Exceedence for Significant Wave Height at points 64–72 of Fig. 7.29.

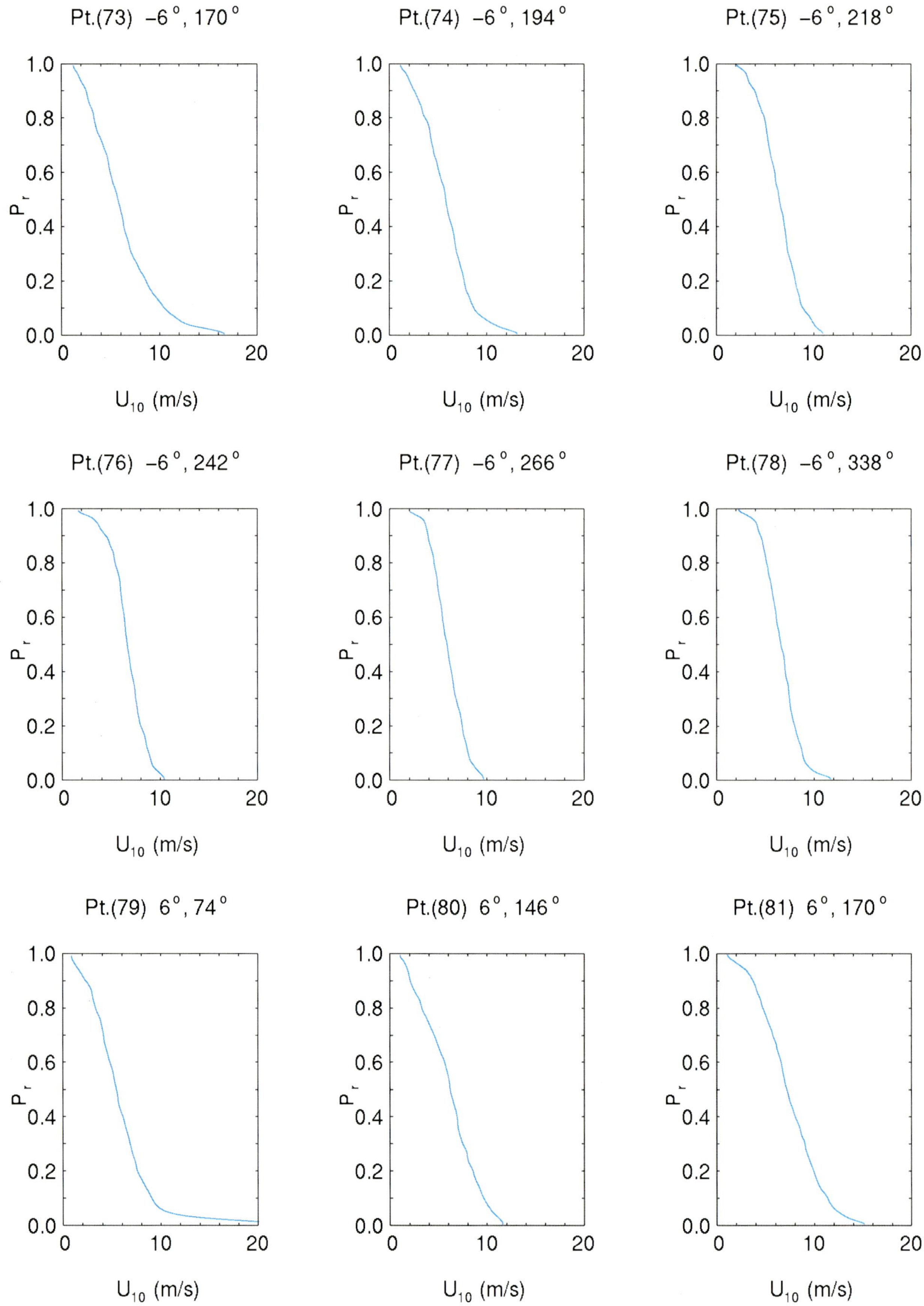

Fig. 7.31(q) Probability of Exceedence for Wind Speed at points 73–81 of Fig. 7.29.

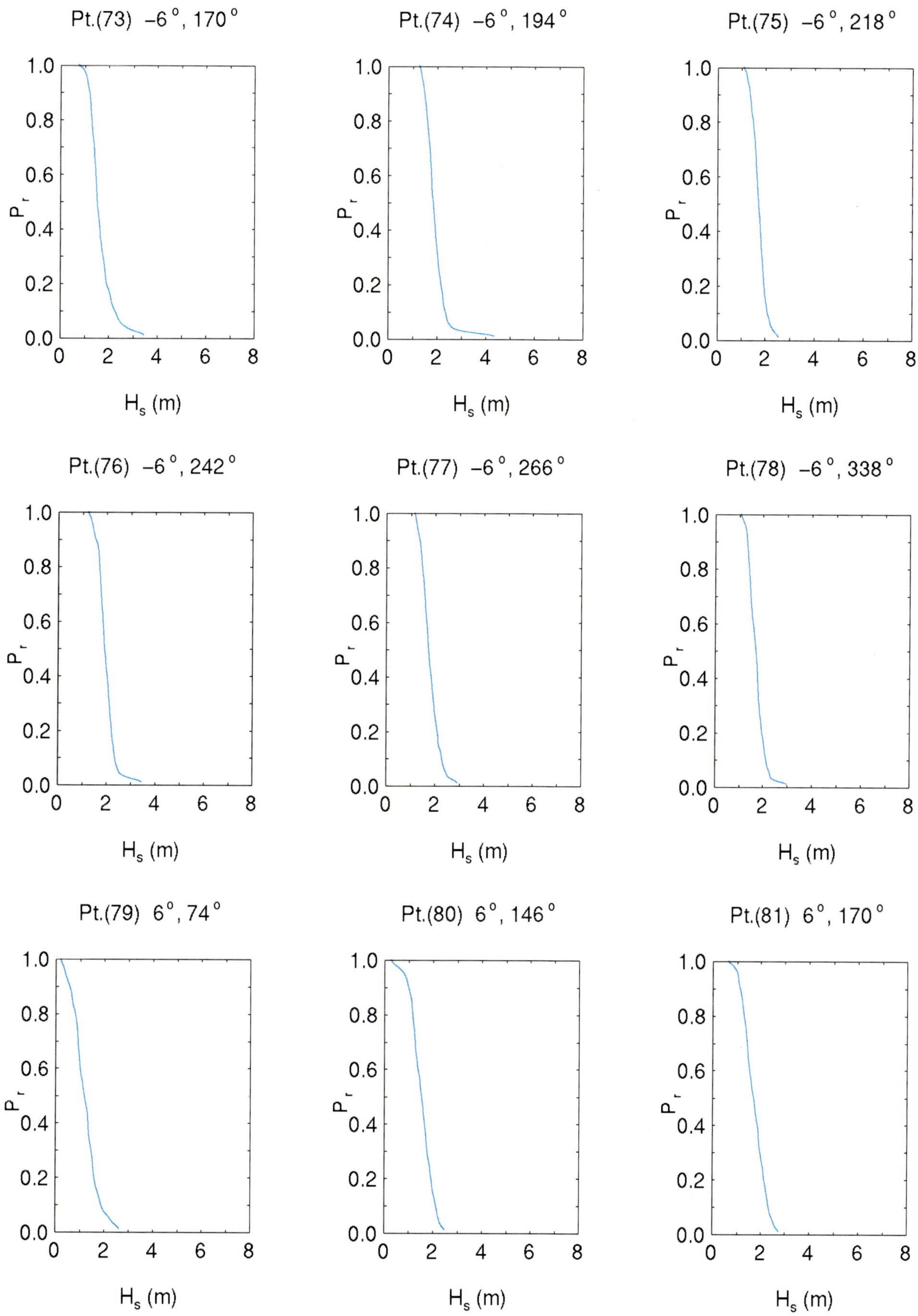

Fig. 7.31(r) Probability of Exceedence for Significant Wave Height at points 73–81 of Fig. 7.29.

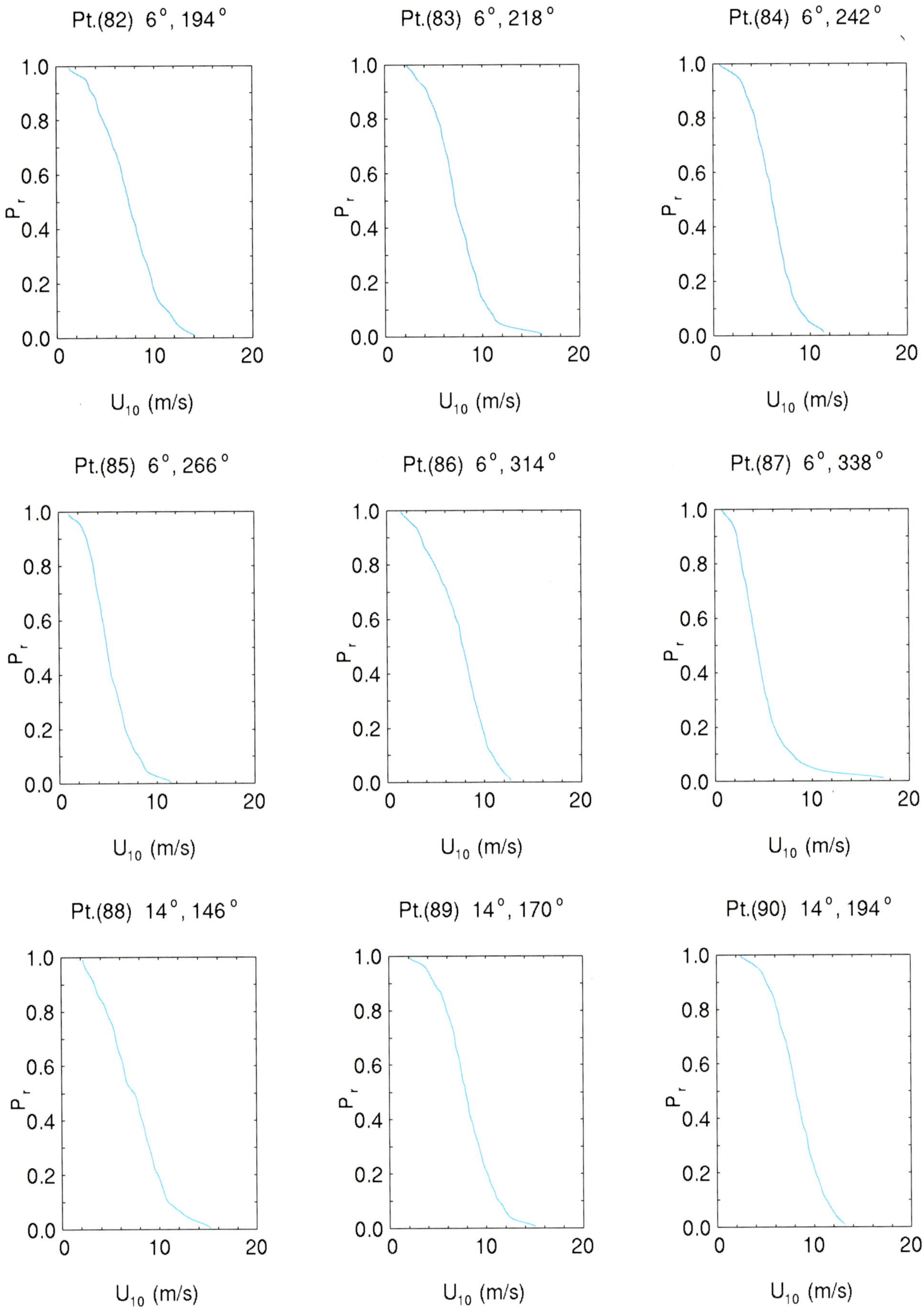

Fig. 7.31(s) Probability of Exceedence for Wind Speed at points 82–90 of Fig. 7.29.

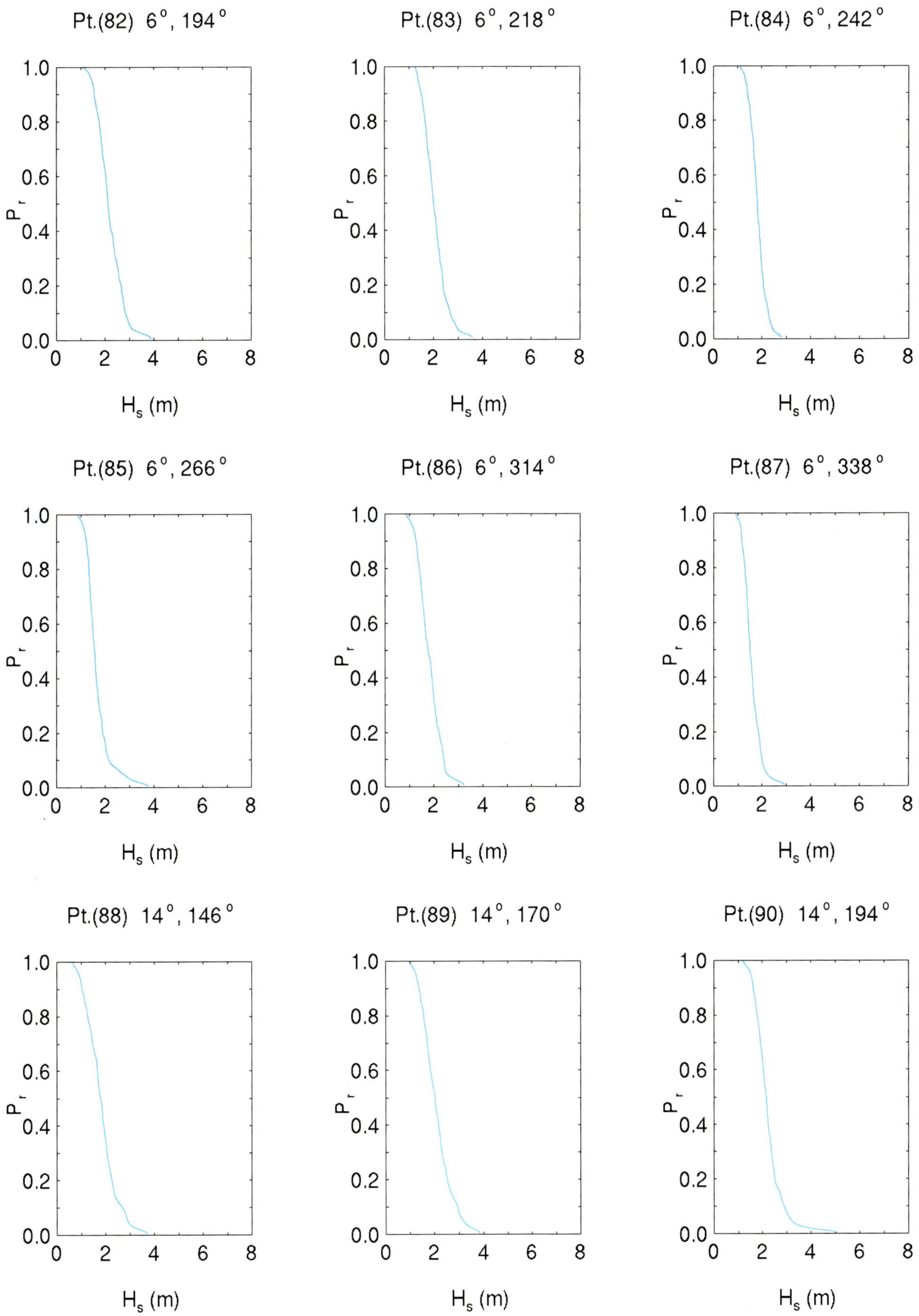

Fig. 7.31(t) Probability of Exceedence for Significant Wave Height at points 82–90 of Fig. 7.29.

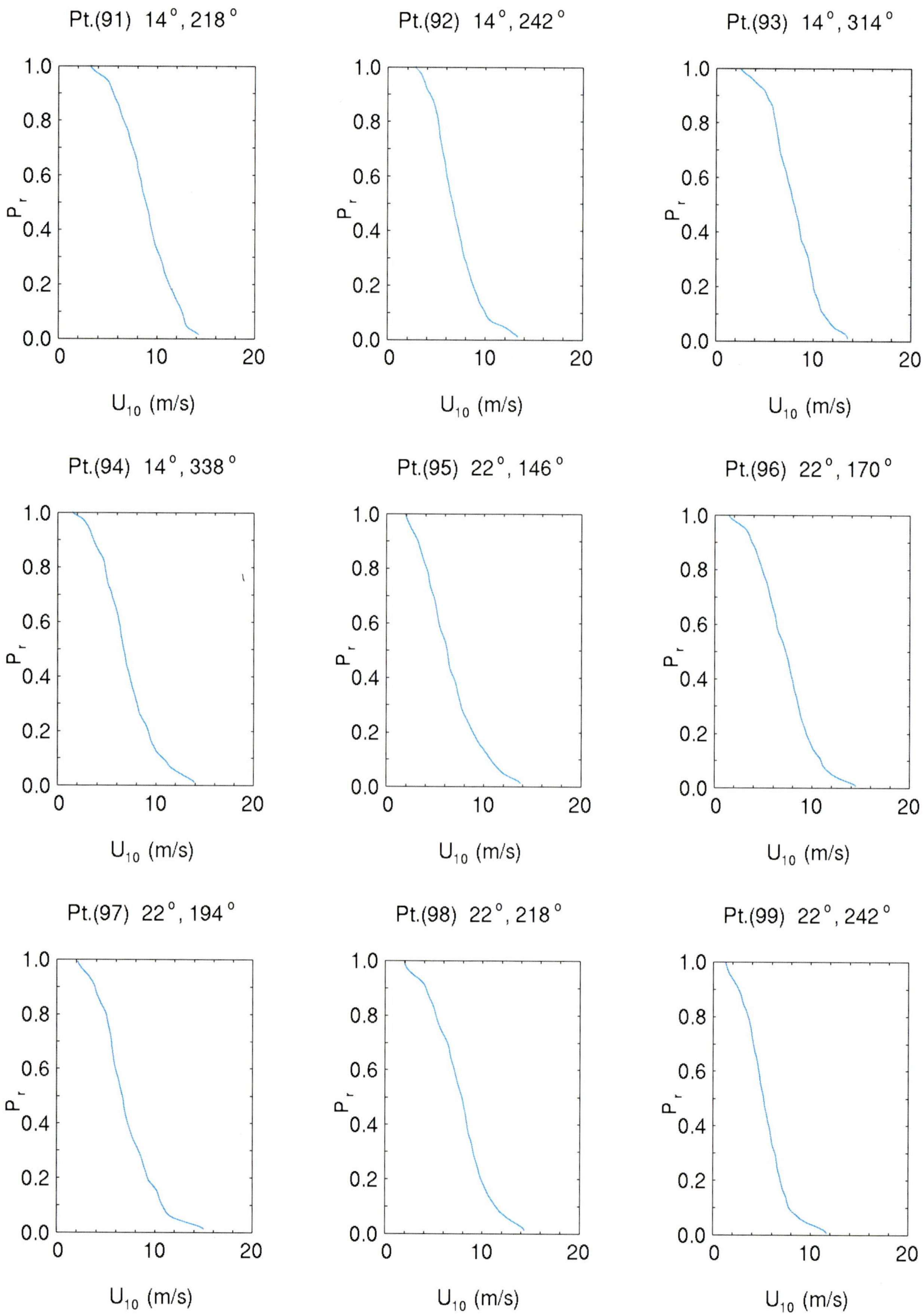

Fig. 7.31(u) Probability of Exceedence for Wind Speed at points 91–99 of Fig. 7.29.

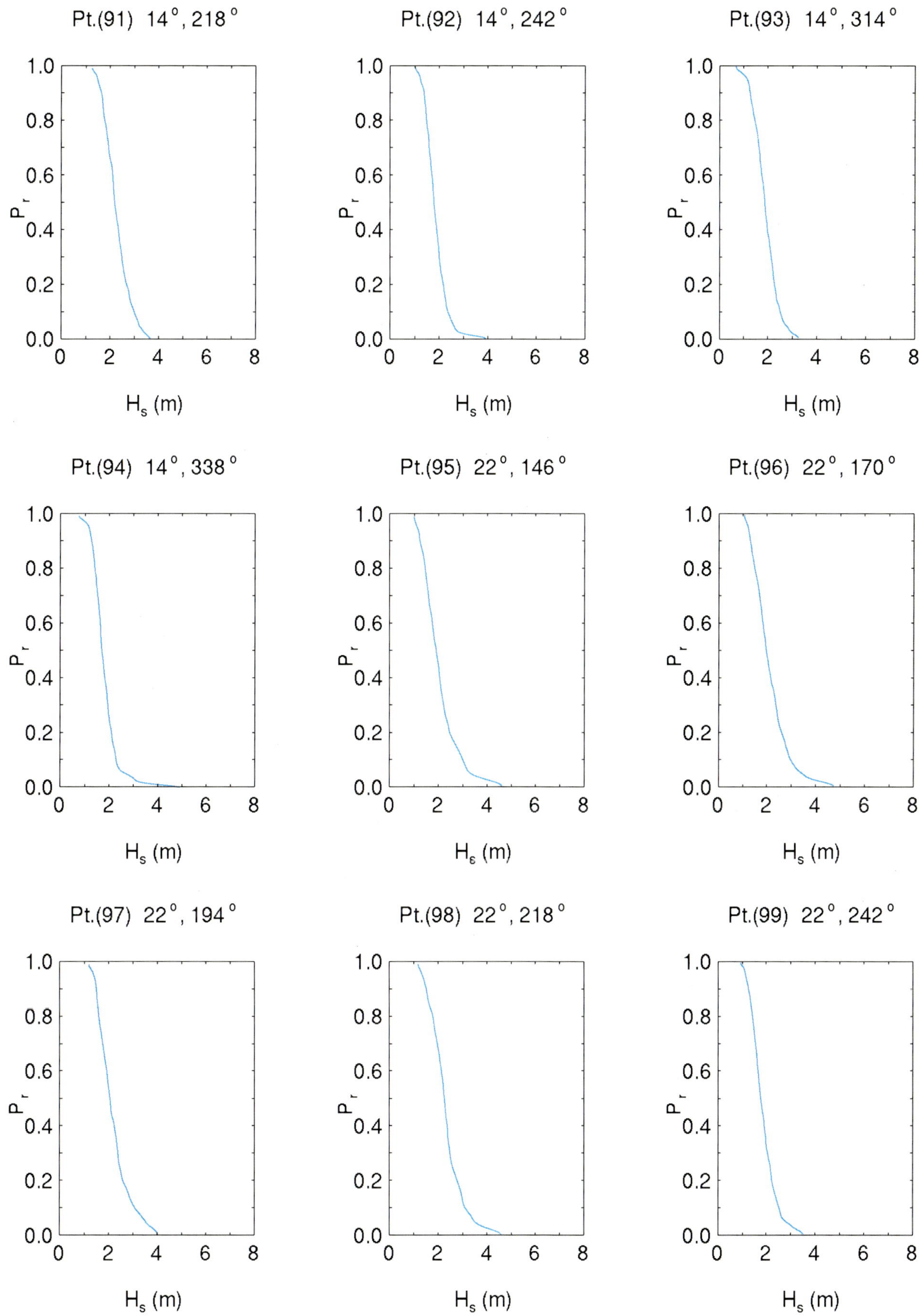

Fig. 7.31(v) Probability of Exceedence for Significant Wave Height at points 91–99 of Fig. 7.29.

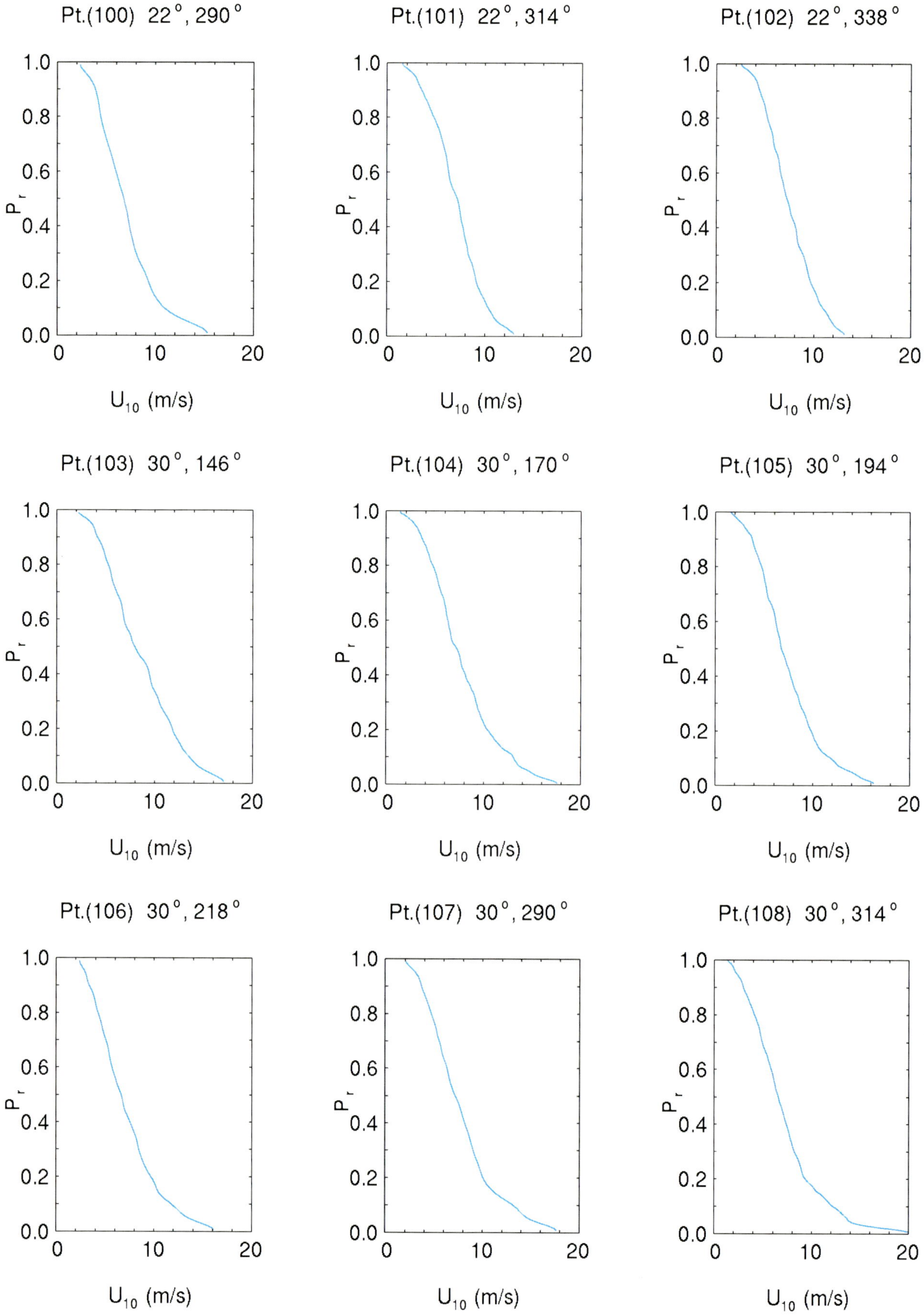

Fig. 7.31(w) Probability of Exceedence for Wind Speed at points 100–108 of Fig. 7.29.

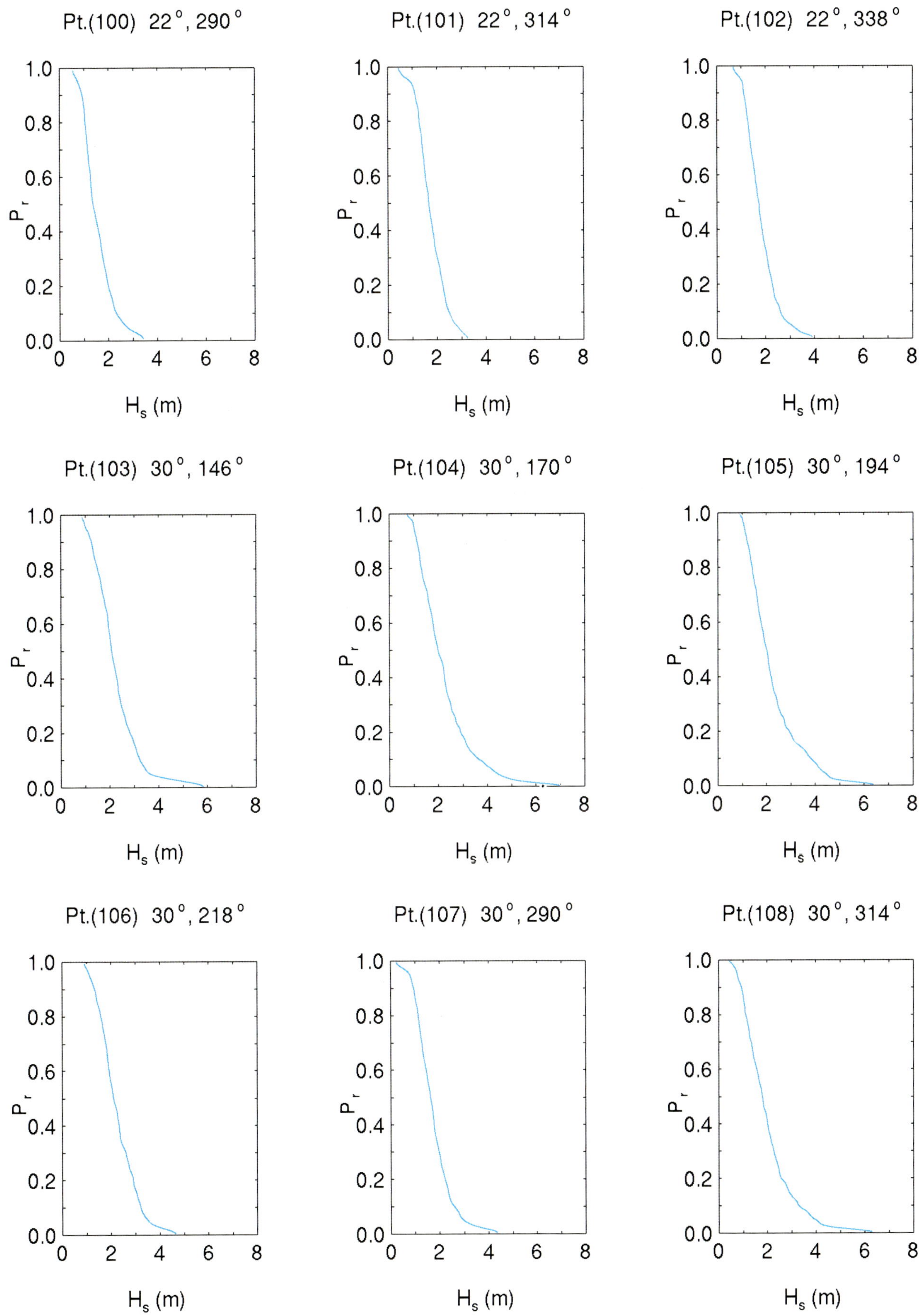

Fig. 7.31(x) Probability of Exceedence for Significant Wave Height at points 100–108 of Fig. 7.29.

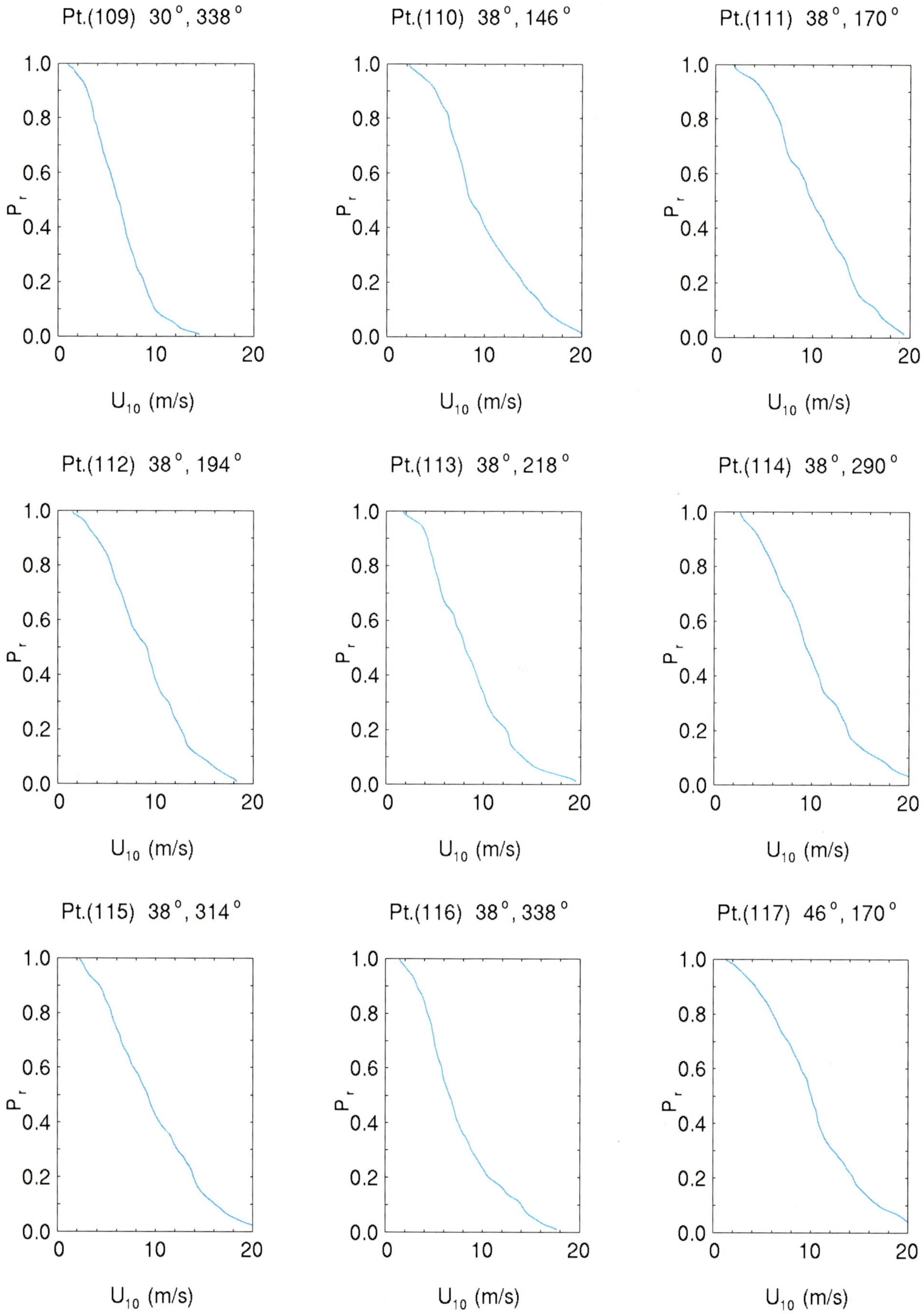

Fig. 7.31(y) Probability of Exceedence for Wind Speed at points 109–117 of Fig. 7.29.

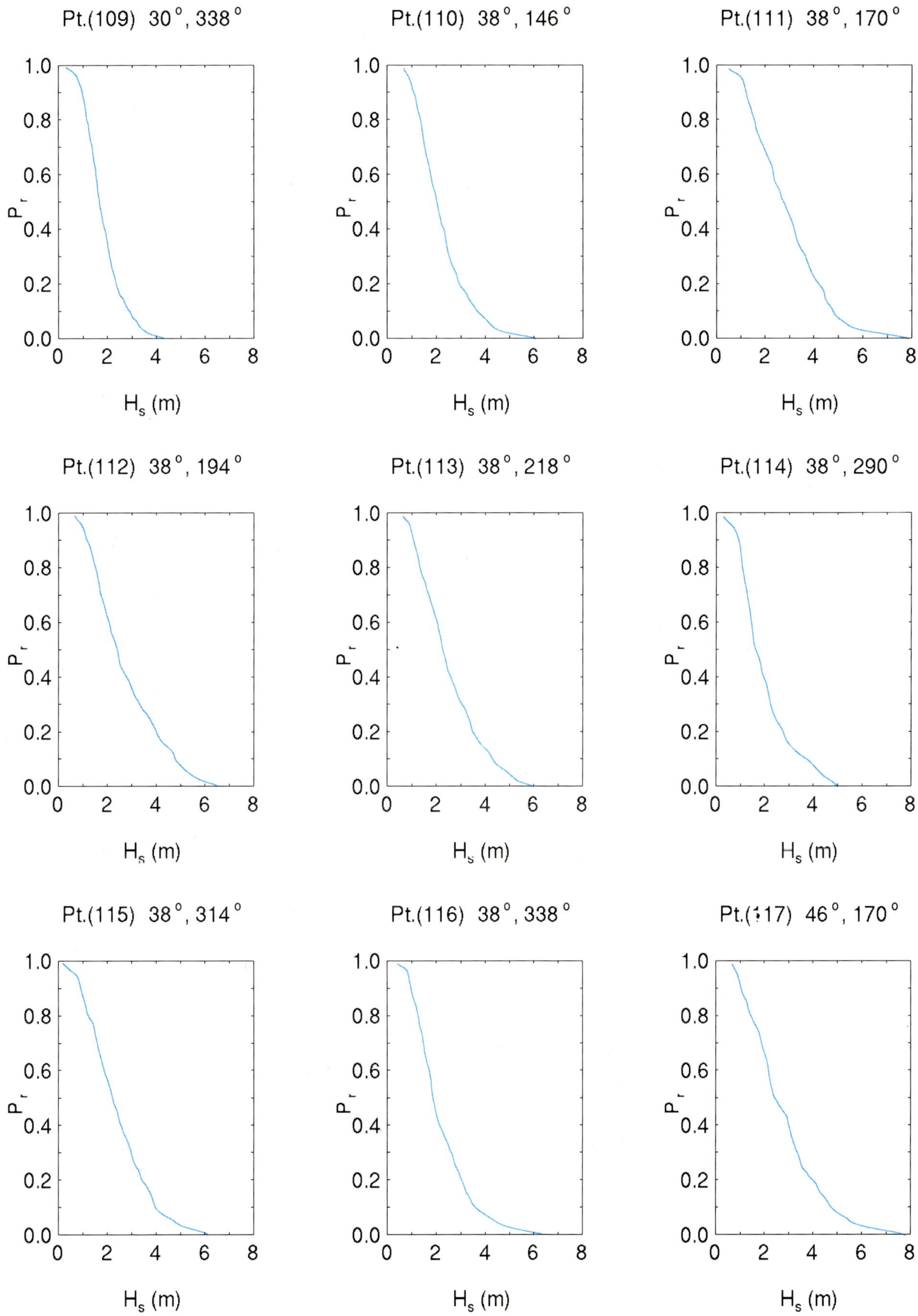

Fig. 7.31(z) Probability of Exceedence for Significant Wave Height at points 109–117 of Fig. 7.29.

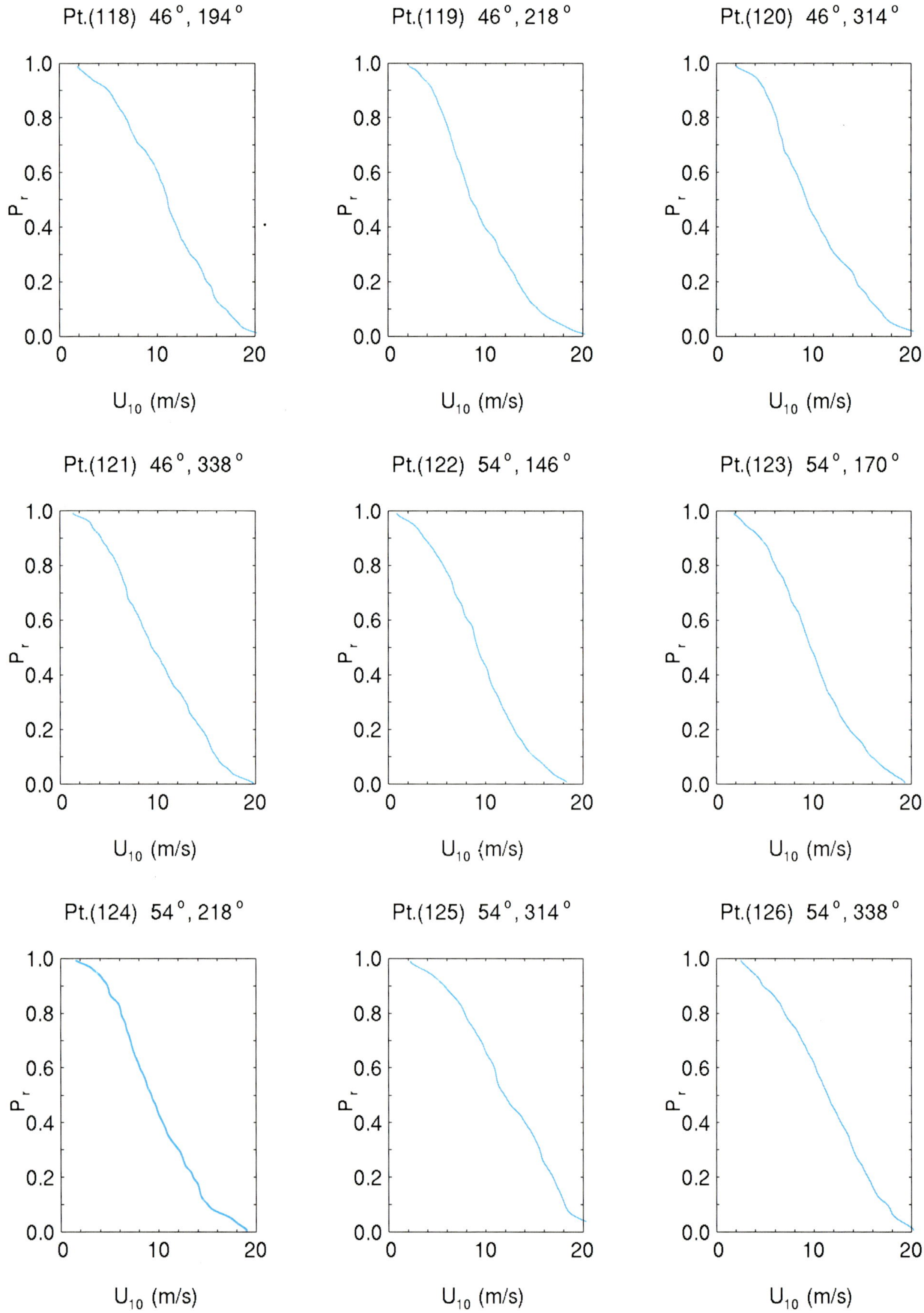

Fig. 7.31(aa) Probability of Exceedence for Wind Speed at points 118–126 of Fig. 7.29.

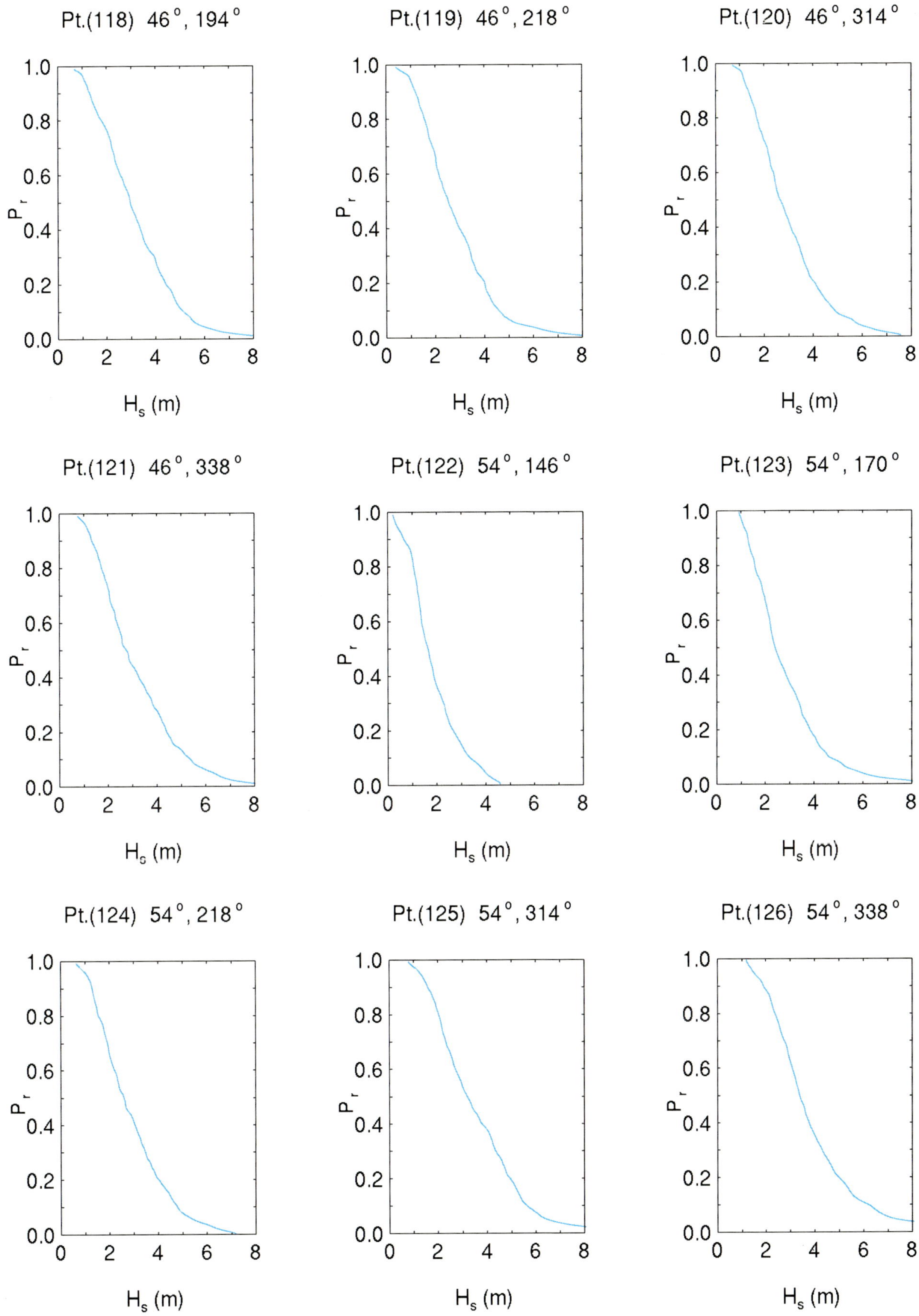

Fig. 7.31(bb) Probability of Exceedence for Significant Wave Height at points 118–126 of Fig. 7.29.

Chapter 8

REFERENCES

Airy, G. B. (1845) *Tides and Waves*. Encyclopedia Metropolitana.

Atkinson, G. D. and Holliday, C. R. (1977) Tropical cyclone minimum sea level pressure maximum sustained wind relationship for the western North Pacific, *Mon. Wea. Rev.*, **105**, 421–427.

Black, P. G. and Holland, G. J. (1995) The boundary layer of Tropical Cyclone Kerry (1979), *J. Atmos. Sci.*, **53**, in press.

Brown, G. S., Stanley, H. R. and Roy, N. A. (1981) The wind speed measurement capability of spaceborne radar altimetry, *IEEE J. Oceanic Engng*, **OE6**, 59–63.

Businger, S. (1991) Arctic hurricanes, *American Scientist*, **79**, 18–33.

Carter, D. J. T., Challenor, P. G. and Srokosz, M. A. (1992) An assessment of Geosat wave height and wind speed measurements, *J. Geophys. Res.*, **97**, 11383–11392.

Challenor, P. G., Foale S. and Webb, D. J. (1990) Seasonal changes in the global wave climate measured by the Geosat altimeter, *Int. J. Remote Sensing*, **11**, 2205–2213.

Chelton, D. B. and McCabe, P. J. (1985) A review of satellite altimeter measurement of sea surface wind speed: With a proposed new algorithm, *J. Geophys. Res.*, **90**, 4707–4720.

Chelton, D. B. and Wentz, F. J. (1986) Further development of an improved altimeter wind speed algorithm, *J. Geophys. Res.*, **91**, 14250–14260.

Cheney, R. E., Douglas, B. C., Agreen, R. W., Miller, L. L., Porter, D. L. and Doyle, N. S. (1987) *Geosat Altimeter Geophysical Data Record User Handbook*, NOAA Tech. Memo. NOS NGS-46.

Cotton, W. R. and Anthes, R. A. (1989) *Storm and Cloud Dynamics*. Academic Press.

Dean, R. G. and Dalrymple, R. A. (1991) *Water Wave Mechanics for Engineers and Scientists*. World Scientific.

DeMaria, M. and Kaplan J. (1995) Sea surface temperature and the maximum intensity of Atlantic tropical cyclones, *J. Climate*, in press.

Dobson, E. B. and Monaldo, F. M. (1996) Radar altimeter wave height measurements. In *Remote Sensing Techniques for Oceanographers* (M. Ikeda and F. Dobson, eds). CRC Press, New York, in press.

Dobson, E., Monaldo, F. and Goldhirsh, J. (1987) Validation of Geosat Altimeter-derived wind speeds and significant wave heights using buoy data, *J. Geophys. Res.*, **92**, 10719–10731.

Ebuchi, N., Kawamura, H. and Toba, Y. (1992) Growth of wind waves with fetch observed by the Geosat altimeter in the Japan Sea under Winter monsoon, *J. Geophys. Res.*, **97**, 809–819.

Emanuel, K. A. and Rotunno, R. (1989) Polar lows as arctic hurricanes, *Tellus*, **41A**, 1–17.

Fedor, L. S. and Brown, G. S. (1982) Wave height and wind speed measurements from the Seasat radar altimeter, *J. Geophys. Res.*, **87**, 3254–3260.

Findlater, J. (1969) A major low-level air-current near the Indian Ocean during the northern summer, *Quart. J. Royal Meteor. Soc.* **95**, 362–380.

Glazman, R. E. and Pilorz, S. H. (1990) Effects of sea maturity on satellite altimeter measurements, *J. Geophys. Res.*, **95**, 2857–2870.

Goda, Y. (1985) *Random Seas and Design of Maritime Structures*. University of Tokyo Press, Tokyo.

Goldhirsh, J. and Dobson, E. B. (1985) A recommended algorithm for the determination of ocean surface wind speed using a satellite-borne radar altimeter. Report JHU/APL SIR-85-U005, Johns Hopkins University, Applied Physics Laboratory, Laurel, MD.

Guillaume, A. (1987) VAG-Modéle de prévision de l'état de la mer en eau profonde, *Note de travail 178*, Etab. d'Etudes et Rech. Météorol., Toulouse, France.

Guillaume, A. and Mognard, N. M. (1992) A new method for the validation of altimeter-derived sea state parameters with results from wind and wave models, *J. Geophys. Res.*, **97**, 9705–9717.

Gumbel, E. J. (1958) *Statistics of Extremes*. Columbia University Press, New York.

Hadley, G. (1735) Concerning the cause of the general trade winds, *Philosophical Trans.* , **39**.

Hasselmann, K. *et al.* (1973) Measurements of wind-wave growth and swell decay during the Joint North Sea Wave Project (JONSWAP), *Dtsch. Hydrogh. Z. A*, **8**, 12.

Holland, G. J. (1980) An analytical model of the wind and pressure profiles in hurricanes, *Mon. Wea. Rev.*, **108**, 1212–1218.

Holland, G. J. (ed.) (1993) *The Global Guide to Tropical Cyclone Forecasting WMO/TD-560*. World Meteorological Organization, Geneva.

Holland, G. J. (1995) The maximum potential intensity of tropical cyclones, *J. Atmos. Sci.*, in press.

Holton, J. R. (1992) *An Introduction to Dynamic Meteorology*. Academic Press, San Diego.

Neumann, C. J. (1993) Global overview. In *The Global Guide to Tropical Cyclone Forecasting*, Chap. 1.1–1.43 (G. J. Holland, ed.). WMO/TD-560, World Meteorological Organization, Geneva.

Philander, S. G. H. (1990) *El Nino, La Nina, and the Southern Oscillation*. Academic Press, San Diego.

Pierson, W. J. and Moskowitz, L. (1964) A proposed spectral form for fully developed wind seas based on the similarity theory of S.A. Kitaigorodskii, *J. Geophys. Res.*, **69**, 5181–5190.

Ramage, C. S. (1968) Role of a tropical Maritime Continent in the atmospheric circulation, *Mon. Wea. Rev.*, **96**, 365–370.

Rasmussen, E. (1979) The polar low as an extratropical CISK disturbance, *Quart. J. Royal Meteor. Soc.*, **105**, 531–549.

Sanders, F. and Gyakum, J. R. (1980) Synoptic-dynamic climatology of the bomb, *Mon. Wea. Rev.*, **108**, 1589–1606.

Sellers, W. D., (1965) *Physical Climatology*. University of Chicago Press, Chicago.

Shaw, N. (1926) *Manual of Meteorology, Vol. 1, Meteorology in History*. Cambridge University Press, Cambridge.

Shuhy, J. L., Grunes, M. R., Uliana, E. A. and Choy, L. W. (1987) Comparison of Geosat and ground-truth wind and wave observations: Preliminary results, *Johns Hopkins APL Tech. Dig.*, **8**, 219–221.

Simpson, R. H. and Riehl, R. (1981) *The Hurricane and its Impact*, pp. 366–372. Louisiana State University Press, Louisiana.

Snodgrass, F. E., Groves, G. W., Hasselmann, K. F., Miller, G. R., Munk, W. H. and Powers, W. H. (1966), Propagation of ocean swell across the Pacific, *Phil. Trans. Roy. Soc. London A*, **259**, 431–497.

Steedman, R. K. (1993) Data report on Southern Ocean surface wave climate adjacent Macquarie Island, November 1988 to October 1989. Steedman Science & Engineering, Report No. R594, Perth, Australia.

Tournadre, J. and Ezraty, R. (1990) Local climatology of wind and sea state by means of satellite radar altimeter measurements, *J. Geophys. Res.*, **95**, 18255–18268.

Twitchell, P. F., Rasmussen, E. A. and Davidson, K. L., eds (1989) *Polar and Arctic Lows*. A. Deepack Publishing, Hampton, VA.

USSR Ministry of Defence (1974) *Ocean Atlas: Pacific Ocean*. USSR Ministry of Defence, Moscow.

Walker, G. T. (1924) Correlation in seasonal variations of weather, X. Applications to seasonal forecasting in India, *Mem. India. Met. Dept.*, **24**, 333–345.

Witter, D. L. and Chelton, D. B. (1991) A Geosat altimeter wind speed algorithm and a method for altimeter wind speed algorithm development, *J. Geophys. Res.*, **96**, 8853–8860.

Young, I. R. (1986) Probability distribution of spectral integrals, *ASCE J. Waterway, Ports, Coastal and Ocean Engng*, **112**, 338–341.

Young, I. R. (1993) An estimate of the Geosat altimeter wind speed algorithm at high wind speeds, *J. Geophys. Res.*, **98**, 20275–20285.

Zambresky, L. (1989) A verification of the global WAM model December 1987–November 1988. European Centre for Medium-Range Weather Forecasts, Tech. Rep. No. 63.

Chapter 9

GLOSSARY

9.1 Common Terms

Baroclinic: Has vertical structure, in the form of variable density, temperature and winds.
Barotropic: Single level, with no vertical structure.
Basin: Either an oceanic region with coherent features (e.g. the north Indian Ocean Basin), or a set of bathymetric data for use in numerical models.
Beaufort Number: Categorization of wind speed based on visual state of the sea or land effects.
Centre of a Tropical Cyclone: Varies according to the analysis method; typically one of geometric centre of eye, minimum pressure, zero wind, or end of a spiral band.
Deep/Shallow Water: Used in the context of gravity waves, water that is several wavelengths deep; thus greater than 100 m constitutes deep water for an ocean swell (wavelength tens of meters), but a storm surge (wavelength several tens of kilometres) is in shallow water anywhere on a continental shelf.
Extratropical Cyclone: A synoptic scale low pressure system which derives its energy primarily from available potential energy in a pre-existing horizontal temperature gradient.
Fetch: The area of sea, especially along the direction of the wind, over which winds are relatively constant and wave generation occurs.
Freak Waves: Single waves that may arise from local amplification of intersecting waves, interactions between waves and currents, or development of solitary waves. Of considerable danger to mariners because of their unexpected nature (often coming from an unexpected direction) and potential for developing to great height.
Fully Developed Sea: Sea in a quasi-steady state in which the energy gained by waves from the wind is approximately equal to that lost to wave breaking and other mechanisms.
Gravity Wave: A perturbation along a vertical density discontinuity, or gradient, which has gravity as its restoring force; for example ocean waves, storm surges. In the atmosphere, gravity waves are generated whenever an imbalance between the mass and wind fields develops.
Inverted Barometer Effect: The uplift of water in the centre of a low-pressure system (such as a tropical cyclone) as a dynamic adjustment to the low air pressure; roughly 1 cm sea level rise for every 1 hPa drop in sea level pressure.
Marsden Square: Any region on earth with sides comprised of equal degrees of latitude and longitude.
Murphy's Law: Anything that can go wrong, will go wrong. There are two useful corollaries: 1. The chance of something going wrong is directly proportional to the amount of damage that will ensue; 2. Murphy was an optimist.
Partially Developed Sea: Sea which has not reached a steady state for the imposed wind conditions and is still developing.
Radar: Radio Direction And Ranging, a conical beam of pulsed electromagnetic energy in the microwave range transmitted outward from an antenna.
Radius of Maximum Winds: The radial distance from a cyclone centre to the mean position of the band of maximum winds.
Solitary Wave: Single waves, or an isolated train of waves that can be of considerable height and very different to the prevailing ocean wave conditions and have been associated with some spectacular maritime incidents (see Mariners Weather Log, Fall 1993). These may take the form of solitons, tsunamis, tidal bores, and tidal races or hydraulic jumps.

Soliton: A single wave, or short wave train that has developed from either a special combination of wind swells or interactions of tides with local bathymetry. On the surface these consist of a well-defined set of wave crests. Solitons also may develop on internal density gradients in the ocean, and are manifested at the surface by a boiling effect.

Spherics: Shortened from atmospherics, the crackle heard on radio from lightning discharges in the atmosphere.

Storm Surge: An abnormal rise of sea water associated with a tropical cyclone. It is described as the still water elevation above the local astronomical tide, uncontaminated by high frequency, short gravity wind waves.

Storm Tide: The total ocean elevation, including the astronomical tide, above or below a standard datum resulting from the passage of a tropical cyclone.

Tidal Bore: Generated in estuaries and rivers with high tidal range and a sharp narrowing and shoaling at the entrance. Manifested as a single wave, which may be breaking, that travels along the river or estuary and is followed by a rise in water level.

Tidal Race or Rip: Strong currents, and in extreme cases an hydraulic jump where the water level sharply changes by up to a metre; generated by high tides interacting with local bathymetry.

Tropical Cyclone: A low pressure system which derives its energy primarily from: 1. evaporation from the sea in the presence of high winds and low surface pressure; and 2. condensation in convective clouds concentrated near its centre. The precise definition varies considerably around the world (Neumann, 1993), but generally a low pressure system over the tropical oceans with sustained winds of at least 17 ms^{-1} is called a tropical cyclone.

Tropical Cyclone Severity Scale: Five categories indicating the damage potential of severe tropical cyclones; the Saffir–Simpson Hurricane Scale (Simpson and Riehl, 1981) was developed to provide a sliding scale of damage potential for hurricanes, including that arising from storm surge. A similar scale, though adapted for local conditions, is used in Australia.

Hurricane or Severe Tropical Cyclone Level	Maximum Sustained Winds	
	Saffir–Simpson 1 min mean ms^{-1} (mph)	Australian Scale 10 min mean ms^{-1} (mph)
1	33–42 (74–94)	17–25 (38–56)
2	43–49 (95–110)	25–35 (57–78)
3	50–58 (111–130)	35–46 (79–103)
4	59–69 (131–155)	46–63 (104–141)
5	$>$ 69 ($>$ 155)	$>$ 63 ($>$ 141)

Tsunami: Ocean wave generated by an earthquake or volcanic explosion; travels through the deep ocean with height of a few cm and a speed of 200–300 ms^{-1}; on reaching shallow coastal regions the wave slows and grows to huge heights of potentially 60–70 m (see Mariners Weather Log, Fall 1993). These are often incorrectly referred to as tidal waves.

Wall Cloud: The region of deep cloud surrounding a mature cyclone eye.

Wave

Height or Amplitude: The vertical distance between a wave crest and trough;

Length: The horizontal distance between two wave crests;

Period: The time taken for two wave crests to pass a fixed point;

Set-Up: Water forced inshore by breaking waves;

Significant Wave Height/Period: The average height/period of the highest third of all waves in a sample of wind-waves;

Standing Men: Steep waves generated when ocean waves propagate against a strong current;

Swell: Smooth, regularly spaced waves that have propagated long distances from their initial generation region;

Wind Waves: Choppy and chaotic waves generated locally by the wind.

Wind

Mean, Average or Sustained: Taken as the 10 min mean wind, except in the U.S.A. and North Atlantic, where 1 min means are used, and in India, where 3 min means are used; the conversion factor from 1 to 10 min mean is 0.871; this varying definition of what constitutes a wind speed is a source of considerable confusion amongst mariners and has led to incorrect assessments of dangerous conditions.

Gust: The highest wind burst, generally taken as the 1 or 2 s value; gust factors vary according to the surface conditions and the type of wind system. Indicative gust factors are provided in the following table. These are defined by the ratio of peak 2 s wind to the mean wind at 10 m elevation and are relevant to wind speeds of at least hurricane force. Parentheses give an indication of the range in gust factors (Holland, 1993).

	Ocean	Flat grassland	Woods or city
1 min mean	1.25 (1.17–1.29)	1.35 (1.29–1.45)	1.65 (1.61–1.77)
10 min mean	1.41 (1.37–1.51)	1.56 (1.51–1.70)	2.14 (1.89–2.14)

9.2 Unit Conversion

Multiply the line unit by the indicated number to get the column unit; for example, 1 mile is 1.61 km. Take care to convert to the degree of accuracy of the original observation only; for example a ship estimate of 20 kt based on sea state should convert to 10 ms^{-1}, a literal transformation to 10.256 ms^{-1} is incorrect.

9.2.1 Length

	km	mi	nm	°lat
km	1	1.61	1.85	111.137
mi	0.62	1	1.15	69
nm	0.54	0.87	1	60

9.2.2 Time

	s	min	hr	day
s	1	60	3.6×10^3	8.64×10^4
min	⋆	1	60	1.44×10^3
hr	⋆	⋆	1	24

9.2.3 Speed

	km h^{-1}	mph	kt	ms^{-1}
km h^{-1}	1	1.61	1.85	3.6
mph	0.62	1	1.15	2.24
kt	0.54	0.87	1	1.95
ms^{-1}	0.28	0.45	0.51	1

9.2.4 Conversion °C to °F

$$F = \frac{9}{5}C + 32$$

9.2.5 Conversion K to °C

$$C = K - 273$$

9.3 The Beaufort Scale

The Beaufort Scale is a commonly used characterization of the wind speed in terms of the visual appearance of the sea surface (see over).

BEAUFORT SCALE OF WIND FORCE

Beaufort number	Descriptive term	Wind speed ms^{-1}	Wind speed knots	Specifications for observations On board ship (open sea)
0	Calm	0–0.2	< 1	Sea like a mirror
1	Light air	0.3–1.5	1–3	Ripples with the appearance of scales are formed, but without foam crests
2	Light breeze	1.6–3.3	4–6	Small wavelets, still short but more pronounced; crests have a glassy appearance and do not break
3	Gentle breeze	3.4–5.4	7–10	Large wavelets; crests begin to break; foam of glassy appearance, perhaps scattered white horses
4	Moderate breeze	5.5–7.9	11–16	Small waves, becoming longer, fairly frequent white horses
5	Fresh breeze	8.0–10.7	17–21	Moderate waves, taking a more pronounced long form; many white horses are formed (chance of some spray)
6	Strong breeze	10.8–13.8	22–27	Large waves begin to form; the white foam crests are more extensive everywhere (probably some spray)
7	Near gale	13.9–17.1	28–33	Sea heaps up and white foam from breaking waves begins to be blown in streaks along the direction of the wind
8	Gale	17.2–20.7	34–40	Moderately high waves of greater length; edges of crests begin to break into spindrift; the foam is blown in well-marked streaks along the direction of the wind
9	Strong gale	20.8–24.4	41–47	High waves; dense streaks of foam along the direction of the wind; crests of waves begin to topple, tumble and roll over; spray may affect visibility
10	Storm	24.5–28.4	48–55	Very high waves with long overhanging crests; the resulting foam, in great patches, is blown in dense white streaks along the direction of the wind; on the whole, the surface of the sea takes on a white appearance; the tumbling of the sea becomes heavy and shock-like; visibility affected
11	Violent storm	28.5–32.6	56–63	Exceptionally high waves (small and medium-sized ships might be for a time lost to view behind the waves); the sea is completely covered with long white patches of foam lying along the direction of the wind; everywhere the edges of the wave crests are blown into foam; visibility affected
12	Hurricane	> 32.7	> 64	The air is filled with foam and spray; sea competely white with driving spray; visibility very seriously affected

INDEX